To Steve and the family
this book and its chapters are about the future. Enjoy it.
I hope that we will be working together and many other things &
lets think about page two about

Best
Wooly
Aug 2014

— Bert

The Next Economics

Woodrow W. Clark II
Editor

The Next Economics

Global Cases in Energy, Environment, and Climate Change

Editor
Woodrow W. Clark II
Clark Strategic Partners
Beverly Hills, CA, USA

ISBN 978-1-4614-4971-3   ISBN 978-1-4614-4972-0 (eBook)
DOI 10.1007/978-1-4614-4972-0
Springer New York Heidelberg Dordrecht London

Library of Congress Control Number: 2012951413

© Springer Science+Business Media New York 2013
This work is subject to copyright. All rights are reserved by the Publisher, whether the whole or part of the material is concerned, specifically the rights of translation, reprinting, reuse of illustrations, recitation, broadcasting, reproduction on microfilms or in any other physical way, and transmission or information storage and retrieval, electronic adaptation, computer software, or by similar or dissimilar methodology now known or hereafter developed. Exempted from this legal reservation are brief excerpts in connection with reviews or scholarly analysis or material supplied specifically for the purpose of being entered and executed on a computer system, for exclusive use by the purchaser of the work. Duplication of this publication or parts thereof is permitted only under the provisions of the Copyright Law of the Publisher's location, in its current version, and permission for use must always be obtained from Springer. Permissions for use may be obtained through RightsLink at the Copyright Clearance Center. Violations are liable to prosecution under the respective Copyright Law.
The use of general descriptive names, registered names, trademarks, service marks, etc. in this publication does not imply, even in the absence of a specific statement, that such names are exempt from the relevant protective laws and regulations and therefore free for general use.
While the advice and information in this book are believed to be true and accurate at the date of publication, neither the authors nor the editors nor the publisher can accept any legal responsibility for any errors or omissions that may be made. The publisher makes no warranty, express or implied, with respect to the material contained herein.

Printed on acid-free paper

Springer is part of Springer Science+Business Media (www.springer.com)

This book is dedicated to my wife, Andrea, and our son, Paxton, who both supported and will benefit from this book's vision for the future.

Foreword

I have known Woody and his work for two decades, since we initially collaborated on United National Intergovernmental Panel on Climate Change assessment reports. Woody was one of the first guest lecturers in my energy courses at the University of California, Berkeley. What I noticed consistently was that Woody was always looking ahead – sometimes far, far head. In *The Next Economics*, his timing could not be more on the money. With the world economies in trouble, in large part due to the failure of the Western economic models, he has provided vision that makes a difference. Central to his analysis of the opportunity for a paradigm shift to a Green Industrial Revolution is the fact that economics itself needs to become a science. As a physicist, I very much appreciate the effort to build a clear analytic foundation for the tools to assess sustainability. And why we need to have these ideas wide spread and included into our programs, research sooner than later.

Professor Dan Kammen
Co-Director, Berkeley Institute of the Environment;
Founding Director, Renewable and Appropriate
Energy Laboratory
University of California, Berkeley

Woody Clark is a prolific author who has been at the forefront of some of the most important issues of environment and the economy confronting the world today. His scope of interest, experience and influence is truly global. As Director of the Institute of Environment and Sustainability (IoES) I have had the pleasure of working with Woody on research and educational initiatives. One recent and exciting aspect of this was Woody's work with Dr. Ren Sun, Director of the Cross-Disciplinary Scholars in Science and Technology program, here at UCLA where Woody developed and taught a course for the IoES in 2012. During my tenure as Director, I have been astounded by Dr. Clark's scholarly productivity. For example, in 2009–2010 he produced and spoke on "Sustainable Communities" with case studies and data from around the world. These are issues of importance to my Institute certainly. I am particularly excited by his new book *The Next Economics* coming out in 2012. This book argues in part how economics, as practiced in the world of policy, needs to become more scientific in approach and further removed from clouding by shifting public opinions and political biases. The world of scholars and decision makers should consider the recommendations for the 'Next Economics" and the need for "Social Capitalism" to help see us through the potentially rough waters of the twenty-first century.

Professor Glen M. MacDonald
UC Presidential Chair and
Director, Institute of Environment and Sustainability;
Distinguished Professor of Geography and Ecology
and Evolutionary Biology
University of California, Los Angeles

Since Woody was a Fulbright Fellow in 1994 at Aalborg University (AAU), Denmark, we have collaborated and been close professional and personal friends. For over two decades, we have taught together, done research and published papers. Among others Woody contributed to my book *Renewable Energy Systems* in 2010. His books on *Sustainable Communities* (2009 and 2010) have included chapters that I have done with my colleagues at AAU. Our work continues today. This book on economics becoming a science is a significant step forward in a world where opinions and political biases tend to dominant and influence the truth. We need to take his ideas and make them into programs, degrees and awards. The implementation of Renewable Energy Systems calls for his insights into the understanding of Economics. We need to act now.

Professor Henrik Lund
Editor-in-Chief of ENERGY – The International Journal
Technology Professor in Energy Planning
Aalborg University, Denmark

A 95 year old Tribal Elder, Archie Mosay, was once asked how he foresaw the future.

With visionary clarity, his answer was that he saw the earth healing itself, green grass, big trees swaying in the wind and the water clear and blue. We live in a sea of energies joining together all that live on the earth. The way of life of Native America has, throughout the centuries, respected the gifts bestowed by Mother Earth, the primary of which are sun, water and wind. Green technologies employing these gifts help nurture the long-term well-being of the earth and its people. It is time all people listened to the earth with their hearts and come to understand that benefits coming from green technologies are of far greater value than the cost/benefit derived from the technologies; it is the healing and preservation of our earth for future generations. Native America stands ready to lead the way toward a new beginning that embraces green energy initiatives benefiting all peoples of the earth.

Woodrow Clark envisions the future the same as our Tribal Elders. Humankind for millennia has been wasting the earth's gifts. In our modern world, Woodrow Clark applies the science of economics to help focus attention and bring about earth compassionate public policies for the benefit of all people of the earth.

Rick Hill
Oneida Tribe

Dave Coon
Lake Superior Chippewa Nation

Jennifer Alekson
Citizen Band of the Potawatomi Nation

Preface

The *Special Issue of Contemporary Economic Policy* (CEP) that I coedited with Professor Michael Intriligator had been over 2 years in the making. CEP is one of two major economic publications from the Western Economic Association International (WEAI). These journals are well known for examining contemporary economic issues and exploring new approaches to them. Mike and I worked hard on the *Special Issue*. With 11 peer-reviewed articles that were also reviewed by the CEP editor, Wade Martin, we were proud of the results: five of the eleven articles will be published in 2013 and are now chapters in this book.

The book takes all these papers and includes a few that provides the framework for discussion of economics which is seen a "field of study," according to a special issue of the Economist (2009) with a picture of the Bible melting stating that modern economic theory is failing, about 9 months after the global recession in the fall of 2008. The basic conclusion from this special issue and a series of other articles that turned into a debate among economists is that "economics is not a science", but needs to become one.

Economics must move "toward a science" was the subtitle of my book with Professor Michael Fast on Qualitative Economics (2008) earlier that same year. This book provides new and creative thinking about the field of economics. A special thanks goes to Wade for his encouragement and very diligent oversight of the entire CEP issue and to Mike for his solid and consistent support of looking into new ways to consider economics scientific in order to solve societal problems.

The background for this book and the CEP *Special Issue* are important. Originally, we all wanted to do the special issue along the lines of a reflection of new thinking within the field of economics. We saw this as a point of departure from the western-developed world today that has energy security issues about its future, especially with the impact on climate change. Defining and exploring the depths of economics is at the core of this book and reflected in every chapter.

I spearheaded the *Special Issue* of CEP because I saw economics as being in serious trouble, even before the economic collapse American economic collpse in the fall of 2008 and the global economic crisis that continues today. A year before global economic collapse, I organized a panel for the annual WEAI conference in Seattle in 2007. The presenters, some of whom contributed to the *Special Issue*, and others that are now in this book were concerned with the "field of economics" in general. They were concerned that it was covering broader societal issues from an economic perspective.

For example, how can communities and nations develop with no political and economic plans and little concern for the environment, people, health, and the climate. Today, America still has no national energy or even mass transportation plans. Yet, every family and business has a plan if not month by month, then certainly an annual one. I have taught business plans and entrepreneurship in graduate business and MBA programs. Every person, group, community, and nation needs a plan. The fact that America today is divided is both a major cause for the nation not to progress and lead what I call in another book with Grant Cooke on Global Energy Innovation (2011) the "Green Industrial Revolution" is destructive to everyone and detrimental to future generations. The problem is today's ideological politicians in every region, state, and country. I had experienced this enormous divide over a decade ago when I was very involved with the UN Intergovernmental Panel on Climate Change (UN IPCC). Nations around the world need to agree upon a plan to mitigate climate change. I personally had to try to get 129 nations to agree upon the executive summary for the third report by the UN IPCC in 1999. While we finally agreed on a report, it took almost another decade to proclaim that climate change was the result of people and that the world needed a plan to stop and reverse climate change. That plan has yet to be done and implemented.

When Al Gore won the Nobel Peace Prize in December 2007, hundreds of us with the UN IPCC shared it with him. However, what was never, even now, really discussed was that Gore identified and dramatically presented in his film, *An Inconvenient Truth* (2007), was that the climate is changing dramatically today. But the UN IPCC did the same with scientific evidence by not providing a plan. The problem of climate change was discussed and proven scientifically. What was not recognized then was that Gore and many members of the UN IPCC, work on the "solutions" to climate change, ranging from sustainable communities, renewable energy, commercialized technologies, and finance.

In my case, I have two books, *Sustainable Communities* (Springer, 2009) and *Sustainable Communities Design Handbook* (Elsevier, 2010) with cases about sustainable communities and how they can be created, financed, implemented, and maintained. In the next year, I am completing a new book on *Global Sustainable Communities Design Handbook* (Elsevier, 2013) with cases of sustainable communities and how they were designed, developed, and planned with resources, finances, and educated workers. The book sets a standard from which a series of books on this topic can be published annually in book, journal, and online formats.

Basically, the problem with the "field of economics" is that for over four decades, it has taken conventional or "neoclassical economic theories" from Adam Smith and tried to apply them. The Smith model for western capitalism, however, was and is today simply a "theory". There have never been actual cases of neoclassical capitalism. For example, these theories depend on "market forces" that are a balance between supply and demand, but never work (ibid., 2008). They never account for key issues facing society, such as social revolutions, economic recessions, and climate change. I experienced this in the role as Renewable Energy Advisor to Governor Gray Davis of California (1999–2003) where there was a need to change economics away from the "market forces" that was created in prior state government administrations with their deregulation of the energy sector.

Governor Davis came into office and was immediately confronted with an energy crisis caused (starting in 2000–2003, but that continues today) by the prior two governors before him, because they argued that "deregulation" of the energy sector in 1996 from public utilities should go to private companies to generate power and supply the state with energy. New companies would be competitive and therefore lower prices for energy to consumers. Just the opposite happened. And without very much oversight in the laws for deregulation, the problem had to be taken on by Governor Davis, after he was elected in 1999.

By spring 2000, California had an energy crisis with rolling blackouts and brownouts even though there was plenty of energy supply. I had warned Governor Davis' senior staff that this would happen 6 months or more before the brownouts started in San Diego. California deregulation was copied in other states and nations which called it "liberalization or privatization." The national utility-controlled energy systems converted from being public-controlled companies to private businesses. The market forces economic model would create competition and hence reduced energy costs, but did just the opposite of that.

The California energy crisis came without warning as the new private energy companies controlled and manipulated prices, with services through their control of energy. The economic model failed in California and other nations as well. There was something wrong. Private companies manipulated the "energy market" and caused severe problems throughout the state. The California energy crisis was just the beginning, because supply and demand did not work when the state was immersed in brownouts and blackouts that threatened businesses and individual health that all needed power for commerce and medical care.

The economists' explanation, issued at one point in a public memo to Governor Davis (Spring 2001), argued that "market forces" would prevail and get the state energy needs back on course. In reality, those market forces were "gaming the energy sectors" with illegal and deceptive accounting. These companies were responsible for conducting fraudulent actions. The firms (Enron and many others) and their accounting firms "verified" the economic energy data as valid, when it was not. The state investigated and took those people and their companies to court, where individuals were convicted and sent to jail (Clark, 2003; Clark and Demirag, 2002 and 2006). Several chapters review and discuss economic models and where or why they have failed. But this book also sets out in a number of chapters to create and inspire new economic models. In particular, it strives to turn economics into a science with examples in the different chapters.

The book was first inspired by other work with economists seeking changes in their field. At another WEAI conference in Honolulu in July 2008, the issue about how precise and accurate economics was raised in a different way by a panel that

I chaired and also presented a paper. The topic was modern economic theory and what was wrong with it. I coauthored with Professor Michael Fast from Aalborg University in Denmark the book *Qualitative Economics* (ibid., 2008) that came out just in time for the conference and was a key part of one session there. By that fall, the global economic crisis hit the USA and went around the world, in which much of modern economic theory came to be questioned by economists themselves. If economics was a science, why was it not able to predict the global economic crisis in 2008?

The time was perfect then for the CEP Special Issue. We felt that the "field of economics" was so vast there needed to be a focus on only a few topics for the *Special Issue*: Global Cases in Energy, Environment, and Climate Change. We decided that these areas were a challenge for economists but needed to be studied.

Finally, there is need to be cross-disciplinary areas in order for a fresh look to be given to economics. These areas and how they interacted are a starter. Based on past economic models, these areas have been lost or not fitted into modern economic theory. Clearly, economics needs to research and probe these areas, as they are major determinants in the economics of the future. The challenge is to explore and look deeply into economics, in order to turn "the field" into a science.

Beverly Hills, CA, USA   Woodrow W. Clark II

References

Clark WW II (2013) Author and Editor. Global sustainable communities design handbook. Elsevier Press, New York
Clark WW II, Michael I (2013) Global cases in energy, environment, and climate change: some challenges for the field of economics, Special Issue of Contemporary Economic Policy (CEP). Western Association of Economics International, Blackwell Publications, Fullerton
Clark WW II, Gant C (2011) Global energy innovation. Preager Press, New York
Clark WW II Author and Editor Sustainable communities design handbook. Elsevier Press, New York
Clark WW II (2009) Author and Editor. Sustainable communities. Springer Press, New York
Clark WW II, Michael F (2008) Qualitative economics: toward a science of economics. Coxmoor Press, London
Clark WW II, Demirag I (2006) US financial regulatory change: the case of the California energy crisis. Special Issue, J Bank Regul 7,(1/2):75–93
Clark WW II (2003) Point and counter-point: de-regulation in America. Utilities Policy, Elsevier, Fall
Clark WW II, Demirag I (2002) Enron: the failure of corporate governance. J Corp Citizenship 8(Winter):105–122
Economist (2009) Modern economic theory: where it went wrong – and how the crisis is changing it. Special issue with cover of Bible Melting, London, 18 July 2009
Gore Al (2007) An inconvenient truth. Paramount Studios. Hollywood, California

Contents

1. Introduction 1
   Woodrow W. Clark II

2. The Next Economics 21
   Woodrow W. Clark II

3. Market Solutions for Climate Change 43
   Malcolm Dole, Jr.

4. Qualitative Economics: The Science Needed in Economics 71
   Michael Fast and Woodrow W. Clark II

5. Energy Planning for Regional and National Needs:
   A Case Study – The California Forecast (2005–2050) 93
   Gary C. Matteson

6. Achieving Economic Gains Through the Setting
   of Environmental Goals: The Case of California 125
   Tracey Grose

7. Social Capitalism: China's Economic Raise 143
   Woodrow W. Clark II and Li Xing

8. The "Cheap Energy Contract": A Critical Roadblock
   to Effective Energy Policy in the USA 165
   Michael F. Hoexter

9. Economic-Environmental Performance of Micro-wind
   Turbine in Mediterranean Area 185
   Nicola Cardinale, Gianluca Rospi, Giuliano Cotrufo,
   and Tiziana Cardinale

10. Energy Conservation for Optimum Economic Analysis 207
    Stephen C. Prey

11 **Blue-Green Agricultural Revolution** 237
    Daniel Nuckols

12 **Going Beyond Growth: The Green Economy
    as a Sustainable Economic Development Strategy** 251
    Laurie Kaye Nijaki

13 **Conclusions: The Science of Economics** 275
    Woodrow W. Clark II

Index 287

Chapter 1
Introduction

Woodrow W. Clark II

Abstract This chapter concerns how economics must change its conventional western- oriented paradigm from Adam Smith and his followers of "the market economy" to a new global economic paradigm that is rooted in societal issues and concerns, ranging from environmental, social, and health issues to more that include the solutions to climate change, medical health, education, and a broad range of concerns for humankind. The problems are vast. However, the solutions often start with innovations and technologies to support and solve societal problems. Hence, there are higher costs as with all solutions to any problem.

However, the basic barrier to stopping climate change rests with economics. The standard reply is that "it" (whatever "it" means) costs too much. The only economic cost-benefit analysis is higher taxes. While that may be true in some ways, according to the standard classical economic model, it raises a fundamental question about economics itself: economics is not a science because no one can predict that higher taxes enhances or hurts economic growth. Almost every scientist agrees with that statement. But most differ to the economic analysis since they are controlled by the corporate CEOs and political leaders, who allocate funds for research and new technologies. Yet as this chapter discusses, economics is not a science, as it cannot predict economic trends or events.

Hence, the chapter discusses what needs to be done to make economics a science, setting the stage for the other chapters in the book with their review of societal issues (primarily environmental) and how economics needs to address the financial costs in different ways ranging from externalities to life-cycle analyses. Each chapter is reviewed and summarized in this chapter to that the reader has a good solid basis for the last chapter that provides a new economic paradigm in detail that is grounded and based on science.

W.W. Clark II, Ph.D. (✉)
Qualitative Economist, Academic Specialist, Cross-Disciplinary Scholars in Science and Technology, UCLA and Managing Director, Clark Strategic Partners, California, USA
Website: www.clarkstrategicpartners.net

Keywords Science • Paradigms • Classical and conventional economics • Case examples

The Next Economics was an idea that I had for a book about 5 or 6 years ago. Then after participating in several Western Economic Association International (WEAI) Conferences, I did one paper and a panel on Status of Economics. When my friend Professor Michael Intriligator was president of the WEAI, I talked to him about doing a special issue for one of the two journals that WEAI published each quarter. After talking with the publisher, Professor Wade Martin, we all decided on Contemporary Economic Policy (CEP) for a special issue to be called "Global Cases in Energy, Environment, and Climate Change: Some Challenges for the Field of Economics."

Professor Intriligator became the coeditor, and we gathered 11 papers covering this topic. In the CEP special issue, only five of the papers are to be published. Two of the papers are changed and published here in The Next Economics, along with the other six that were not accepted for the CEP. The editorial opinion about the six papers left out of the CEP, but published here, was basically that these six papers were not traditional papers in economics. Hence, the purpose of this book is to expand traditional economics by examining and providing cases of economics as this field, but applied to environment, energy security, and climate change topics. The issue for everyone today is that the costs for saving the environment and solving climate change are unknown and often given as an excuse to do nothing. The most common comment from traditional neoclassical economic paradigm is that the "market" will find solutions. This book directly counters that assumption with solid data and findings in order to create economics based on science, rather than politic opinion and unfounded policies.

Without doubt, economics needs to be applied to these global societal areas. What was not done in the CEP was an analysis of the philosophical history of economics in terms of how it has impacted global environment and climate change issues. The paper that addressed that issue from Professor Michael Fast and myself is revised as a chapter in this book and argues for a looking at economics in a far different manner and theoretical paradigm than the field has had over the last two centuries and especially the last 40–50 years. Therefore the next step needs to be a far more comprehensive examination of what, where, and why economics has come only one western philosophical paradigm base and not others. Adam Smith was not the only economics philosopher over 200 years ago. There were others. And there were many economists who were from different cultures and wrote in different languages with positive results from their approaches to economics. Japan and now China are cases proving that point.

What is more concerning, however, is that Adam Smith and his classical economics became propagated as an ideology, from only one particular perspective and point of view, especially in the last four decades. Yet while Adam Smith prided himself in taking ideas and concepts from physics and mathematics, he did not use either science as the base in calculations and formulas for economics. Instead, Adam

Smith used the concept of a balance between physical forces, which in the end would work in field of economics as the balance between supply and demand. However in reality, physics is not only about balance or only as mathematics. What the world has witnessed in the last decade, is that market forces as western capitalism, certainly does not demonstrate such as a balance. The global economic collapse in the fall of 2008 continues today and is documented proof of that. Science is far more than the balance of physical forces. Nor for that matter, any science, be it chemistry, engineering, mathematics, or others, is not based on balances alone. And clearly, no science has an invisible hand. The fact is that science historically demands heavily on government from research funds to tests and technological applications.

So what is science? Consider "Physics (which) is often described as the fundamental science, as it seeks to understand the 'rules' or 'laws' by which the universe operates" (Perkins 1996). What The Next Economics (NE) does is to set the stage for a far more in-depth and global investigation of economics than the CEP special issue did. Consider now a summarization of the issues in each chapter of NE so that they can provide background and guidance to further study, research and tests for analysis in order to create a science of economics.

Let us start with Rifkin, who is a well-known environmental Economist in his book, Entropy (1980) which discusses the extension of economics beyond the neoclassical theoretical paradigm of Adam Smith to include social issues such as the environment. Perkins, a physicist, notes that "entropy" is a scientific way to describe how "the universe is running down and getting more disordered" (op.cit., Perkins, p. 3). That concept might apply to neoclassical economics but certainly does not provide a direction in which economics must move, due to its consistent failure over the last few years to become a science.

In Rifkin's last book, The Third Industrial Revolution (2011), he tries to make the case that economics needs to be connected to climate change. And in order to do that, he argues that thermodynamics is the key as it was the basis for making neoclassical economics into a science. Rifkin describes then how thermodynamics explains the balance between inputs and outputs as that can apply to the environment and other externalities, which make up the Third Industrial Revolution (TIR).

While this is an attempt to explain economics as a science, and in particular to address the concerns for the environment and climate change, the arguments fall short. Four basic issues remain with economics which Rifkin and others fail to address. The primary one is the acceptance of Adam Smith's theoretical basis for economics being scientific. The theories of the seventeenth and eighteenth century are dated due to the use of science in a limited manner and rooted in this historical century knowledge of science. Since then, science has developed and expanded with new theories and extensive research.

Second is the focus on science in a limited manner. While the traditional link between Adam Smith and Sir Isaac Newton remains the basic barrier for revolutionizing economics, it is the wrong approach rooted in the wrong assumptions. What is wrong are the particular and limited aspects of science that Smith used from Newton. In other words, Smith and his economic paradigm was based on his creation of an ideal world that never existed then; nor does it today.

In short, science is much more than thermodynamics, and the balance between forces with what economics has come to label the "invisible hand". In science, there is no invisible hand. In order for something to be science (like economics should be), everything needs to be known and accounted for, repeated, and then predictable. Economics fail to do that. As Perkins puts it, "the vast majority of these physicists had in common a few essential things: they were honest; they actually made the observations that they recorded, and they published the results of their discovery of the "rules" in a form that others could duplicate and confirm. This is the basis of physics as a science" (op.cit. p. 5). That is not what economists do today or for more than half a century now. For most of them, they play the role of the invisible hand becoming visible with offering opinions, political conclusions, and even plans which are never based on repeated analyses, data, tests and predictions.

The problem with economics moves into the third issue directly. What "role model" is there for the field of economics to become a science? Economics is not a science. But there is one area of the "social sciences" that is scientific, linguistics. Above all, economics needs to be modeled on linguistics. In Qualitative Economics, Clark and Fast (2008) make this point. Linguistics, particularly through Noam Chomsky and his transformational grammar work over the last 50+ years, has made linguistics the model for science in an area of research that was once considered "social science."

See Chomsky, Reflections on Language (Pantheon Books 1975) and Syntactic Structures (The Hague: Mouton and Co. 1957), among other books and articles for how linguistics is a science that economics should follow. For example, Chomsky's first book in 1957 was first published in The Hague because no US or English language publisher wanted to print a book that talked about linguistics becoming a science. Yet Chomsky's arguments and work since then turned a corner for the scientific study of languages. The point is that Adam Smith and those who interrupt his work today are still using the scientific philosophy and knowledge from hundreds of years ago, while there are more recent breakthroughs in science within the last 50 years. Linguistics is the outstanding example.

Chomsky asks, "What is the 'science-forming capacity' that enables us to recognize certain proposed explanatory theories as intelligible and natural while rejecting or simply not considering a vast array of others that are no less compatible with evidence?" (Chomsky 1980: 250). Basically, science must describe, explain, and predict phenomena. Scientific statements are only valid if they can be replicated and proven through predictions. Modern linguistic theory led by Chomsky has been able to do just that in a nonphysical and natural science environment. The key is to extend the construction of scientific theory for languages beyond the descriptive phase and into an explanatory and predictive phase.

As described in Qualitative Economics (Clark and Fast 2008) from Chomsky:

> by way of example, considers a typical linguistic situation. A sentence (S) contains a noun phrase (NP) followed by a verb phrase (VP) or in symbols, represented as (NP VP ---> S), where: among the categories that figure in the categorical component are the 'lexical categories,' noun (N), verb (V), adjective (A), and others. (Chomsky, op. cit. 1980: 80)

1 Introduction

Therefore, the representation of the parts of the sentence is made into symbols that allows the surface structure (spoken language) to be broken down into components. An arrow denotes the horizontal and vertical transformations (---> or $_v$) from the deep structure (NP --> VP) into the surface structure sentence (S) and its "deep structure" or the meanings of the words, phrases, and sentences.

Today, Chomsky remains a leader in scientific thinking through linguistics, as it needs to apply to societal and economic issues (Chomsky 2012). Concerning "the environmental catastrophe, practically every country in the world is taking at least halting steps toward trying to do something about it. The United States is also taking steps, mainly to accelerate the threat. It is the only major country that is not only not doing something constructive to protect the environment…" (Chomsky 2012: 5).

Finally mathematics is the other scientific model for economics in both formulas and processes, including definitions of numbers, symbols, and their results. As Perkins (op. cit. 1996: 1) puts out, "An associated feature of this simplicity or beauty is that the 'rules' can be written down very elegantly in mathematical form. This can be a problem when we first learn physics because our mathematical skills are not usually sufficiently developed."

Consider the definition of mathematics in Wikipedia (2012):

> Mathematicians seek out patterns and formulate new conjectures. Mathematicians resolve the truth or falsity of conjectures by mathematical proof. The research required to solve mathematical problems can take years or even centuries of sustained inquiry. Since the pioneering work of Giuseppe Peano (1858–1932), David Hilbert (1862–1943), and others on axiomatic systems in the late 19th century, it has become customary to view mathematical research as establishing truth by rigorous deduction from appropriately chosen axioms and definitions. When those mathematical structures are good models of real phenomena, then mathematical reasoning often provides insight or predictions about nature.
>
> Through the use of abstraction and logical reasoning, mathematics developed from counting, calculation, measurement, and the systematic study of the shapes and motions of physical objects. Practical mathematics has been a human activity for as far back as written records exist. Rigorous arguments first appeared in Greek mathematics, most notably in Euclid's *Elements*. Mathematics developed at a relatively slow pace until the Renaissance, when mathematical innovations interacting with new scientific discoveries led to a rapid increase in the rate of mathematical discovery that has continued to the present day.

Part of this definition of mathematics includes some discussion of the "field of mathematics" and how it became scientific. "Mathematics can, broadly speaking, be subdivided into the study of quantity, structure, space, and change (i.e., arithmetic, algebra, geometry, and analysis). In addition to these main concerns, there are also subdivisions dedicated to exploring links from the heart of mathematics to other fields: to logic, to set theory (foundations), to the empirical mathematics of the various sciences (applied mathematics), and more recently to the rigorous study of uncertainty."

Perkins summarizes well the relationship between rules and mathematics (the core of implementing the scientific process) as applied to energy:

> However, knowing sufficient mathematics enables us to write the rules down very simply; and very importantly, allows us to calculate many useful things. For example, knowing the rather simple mathematics of electricity allows you to calculate the rating of the fuse

required in your hi-fi. An alternative is to pay an electrician $75 to do it for you! Again, knowing some of the mathematical expressions of nuclear physics permits us to calculate the electrical energy that can be generated from 1 kg of uranium. The answer is about the same as you would get from 1000 tons of coal! (ibid. p.3).

In this introduction to NE, I quote and cite each chapter to make the point about how economics can move from a field of study to become a science. The idea is to give the reader an overview of the main points in the chapters and the book itself. Hence, only the salient points and issues are covered that need more economic study. In particular, the goal is to stimulate further research and review of the field of economics. The time has come for just that. The purpose, in short, is to conclude what has become a pattern in economics, which leads to a global economic paradigm, if not total philosophical change in the field of economics itself. In short, The Next Economics helps make economics a science.

For example, recently an article was published (Simmons et al. 2011) on "False-Positive Psychology" about some research and a paper published by colleagues at an University in Netherlands that showed how research results were questionable but even more significantly were used as false data for a peer reviewed published paper. As Simmons and his colleagues put in their paper, "Our job as scientists is to discover truths about the world. We generate hypotheses, collect data, and examine whether or not the data are consistent with those hypotheses" (ibid., p. 1). While this is correct, science is also about replicating the hypotheses and thus providing predictions.

Thus, when economic history examines the struggles over energy, not only today in the Middle East with the USA now engaged in the longest and most costliest war in its history, but also to examine USA energy security in terms of other world wars in the past two centuries, the problems in terms of people, communities, and economics are staggering. There are no calculations or numbers that even come close to the results of these conflicts. What we can do, and have tried to provide some examples and cases, is ask questions and provide economic data that provides guidance and thought provoking debate about economics itself.

Now, today, economics needs to add societal issues such as health and environmental impacts into it numbers and formulas, not only locally but also internationally to understand and change the damages from climate change. The world is round. Hence, vast areas of land are damaged, the air is toxic, and water is polluted due to ocean and atmospheric changes from one region to another. Therefore, the numbers and calculations reported cannot even come close to being accurate. The economic costs must be comprehensive, based on hypothetical cases that are observed, examined, and redefined and tested in order to be scientific, hence, accurate and predictive.

What was done in the CEP special issue is a start to make economics a science. This book completes that task.

Economics is going through an enormous paradigm shift as it develops into a science from just being a "field of study" especially now with a focus on climate change that could help be illustrative of the entire field. In fact, applying some of the philosophical roots, analytical tools, formulas, and methodologies from the collection of these papers should provide a road way and map into the future of economics. Certainly, there will be twists and turns, and even a few dead ends and off ramps, but economics needs just that.

Without a doubt, economics is in a quandary while it tries to find its direction and even a new paradigm. What we have done here is to provide the basis for that discussion and examination into the field itself. So what is the new paradigm? Where is it going, and what will it do? We have a few initial thoughts and ideas about that and the challenges that go with it.

For example, Bailey and Wolfram (2012) in the Wall Street Journal, published an article about how energy has become a key part in addressing what communities and countries can do about climate change.

In the article, the concern is how renewable systems were only available to the rich due to their "hefty cost" which thus restricted and limited the "market for residential solar installations (for example) to cash-rich homeowners, restricting the potential for growth" (ibid 2012). Yet this "innovation," like many others, needs to be short term and tightly controlled.

The basic problem is that leasing contracts are stranded costs in that they must be paid back over a long period of time, with usually an additional purchase price. And the financing is higher due to third parties providing the financing at a price above the actual purchase costs to the person leasing the energy system. What is more useful to do is to think and consider other finance models, for example, the use of mortgages for including energy (renewable power, storage, refueling, gird models, etc.) just as water, waste, electric along with air conditioning, and heating are done today within the costs for most buildings. Such financial innovations are now being included in current laws under considered in California. Now, the costs for renewable energy are another asset valued in a building that are bought and sold with that building.

One of the significant factors in looking at climate change is the role of the continuing role of the UN and all nations to find resolutions. The fact that yearly meetings are being held to find answers to climate change had not resulted in any significant global actions. Part of the reason concerns the nations involved. Another key part are the politicians and to whom they report when they are back home. But in the end, it is economics. For example, the UN IPPC (Intergovernmental Panel on Climate Change) has been debating if climate change is a result of humankind or the natural evolution of our planet. In 2007, the UN IPPC stated clearly the problem was humankind. However, in that debate, was the underlying conflict and theme rooted in economics: Can humankind adapt (that is, live with) or mitigate (that is, stop) climate change? In short, what are the basic economics and costs associated with one or the other of these areas?

Now let us move into the core issues surrounding climate change: economics and the people who control or oversee groups, companies, and countries who benefit either through adaptation or mitigation or both. That core basis for debate within economics leads to the chapters. Consider, however, the important philosophical perspective above which is about how economics sets the forum for a dramatic shift in economics which would lead the field of economics to becoming a science.

According to Adam Smith, the founder and most well-known creator of modern economic theory who is often noted as being founder of "classical economics", the use of taxes for both controlling consumption and considered financial resources can be "Sugar, rum and tobacco are commodities which are nowhere necessaries of life, [but] which are ... objects of almost universal consumption and which are therefore extremely proper subjects of taxation" (Toedtman 2012). The application of this basic neoclassical principle that is never mentioned in the modern economic theory over the last four decades is applicable to the economics of climate change.

The first chapters therefore discuss the "basic problems in the field of economics" as it impacts externalities such as climate change, environment, and national security. In that regard, one of the fundamental problems rests with how to analyze economics and the environment. Dole points out several economic approaches that are meant to solve environment problems. He concludes after reviewing cap and trade economics which has become popular with environmental policy makers, numerous NGOs, and businesses, does not work. Part of the issues facing President Obama with his attempt to create a national energy program was over cap and trade economics. Dole notes in "market solutions for climate change" (CC) that command and control (CAC) works in many ways. CAC is fairly successful in Nordic countries and China where government oversight and even control of economics are accepted and successful.

However, as Dole points out, CC is questionable in the USA, primarily for political reasons. After reviewing market solutions to CC and especially the current focus in California on cap and trade (CAT), Dole argues that the only economic way to mitigate climate change is through a carbon tax. Adam Smith would agree with that, but today's political interruption of his economic paradigm disagrees with that conclusion. Hence, in the classical economic paradigm, laws need to be created and managed which make individuals and companies pay for their negative impact on the environment through pollution and emissions. The issue of economics becoming a science is not solved, but the classical economic paradigm of Adam Smith would agree with that tax policy.

Then Li and Clark look into what these new approaches to economics might actually do on the microlevel, but also how to characterize the new emerging economic paradigm through national planning such as "energy economics in China." Both authors have worked in China and find that their way to characterize all these concerns over the environment and economics can be seen and described as "social capitalism." Below are some social issue areas (climate change and the environment) and economic applications for financing renewable energy systems in order to make them affordable. The key issue is how can renewable energy, one of the more significant, concrete ways to solve and slow climate change, be economically viable for people in their homes, work, and public service roles. In short, to mitigate climate change, affordable technologies and systems must be available today and not just in 10–20 years.

The second section of articles focuses on the "Next Economics" starting with an overview of what this means for moving economics into a science through different historical and philosophical perspectives that apply to the societal issues of environment, energy, and climate change. The other chapters in this section discuss issues and ideas in this area, but there are missing pieces. The critical one is the health costs to people with climate change impacting everyone's daily life.

Then as Clark presents in the "Next Economics" chapter, there is a Green Industrial Revolution already starting around the world. Here, the difference in the new economic paradigm is that more than economic numbers must be placed the equation. One of the key concepts concerns communities becoming sustainable. More will be said about sustainable communities below (Clark 2009, 2010). But at this point, the key issue is that communities of every kind are configured through infrastructures. Today, most economic analyses are focused on one area or another, like energy, transportation, waste and water. Few economic analyses study the overall integration of the entire community infrastructure systems. Even more significant are the needs of areas not usually considered in economics, like the environment, climate change, and health issues.

Awareness of these "externalities" to economics has become significant with the reports from the UN IPPC and debates among nations over the Kyoto Accords. What has become obvious in all of these discussions about the environment is that the western nations are dependent on neoclassical economic theories. However, these economic theories fail to be able to take into account the perspectives of all western nations and certainly leave significant economic scientific questions in others.

Fast and Clark next present a new philosophical approach to economics altogether. In their book, Qualitative Economics (QE) 2008, Clark and Fast provided a challenge to economics from the basic ground level: philosophical roots to western economics. In that book and this chapter, the fact is that there are other western philosophical theories, some of which are close to Asian theories and philosophers, that challenge the roots of neoclassical economics. However, moving into the modern twenty-first century, the legacy of the twentieth century economic theories is questionable. The subtitle of QE, originally drafted and written in the late 1990s and revised in the mid-2000s, is "toward a science of economics," exactly the same concern of the Economist special issue (July 2009) with the Bible melting on the cover, about "modern economic theory" failing by it documenting significant concerns over the global economic collapse nine months earlier.

If indeed economics is not a science and perhaps an art or social study, then how can it become a science? As the QE points out and documents at length, science is a not only a matter of numbers and formulas. Instead, science of any kind concerns creating and testing hypotheses, one after another. The use of numbers and statistics is not as significant as the mathematics and philosophical basis on which the hypotheses and data are centered. Hence, the science of economics does not exist.

Also if that is the case, how can economics become a science? This is the challenge. What QE did was look at various sciences and social "sciences" to which this chapter also cites as there being only one, linguistics. The creation and growth of modern linguistics can be directly attributed to Noam Chomsky at MIT. His work for almost half a century has made linguistics a science – one in which the field of economics needs to research and learn from. The chapter presents some of those structural forms of linguistics that make it a science with some case studies about these applications from linguistics to create the science of economics. Clark and Lund (2006) looked at economics in the context of sustainable global communities and then Clark (2007) writes how a new economics can be focused on the environment and energy to emerging Asia nations and China in particular (Clark and Isherwood 2007, 2009 and 2010).

Each of these chapters and many more now, as well as those in the three sections of the book, covers important economic topics in climate change. What they all conclude is that economics cannot be narrow and hence limited. Science is certainly not that way. The focus on sustainable communities (Clark 2009, 2010) provides a basis for cases and baseline data from which further hypotheses can be made, data gathered, and concrete scientific economic data derived, tested and replicated. The result is that sustainable communities are one basis for the economic understanding of climate change. If human kind is to resolve climate change and go beyond adaptation to it, then there must be hard scientific economic ways to mitigate and stop climate change.

Critical to understanding energy are its economics. As Matteson in this chapter takes the topic on a regional level, using California as a case, he then moves into the national needs for the USA to produce a national energy plan. His data provides a basis for prediction. And the results are concerning as they provide comparative data on the impacts of energy production to the environment. Professionally, Matteson spent most of his career managing the energy system for the University of California, Berkeley, and then the entire University of California system with its ten campuses. He was able to see what the energy demands were for the campuses and then for the state of California.

What is particularly concerning are those energy needs and the sources for the energy as the shift from coal (plants and power transmitted from long distances in other states, since California has no coal resources) to natural gas, now unfortunately the predominant energy resource, to the current growth of renewable energy power from solar and wind. However, there are significant barriers to the consumption of renewable energy sources due to the control of the power being transmitted primarily through central energy producing systems controlled by a few large power utilities.

While Matteson had been an advocate of energy deregulation power generation, he saw what happened to California in the early part of the twenty-first century. The solution for the future, from his data, was not just centralized power generation from a small group of utilities, either as government or private businesses, but on-site power for buildings, residents, complexes like shopping malls, offices, and academic institutions. The near and long-term future energy systems need to be planned for this shift from central power plants to local power generation. This was the basic conclusion for agile energy systems (Clark and Bradshaw, 2004) about the need to have a combination of both current central power plants and on-site power systems.

Grose then outlines and discusses in detail the chapter on "Achieving Economic Gains"; the "core green economy" needs to be adaptive and functional within the rest of the economy. The chapter reviews conventional economic issues (15 in all) as segments to the "core green economy" ranging from efficiency, transportation, and infrastructures to water, waste, and agriculture to advance materials and buildings but not including renewable energy generation. The analogy to the IT industry for the green economy is similar to Clark (2011) Rifkin (2011). And some of the ideas including the need for public policy specifically refer to the case of California.

But the key issue in her chapter is the link also to achieving these economic gains with "The Setting of Environmental Goals." The case of California in its energy crisis at the turn of the millennium is important. Providing public policies on conservation and efficiency along with setting an RPS (Renewable Energy Portfolio Standard) and Clean Car Act among others is important. However, public policy(s) needs more as the last decade has demonstrated: financial and monetary support. In short, the policies need money in order to implement them. This does not have to be public finances and/or incentives, but actions that correct the problem, as California had, in the energy sector, such as the public policy for "deregulation." However, de-regulation caused a major economic crisis in California that continues today and now the USA and other nations (Clark and Bradshaw 2004).

The issue of adaptation is the key. Does it do enough? When Rifkin uses entropy in that way to relate economics from science to the environment and climate change, it too does not go far enough. While the UN debates, even today, what to do about climate change, the basis problem is between those nations who want to adapt to it against others who argue that climate change must stop or mitigated and be reversed. The latter view of global climate and environmental perspectives sets the stage of what must be done since vast regions of the world are continuing to experience severe atmospheric and ocean changes.

Adapting to these changes needs to be done for survival, but mitigating and stopping them are the only real solution. This conflict in philosophy is the same as the consideration of economics today in areas that are new. However, if the more comprehensive and global perspective is taken in economics, then economics can become a science. It must be scientific to get results, replicate them, and construct rules that are measurable and evaluated again and again and again. This science of economics brings into it other critical areas and the actual solutions to climate change. That is, what can people, governments, and businesses must do about it. There is no "invisible hand" that applies to science or the real world.

As Clark and Li provide in their chapter on "social capitalism," there is a need to think of basic economics in a new and different way. Social capitalism is part of that way of thinking and how to make economics scientific. In other words making money is fine within economics, but economics still needs to connect to society and social issues like the environment. Interesting enough, the Economist appears to be interested in these more historical models of economics as well. In a special issue on state capitalism (Economist, January 12 2012), the Economist talked about how emerging nations were combining government and market economics in order to construct and build their economics.

While the Economist did not agree with that economic model and alto it had been done historically in all western nations, the conclusion was that it worked in those nations today, known as BRIC (Brazil, Russia, India, and China). In fact, the Economist added Chile, Argentina, and some eastern European nations as well. Each of these nations is rapidly becoming powerful economic powers that challenge the past 100 years of both the EU and North America domination.

China has become the most significant and outstanding case in point, as Clark and Li discuss China, but also as Clark and Cooke (2011) provide a global perspective on energy innovation with China "leapfrogging" into the Green Industrial Revolution" while the USA lags behind. The Chinese 5-year plans are examples of what they do as a nation implementing social capitalism. In short, over 1.3 billion people have a national and then local plan to use as a roadmap. In modern twenty-first century China, those 5-year plans are indispensible for government, businesses and international policy and corporate leaders as along with a plan, they provide billions of US dollar equivalent in funds, finance and investment for the entire country.

So what are the economics that need to be examined?

Many communities, cities, and other public organizations such as academic institutions along with the private sector business recognize the need for policies that direct their facilities and infrastructures to be "green" based upon some criteria. For example, the US Green Building Council (USGBC) certification for achieving LEED (Leadership in Energy, Environment, and Design) provides basic criteria to higher standards.

Individual buildings are to be "net-zero" carbon emissions. Secondly, organizations are seeking to make their entire facilities "Energy Independent and Carbon Neutral." Since June 2007, the USGBC has created "community" or LEED Neighborhood standards. This set of criteria reflects the broader concerns for clusters of buildings with designs that are integrated with basic infrastructure needs.

Developing dense, compact, walk-able communities that enable a range of transportation choices leads to reduced energy consumption. "Communities" thus have a broad definition because they can range from college campuses to cities, towns, and villages that are self-sustaining and provide for multiple uses ranging from housing, education, family events, and religion to business complexes, shopping streets, malls, and recreational activities. Thirdly, a sustainable smart community is a vibrant, "experiential" applied model that should catalyze and stimulate entrepreneurial activities, education, and creative learning, along with research, commercialization, and new businesses.

Communities can be sustainable in that new economic and social programs can be created or recreated. However, the difference today is the need for sustainability to go beyond conservation and efficiency to include renewable energy resources such as wind, solar, biomass, ocean, geothermal, and "run of the river" (not large hydroelectric dams or nuclear energy plants) energy sources that do not impact the environment negatively but instead are green and clean. Therein lies a "paradigm change" to an agile energy system (2), which combines local renewable power and fuel resources with grid connected ones.

Because of global concerns, many nations and now regions, states, communities, and cities have developed their own policies to increase renewable energy power generation as part of the solution to respond to the threat of climate change. Since the primary infrastructure sectors that impact global warming are energy and transportation, they must be examined in order to find ways to reverse the warming of the earth. One key element in achieving such goals is to consider how renewable energy can impact and change the transportation sector to be more environmentally sound and sustainable. Several different technologies have been put forward, but in practice, no single technology can solve the problem on its own. Many different contributions have to be combined and leveraged to coordinate with parallel activities in the energy sector. The Pew Charitable Trusts reported that since 2005 (by 2009) there was a 230% increase in clean energy investments (Lillian 2010, p. 4).

On the local or regional level, sustainable and smart communities must have three components: 1) need for a master strategic plan for infrastructure that includes energy, transportation, water, waste, and telecommunications, along with the traditional dimensions of research, curricula, outreach, and assessments; 2) array of issues pertaining to the design, architecture, and sitting of buildings and overall facility master planning; and 3) an perspective of "green," energy, efficient orientation, and be designed for multiple-use by the academic and local community.

The California energy crisis was the tip of a much larger problem in California, as well as in the USA and other industrialized nations. While private companies took over much of the state's energy generation capacity, related and similar issues confront California and other nation-states, such as infrastructures for water, waste, and transportation that are separate but interconnected sectors (op cit. Clark and Bradshaw 2004). The point is that the private sector is not interested in the public good unless it makes them money. Hence these privatized or de-regulated sectors remain in a crisis mode but are ignored and unattended since there have been few visible crises in them. However, the impact of hurricanes and storms as a result of climate change on communities in the southern part of the USA and globally has begun to make the public more aware and ready to take constructive actions, starting with the need to conserve and use energy efficiently.

Even though California is one of the few states that uses the least amount of electricity per capita, the dominant use is from fossil fuels (about 58% from coal and natural gas) that negatively impacts the environment and pollutes the atmosphere. It is far a broader topic of concern than energy deregulation. The simple policy of deregulation is just the micro tip of an incredibly complex series of issues about global warming, waste, and misuse of natural resources, etc. all within the common economic concerns of companies and government.

A part of the solution which came from the California energy crisis (2000–2003) is the creation of what I have called "agile energy systems" in which communities have clusters of buildings, like colleges, local governments, residential divisions, shopping malls, and office buildings that have their own "on-site power generation systems" (ibid 2004). There is still the "central gird" that often depends heavily on fossil fuels like oil, gas, and coal as well as nuclear to generate central power, however on-site power best comes from renewable energy sources, including solar, biomass, wind, and other sources (5). But the agile system policy, which has become reality throughout California today, is to have a combination of local on-site energy generation (e.g., solar systems, combined heat and power, use of biomass, and other renewable sources for energy), along with the central-grid power generation that also needs to move rapidly to renewable energy power systems too.

For example, some clusters of buildings, like colleges, office buildings and shopping malls, use solar power during the day but on non-sunny days and at night-time, the central grid becomes the power source like a battery back up system for the community. A key component to buildings today is their design (such as LEED standards) so that they are environmentally sound. The design and construction of buildings and clusters must be addressed as they shift from a centralized to a decentralized or a combination of energy production.

The place to start is with small, relatively self-contained communities or villages within larger cities and regions. The issue is to get communities off their dependency on central-grid connected energy since most of these power generation sources come from fossil fuels like coal, natural gas, and nuclear power. Local on-site power can be more efficiently used and based on the region's renewable energy resources such as wind, solar, biomass, among others. This model is now being realized in Denmark where many communities are generating power with wind and biomass, combined to provide base load. Denmark has a goal of 50% renewable energy generation (primarily from on-site and local resources) by 2020 (Clark 2009).

The country is well on its way to meeting and perhaps exceeding that national goal (Lund 2009).

For example, the EU has committed EURO two billion with a matching EURO three billion over the next 5 years as a "challenge" to have the public and private sectors become partners. These "civic markets" must collaborate in developing sustainable and smart communities. Sustainable communities include the facilities, land, and the infrastructure sectors that intersect, such as energy, water, IT, waste, and environment, which must have public sector involvement and oversight to set goals, policies, and programs, and provide finance.

The basic issue always is money and finance = economics. Creating sustainable communities of any kind means finding the funds to pay for the technologies, systems, and operations. These areas are skill sets that must be included in the development of any sustainable community (Clark 2010). While articles in this book address various levels and approaches also to the economics of sustainable technologies, they fall into some methods set by industry to be paid for from products like renewable energy. Solar panels are a good example.

In the USA, for example, two pathways have been set to improve energy efficiency since the residential sector is 36% electricity consumption. One is that the government creates demand-side management (DSM) to (1) increase adoption of efficient appliances (e.g., Energy Star program) and (2) building practices (e.g., Leadership in Energy Environmental Design or LEED programs). In 2005, the US Congress placed an emphasis on more efficient lighting which provided a base to the use of such as products such as LED (light emitting diode) lights.

Second is to reduce load on the grid through better monitoring and feedback from meters. Smart grid etc. (2010) monitor and control energy use which reduces and conserves energy use, thus reducing the load demands. In 2007, the US Congress passed and established the official American national policy for the smart grid. In California, enacted a three later (ibid., 2010) a law that ordered the state's investor owned utilities to develop Smart Grid Plans by the summer of 2011. However, now these plans are barely being implemented.

The new economic paradigm follows the critical pathway to the reduction, and efficient use of energy as it impacts personal behavior. The old paradigm did not connect individual behaviors with energy use. Today with smart grids and their impact on personal behavior well documented, there is a need to connect the energy savings from meters, and other energy efficient resources with utilities and policy makers who "will need better information so that these savings" will have more beneficial financial return for the overall economics of central-grid energy demand.

In section three on practical applications of economics to climate change problems, Hoexter discusses the feed-in-tariff (FiT). This approach to financing renewable energy, especially large systems, has worked well in Germany. Now Spain, Japan, Canada, and other nations have created FiT plans as well. The USA is behind. Some cities in the USA have them and California adopted a mild form of the FiT in the fall of 2009. But in general, the USA is far behind in part because such programs have been the most successful for large arrays of solar concentrated systems and wind farms. These large systems transmit power and hence threaten the power generation from energy utility companies. Smaller economic programs and hence renewable energy systems for homes, buildings, and communities such as colleges, resorts, shopping areas, and retirement complexes are needed.

This book also covers some new economic studies in the EU. Italy became a good place to report on wind energy in particular as there are many wind systems being installed today with more advanced and detailed information. N. Cardinale, Rospi, Cotrufo, and T. Cardinale present a study of wind energy in Matera City for its Materana Murgia Park in southern Italy between Basilicata and Puglia as the "economic-environmental performance in the Mediterranean area." The authors review the history of wind turbines which dates back to sailing in 2500 BC and windmills first built during the seventh century in Persia and coming to Europe 500 years later. However, it was not until windmills moved from the Mediterranean to the northern European regions that the technology expanded and grew.

Wind turbines have now spread around the world, and Denmark has become the recognized leader in the technology, industry, and environmental impact. Lund and Ostergaard (2009) provide an extremely good case of a town in Denmark (Frederikshavn) using 100% renewable energy by 2015. With only a few years to go, they are over half way there. Wind is one of the core renewable technologies accounting for over two-thirds of the power generated. As N. Cardinale et al. put it, "the wind energy represents, among renewable sources, the one with the highest potential of use, as it is an absolutely free resource exploitable by using a simple turbine without the high cost of installation."

In Italy, there was a 40% increase in wind power from 2008 to 2009 generating over 4,898 MW of power. Southern Italy represented 88% of the national installed capacity. In the study of four wind turbines for the Park in Matera, the authors found that in a 20-year period of time, the life cycle of the wind turbines is best suited for "micro-wind turbines" since the wind speeds are not as strong as those needed for standard large wind turbines. The micro wind turbines with horizontal axis and a 6-kW power have a cost that is "close to thermal power plants and nuclear third-generation power plants with the (same) environmental benefits." Such results for regions and areas that need power systems which can use renewables like wind (and solar) are good for the environment and far less costly, especially when integrated, than other power systems such as fossil fuel and nuclear power plants.

The Chapter by Prey concerns what government systems can do to lead and implement energy conservation and reduction. In California, for example, the Caltrans (California State Transportation Authority) has the responsibility for buildings and roadways. Prey has discussed the research in LED lights since 1991 and then the extensive use and results of LED lights for streets lights and bridges over a decade ago (1998) that set the specs for the LED manufacturing industry (National Product Specifications) and the standard for conservation and efficiency throughout California and today, the entire USA.

In the early part of the twenty-first century, "the energy/carbon footprint for the statewide owned/maintained traffic signal upgrade exceeded 93% reduction from the incandescent baseline." For Caltrans alone that meant a 13-MW grid load was eliminated very day. And the return on investment was within 3 years. This leads to California, other states and finally the US Congress in 2005 to pass the National Energy Act banning incandescent traffic lamps throughout the USA. By 2011–2012, the California went even further and "moved the development and discussion phase to that of a deployment phase."

In this same regard, Prey argues that states, like California, can often take the initiative in terms of innovations and program implementation for other environmentally sounds and economical technologies such as HVAC (air conditioning) systems from high electricity based to thermal-based absorption ones that use solar thermal pre-temp to reduce the need for carbon-based fuels. The same can be done, argues Prey, with hydrogen for transportation by using ammonia compounds to transport hydrogen for use in energy centers and vehicles. In short, the numbers work for even a positive (short-term) return on investment (ROI) and for reducing carbon and other emissions.

Nuckols in his chapter then emphasizes the need for coalitions both within the USA and internationally as the need to address environmental, land, and economic issues has expanded greatly. He notes, for example, the Apollo Alliance founded in 2003, which has evolved into the blue-green partnership, that brings together traditional labor unions and environment groups. From that years later, a blue-green alliance emerged with blue-collar labor unions and the green environmentalists actively working together. By 2011, both groups merged to "continue to partner in battling for jobs in the green economy."

One area of focus as Nuckols points out for these and related alliances is agriculture. One leading concern today is the economics of agriculture due to increased competition. Yet the concern for the environment and pollution has become more dominant and costly. While some of the changes are needed in particular for large agribusinesses, it seems that the environmental concerns and even costs are borne by the smaller ones through alliances in order "to generate cooperative dialogue that leads to a reduction in the exposure of toxic elements, both in global communities and particular worksites."

The international social movements have also been playing a significant role since the concern over ethical issues and relationships are predominant. "These activists feel strongly that a healthy community and workplace sets the stage for resilient job creation and sustained economic growth." Schumacher, as Nuckols points out, was "divergent" and only solved "higher forces of wisdom, love, compassion, understanding, and empathy." In short, a higher level of concern must be converted into actions. Science, as Nuckols concludes, in our world today cannot be significant on its own because "science on its own can give no reasons for sustaining humankind."

The last chapter is from Nijaki who talks about a "green economy as sustainable" but that goes beyond traditional concepts of development and growth. In short, she argues that "aggregate economic growth alone (measured as GDP) in terms of measures of productivity, skills, and wealth may be an oversimplification of true economic development." Basically, she defines growth as both a distribution issue of the benefits and costs of growth and also the currently more extensive definition to increase the quality of life. This chapter covers a broader definition of economics that helps to frame "green" growth.

Nijaki reviews conventional and standard economic theory, especially as it applies to development and growth. Some Economists even took the classical economic theory a bit further to argue the need for technological innovation and change to stimulate and push development and growth, but expanded the concept into new products, human capital, and production methods. In the last four decades, the emphasis that placed demand-side economics at the center of all growth was replaced by Milton Friedman and his supply-side economics. Based in neoclassical economics, Friedman's economic assumptions from Adam Smith still holds the control of economic teaching, research, publication, and even public policy through Prime Minister Thatcher and President Reagan about the same time in history 30 years ago. However, Thomas Friedman then pushed the three phases of globalization, which morphed from his seeing a world that was flat in to a round world a few years ago due to climate changes in the atmosphere, oceans, and even land masses.

Thus, "economic growth may sometimes be in conflict with measures that protect equity considerations and quality of life goods." But in the end, the "green economy aims to widen the view of economic growth or progress through an integration of environmental considerations in the development process. It reframes growth as "green growth" and thus limits development by taking into account quality of life considerations that are hinged on environmental quality today and into the future. In this way, the metrics for evaluating development choices and their successes are changed to one that seeks to reference the long-run environmental effects of economic action and inaction."

The key is how "green economy" is defined, measured, and scientific. Companies today are seeking to be labeled "green" while they and consumers are looking for green products and services. Naijaki provides a framework for green growth as part of the Next Economics covering four areas in some detail: (1) diversity in sectors, (2) practitioners versus producers, (3) regulation centric, and (4) small and startups. In the end, not all growth and jobs will be green, which she attributes to local and regional government policy makers.

Government involvement is only part of the need in economics for (2005) it to be scientific and concerned with societal issues. And in fact, government is really a major part of the solution. The supply-side economics argues for "market forces" and related neoclassical economic theories with an invisible hand, called government. This approach to economics is not scientific since there factors that create hypotheses, observations, data, and repeated experiments in order to set rules for future measurement, analyses, and evaluation. Government needs to be present for the objective oversight of the economics of science much like a physicist would be in and out of the laboratory.

The concluding book chapter, presents an overview on how economics can become a science with the case of a pending patent for the economics of energy conservation and efficiency through the comprehensive installation of LED lights. This is just the beginning of the Next Economics. There is a lot to do. Join us.

References

Bailey EM, Wolfram C (2012) A whole different kind of innovation. The Journal Report: innovations in energy, Wall Street Journal. www.wsj.com. Accessed 18 June 2012

Chomsky N (1957) Syntactic structures. Mouton, The Hague

Chomsky N (1975) Reflections on language. Pantheon Books, New York

Chomsky N (1980) Rules and representations. Columbia University Press, New York

Chomsky N (2012) Plutonomy and the precariat: on the history of the US economy in decline, pp 1–5. TomDispatch.com. Accessed 8 May 2012

Clark WW II, Bradshaw T (2004) Agile energy systems: global lessons from the California energy crisis. Elsevier, London

Clark WW II, Cooke G (2011) Global energy innovation. Praeger Press, Global Energy Innovation. Preager Press/ New York, NY

Clark WW II, Fast M (2008) Qualitative economics: toward a science of economics. Coxmoor Press, Oxford, UK

Clark WW II, Isherwood W (2007) Energy infrastructure for inner Mongolia autonomous region: five nation comparative case studies. Asian Development Bank/PRC National Government, Manila/Beijing

Clark WW II (2007) Eco-efficient energy infrastructure initiative paradigm. UNESCAP, Economic Social Council, Asia. Bangkok, Thailand

Clark WW II (2008) The green hydrogen paradigm shift: energy generation for stations to vehicles. Utility Pol J, Elsevier Press, New York, NY

Clark WW II (2011) The Third Industrial Revolution. In: Scott M, James H, George B. Sustainable business practices: challenges, opportunities, and practices. ABC-CLIO, New York, NY, pp 263–278

Clark WW II, Lund H, Co-Editors (2006) Special issue on sustainable development: the economics of energy and environmental production. J Clean Prod

Clark WW II, Editor and Author (2009) Sustainable communities, Springer Press

Clark WW II, Editor and Author (2010) Sustainable communities design handbook, Elsevier Press, New York, NY

Clark WW II, Isherwood W (2009) Report on energy strategies for inner Mongolia autonomous region, Utilities Pol. doi:10.1016/j.jup.2007.07.003 In: Clark WW II, Isherwood W (Authors and Co-Editors). Special Issue of Utility Pol J: China: environmental and energy sustainable development, Winter 2010

Friedman, Thomas. The World is Flat. Farrar, Straus & Giroux, 2005

Jin, J (2010) Transformational relationship of renewable energies and the smart grid. In: Clark WW II (Editor and Author) Sustainable communities design handbook. Elsevier Press, New York, NY, pp 217–232

Lillian J (2010) "New & Noteworthy", Sun Dial, Solar Industry, May 2010, pp 3–4

Lund H, Ostergaard PA (2009) Sustainable towns: the case of Frederikshavn – 100% renewable energy. In: Clark WW II (Editor and Author) Sustainable communities. Springer Press, New York, pp 155–168

Perkins LJ, Senior Physicist (1996) What is physics and why is it a 'science'? Lecture at University of California Physics Seminar, University of California, Berkeley, pp 1–3

Rifkin J (1980) Entrophy. Palgrave Macmillan, New York

Rifkin J (2011) The third industrial revolution: how lateral power is transforming energy, the economy and the world. Palgrave Macmillan, New York

Simmons JP, Nelson LD, Simonsohn U (2011) False-positive psychology: undisclosed flexibility in data collection and analysis allows presenting anything as significant. Association of Psychological Sciences, Sage, pp 1–8

Toedtman J (2012), Editor's letter, Tax attack: is it time for a sugar tax? AARP Bulletin, 30 May 2012, website reference

Wikipedia. Mathematics. en.wikipedia.org/wiki/Mathematics, July 2012

Chapter 2
The Next Economics

Woodrow W. Clark II

Abstract This chapter explains what The Next Economics is about with some specific examples and cases that are expanded upon in other chapters by other authors. The focus is primarily upon the green industrial revolution (GIR) which is the topic of another book that Clark and Cooke discuss in their book, Global Energy Innovation (Praeger Press 2011) and will be a book itself due out in 2013. Certainly, there is also a blue industrial revolution (BIR) as one of the chapters in this book illustrates. The point of the GIR (and BIR) requires new way of thinking about a economic paradigm. Clark (2013) discusses some of that in an article that is part of a special issue for the Contemporary Economic Policy journal.

Below in this chapter, the basic areas and countries where The Next Economics has been done successfully are referenced with some examples. The case that stands out the most is China which appears to be addressing economic reform moving from the extremes of Communism and Capitalism to a new paradigm while focused on social issues ranging from the environment, climate change, pollution and carbon emissions to health and medical care, aging population, and the continued growth of communities in order to make them sustainable in terms of strong environmental and emissions standards. This chapter sets the stage for other chapters related to a new economic paradigm called "social capitalism."

Keywords Green Industrial Revolution • Sustainability • Social capitalism

Clark thanks Jon McCarthy for both economic details and edits as well as Lucas Adams, Kentaro Funaki, Claus Habermeier and Russell Vare for international perspectives and data. Namrita, Singh and Jerry Ji checked on data contributions and review of this chapter which is derived from one published in CEP, Winter 2013

W.W. Clark II, Ph.D. (✉)
Qualitative Economist, Academic Specialist, Cross-Disciplinary Scholars in Science and Technology, UCLA and Managing Director, Clark Strategic Partners, California, USA
e-mail: www.clarkstrategicpartners.net

Introduction

A Green Industrial Revolution (GIR) or Blue-Green Industrial Revolution of renewable energy, smart green sustainable communities, water and waste along with advanced technologies has started in China and taken the USA by surprise. The EU, South Korea, and Japan had started a GIR over two decades ago (Clark and Cooke 2011). The GIR is the significant paradigm change from the fossil fuels and nuclear power plants of the Second Industrial Revolution (2IR), which has dominated global economics since the late 1890s, to renewable energy in the late 1990s and growing at an extraordinarily rapid rate in the twenty-first century. While the USA had invented and even began to commercialize many of the technologies developed into mass markets by the EU and Japan, it failed in the last two decades to move ahead of corporate interests, while at the same time recognizing the growing importance of climate change for the future (Chomsky 2012).

Consider China which has twelve 5-year plans and is ready to start its thirteenth in 2014. Each plan provides clear and formulated policies, with budgets, to address national, international, environmental issues and their solutions. China has "leapfrogged" into the GIR in order to avoid the mistakes of the western developed nations in a variety of infrastructure areas (Clark and Isherwood 2008 and 2010). Also the USA must look comprehensively into the corporate and political reactions to the 2011 Japanese tsunami and ensuing nuclear power plant explosions in Fukishima, as well as the 2010 BP oil spill in the Gulf of Mexico off Louisiana. The USA and other countries cannot ignore the environmental consequences and economic costs of the 2IR that have handicapped it moving into the GIR. The end result is not good for the American people, let alone the rest of the world.

The deregulation of industries starting in the Reagan and Thatcher eras was a mistake and a completely naïve view of reality from the neoclassical economics of Adam Smith. There has never been a society or area in the world in which the principles of capitalism have been proven to work in reality. Instead just the opposite has been the reality. Chomsky (2012) looks at the history of economics in far more concrete manner. Even the economist in two special issues labels modern economics as "state capitalism" (January 23 2012) and another, soon after that, as the Third Industrial Revolution (April 2012), a theme from Jeremy Rifkin (2004) and his book with that title in 2012. Clark has published several articles and given numerous talks about the Third Industrial Revolution or 3IR (2008, 2009, 2010, and 2011) but prefers to think of it as the Green Industrial Revolution or GIR (Clark and Cooke 2011). Basically, the GIR concerns renewable energy, smart green communities, and advanced technologies that produce, store, and transmit energy for infrastructures while saving the environment.

The point is that the development of the USA into a powerful world leader had a lot to do with its military strength, but also its economic development for over a century in the Second Industrial Revolution (2IR) in which fossil fuels, combustion engines, and related technologies including atom bomb and nuclear power dominated (Chomsky 2012). The growth of the USA started over a century ago with businesses and their owners who control today the economy. There was little or no competition. But even more significant is that the basis for this wealth is in fossil fuels and continues to be there. Hence, the environment is continuing to be damaged in order to produce more and oil and natural gas causing climate change. But this 2IR retards and places the USA back decades when compared to emerging economics and even some other western developed nations.

As historians have documented the development of the 2IR in the USA, this too was primary based on "state capitalism" since oil companies got land grants, funding, and even trains, transmission and pipelines for transporting their fossil fuels. That governmental support continues today. Consider the issue of the USA getting shale oil from Alberta, Canada and the massive pipelines installed throughout the USA to get the oil processed and distributed. Furthermore, these same companies get tax breaks and credits such that their economic responsibility to the USA is minimal. The argument that America will be "energy independent" with these fossil fuels is false. The USA needs to stop getting its energy from fossil fuels anywhere in the world, including domestically or from its neighbors.

Hence, the argument is that China will buy oil from Canada. Basically, Canada (and the USA) should not even extract oil from the ground, which permanently destroys thousands of acres of land, making them impossible to repair or restore. There are far more and better resources from renewable energy like sun, wind, geothermal, run of the river, and ocean or wave power to provide energy for central power and on-site demands.

Introduction and Background

A Green Industrial Revolution (GIR) emerged at the end of the twentieth century due, in large part, to the end of the Cold War that dominated the globe since the end of World War II. The Second Industrial Revolution (2IR) had dominated the twentieth century because it was primarily based on fossil fuels and technologies that used primarily mechanical and combustion technologies. On the other hand, the GIR is one of the renewable energy powers and fuel systems and smart "green" sustainable communities that use more wireless, virtual communications and advanced storage devices like fuel cells (Clark and Cooke 2011). The GIR is a major philosophical paradigm change in both thinking and implementation of environmentally sound technologies that requires a new and different approach to economics (Clark 2011).

The USA lived in denial about the world "being round" during the 1970s and then again since the early 1990s, which became apparent for both Democrat and Republican Presidential Administrations in their lack of proactive polices globally through the Kyoto Accords and most recently the UN Intergovernmental Panel on Climate Change (UNIPPC) Conference in Kopenhaven (December 2009) and Cancun (2010). On the other hand, in the early 1990s, economic changes in Europe and Asia were made due to the end of the Cold War to meet the new global economy. The Asian and EU conversions from military and defense programs to peacetime business activities were much smoother than that of the USA. Environmental economist Jeremy Rifkin recognized this change and developed the concept of the "Third Industrial Revolution" in his book, The European Dream (2004). According to Rifkin the 3IR took place a decade earlier in some EU countries. He did not recognize that Japan and South Korea had been in a GIR even decades before that (Clark and Li 2004).

At the same time, Clark and Rifkin et al. (2006) published a paper on the "Green Hydrogen Economy" that made the distinction between "clean" and "green" technologies when related to hydrogen and other energy sources. The former was often used to describe fossil fuels in an environmentally friendly manner, such as "natural gas" and "clean coal." Green, on the other hand, means specifically renewable sources such as the sun, wind, water, wave, and ocean power. In short, the paper drew a dividing line between what technologies were part of the 2IR (i.e., clean technologies such as clean coal and natural gas) and the GIR (solar, wind, ocean, and wave power as well as geothermal). The GIR focused on climate change and replacing the technologies and fuels that caused it; or at least mitigate and stop the negative pollution and emission problems that impacted the earth.

Clark and Fast (2008) in founding the science of "qualitative economics" made the point about economics that definitions are needed to define ideas, numbers, words, symbols, and even sentences. Therefore, due to the misuse of "clean" to mean really fossil fuels and technologies clean technologies were not good for the environment. Tickell's documentary film, Fuel (Tickell 2009), made these points too, as it told the history about how "clean" was used to describe fossil fuels like natural gas in order to placate and actually deceive the public, politicians, and decision makers. For example, Henry Ford was a farmer and used biofuels in his cars until the early 1920s, when the oil and gas industries forced him to change to fossil fuels.

Hawkins et al. (1999) refer to the environmental changes as the beginning of "the Next Industrial Revolution." This observation only touched the surface of what the world is facing in the context of climate change. And the irony is that China has already "leapfrogged" and moved ahead of the USA into the GIR (Clark and Isherwood 2008 and 2010). While China leads the USA now in energy demand and CO_2 emissions, it also is one of the leading nations with new environmental programs, money to pay for them and their installation of advanced infrastructures from water to high-speed rail systems.

These economic changes came first from Japan, South Korea, and the northern EU nations. Rebuilding after WWII from the total destruction of both Asia and Europe meant an opportunity to develop and recreate businesses, industries, and the commercialization of new technologies. The historical key in Japan and then later in the EU was get off dependency on fossil fuels for industrial development, production, and transportation. For Japan, as an island nation, this was a critical transformation for them when in the mid-nineteenth century with the American "Black Ships" demanding that Japan open itself to international, especially American, trade. However, as recent events testify, Japan made the mistake of bending to the political and corporate pressures of the USA to install nuclear power plants despite the atomic bombings of two of its major cities in WWII. The final results of tragedies from the 9.0 earthquake in 2011 are not final yet in terms of the nuclear power plants in Fukushima and its global impact on the environment, let alone in Japan and the immediate region of northern Asia.

Soon after the end of the Cold War in the early 1990s, the GIR become dominant in Japan Nordic countries and spread rapidly to South Korea as well as Taiwan and somewhat to India. China came later when it leapfrogged into the twenty-first century through the GIR. Germany, Japan, and S. Korea took the lead in producing vehicles that required less amounts of fossil fuels and were more environmentally "friendly," often called "clean tech." by mistake but due to pressures from the oil and gas industries. Hence, their industrial development of cars, high-tech appliances, and consumer goods dominated global markets.

America ignored the fledging technological and economic efforts in the EU, South Korea, and Japan as the nation tilted into a long period of self-absorption, bubble-driven economic vitality driven by the false economic premises of the western real estate and financial markets. The nation had a history of cheap fossil fuels primarily from inside the USA and given high tax breaks and incentives (op. cit. Tickell 2009). The 2IR also had survived WWII successfully. Furthermore, the end of Cold War meant to Americans that their 2IR was to dominate and in control of global economic markets. The Soviet Union had failed to challenge them. Then came 9/11 and its aftermath along with the longest continuous war in American history as well as the battle with fundamental Islamic terrorists. With its own unique and fractured political debate and power struggles, America labored to make sense of a post-Cold War era where special interests replaced reason and any movement toward a sound domestic economic policy.

Instead, the American ideological belief in a "market economy," entrenched in the late 1960s to mid-1970s, replaced the historical reality of how government and industry must collaborate and work together. The evidence of the problems and hardships from "market forces" came initially from a convergence of events in the early part of the twenty-first century, including a global energy crisis, the dot.com collapse, and terrorist attacks. Spending and leveraging money into the market caused the global economic collapse almost a decade later in October 2008.

The Economist even characterized the basic economic problem the best when in mid-2009, a special issue was published under the title "Modern Economic Theory," superimposed on the Bible melting (Economist 18–24 July 2009). Basically the case was made that economics is "not a science" in large part because its theories and resultant data "did not predict" the global economic recession that started in the fall of 2008 and continues today. From that special issue of the Economist in the summer of 2009, an international debate about conventional modern economics started and continues today.

The Green Industrial Revolution impacts America in a completely different perspective and rational at the local level than at the regional, state, or national levels. Infrastructures of energy, water, waste, transportation, and IT among others and how they are integrated are the core to the GIR (op.cit. Clark and Cooke 2011). These infrastructure systems need to be compatible yet integrated with one another. For example, renewable energy power generation must be used in homes, businesses, hospitals, and nonprofit organizations (government, education, and others) that are metered and monitored as "smart on-site grids" and also used for the energy in vehicles, mass train, and buses among other transportation infrastructures (Knakmuhs 2011). Such "agile energy" or "flexible systems" (Clark and Bradshaw 2004) allow people to generate their own power while also being connected to a central power grid. However, both the local power and central power in the GIR need to be generated from renewable energy sources, with stand by and back up storage capacity.

There are five key basic elements for the Green Industrial Revolution: (1) energy efficiency and conservation; (2) renewable power generation systems; (3) smart grid connected sustainable communities; (4) advanced technologies like fuel cells, flywheels, and high-speed rail; and (5) education, training, and certification of professionals and programs. First, communities and individuals all need to conserve and be efficient in the use of energy as well as other natural resources like land, water, oceans, and the atmosphere. Second, renewable energy generated from wind, sun, ocean waves, geothermal, water, and biowaste must be the top priority for power on-site and also central plants.

The third element is the need for smart green girds on the local and regional levels in which both the monitoring and control of energy that can be done in real time. Meters need to establish base load use so that conservation can be done (systems put on hold or turned off if not used) and then renewable energy power is generated when demand is needed. The fourth element needs to be advanced storage technologies such as fuel cells, flywheels, regenerative brakes, and ultra-capacitors. These devices can store energy from renewable sources, like wind and solar that produce electricity intermittently, unlike the constant supply of carbon-based fuel sources. Finally, the fifth element is education and training for a workforce, entrepreneurial, and business sector that is growing and provides employment opportunities in the GIR.

In general, the GIR must provide support and systems for smart and "green" communities so that homes, businesses, government, and large offices and shopping areas can all monitor their use of the natural resources like energy and water. For example, communities need devices that capture unused water and that can transform waste into energy so that they can send any excess power that is generated to other homes or neighbors. Best cases from around the world of sustainable communities that follow these elements of the GIR exist today (Clark 2009, 2010).

Essentially the GIR was started by governments who were concerned about the current and near-future societal impact of businesses and industries in their countries. The EU and Asian nations in particular have had long cultural and historical concerns over environmental issues. The Nordic nations and Singapore, for example, have started Eco-Cities as well as reuse of waste for more than three decades. Sweden, Denmark, and Norway either have all eliminated dependency on fossil fuels now for power generation or will be in the near future. All but Finland are shutting down nuclear power plants for their supply of energy as well. The same since the 1980s has been true in most other EU nations, except France which is over 75% dependent on nuclear power.

However, the key factor in the EU and Asia have been their respective government leadership in terms of public policy and economics. Consumer costs for oil and gas consumption are at least four times that of the USA due to the higher taxes (or elimination of tax benefits) to oil and gas companies in these other nations. The EU has implemented such a policy for two decades, which has also motivated people to ride more in trains and take mass transit or ride their bikes rather use their individual cars. The USA on the other continues to subsidize fossil fuels and nuclear power though tax incentives and government grants. Not so in the EU and Asia. The impact of fossil fuels on climate change was the basis for changing these policies and financial structures over two decades ago. Today in the USA, unlike northern EU, the impact on the environment has become severe, and thus, it is even more significant for future generations around the world.

The historical difference has been the American contemporary economic ideology of market forces to simply have a balance of supply and demand so that these market force of businesses can thrive and prevail. This neoclassical economic model has failed for many reasons, especially due to one of the two key issues presented in the economist special issue (July 2009) that points out that economics is not a science. This is important for a number of reasons, but the basic one, which pertains to the GIR, is that contemporary economics does not apply to major industrial changes, such as the GIR, let alone the beginning of the 2IR Clark (2013). For most economists to be confronted with a challenge to their field not being a science is disturbing. The "dismal science" may be boring with its statistics, but is not a science at all, since it fails to question the entire contemporary field of economics and its future.

The debate is over how does a community or nation change? Economics is one of the key factors. The issue is, are "market forces" the key economic change factor? The 2IR discovered that market forces or businesses by themselves could not get fossil fuels and other sources of energy into the economy at reasonable economic costs. It took time, government support, and policies that provided the market with capital and incentives. Additionally, a GIR economic paradgm includes economic externalities such as the environmental and health costs.

In short, the "market force" neoclassical paradigm represented American economic policies (and also the UK) for over the last four decades when Prime Minister Thatcher and then President Reagan were the embodiment and champions of this economic paradigm derived from Adam Smith (Clark and Fast 2008). Market force economics had some influence on the EU and Asia but then demonstrated its failure in October 2008 with the global economic collapse that started in the USA on Wall Street. That failure meant some of the government programs in the EU and Asia, which had succeeded, now needed to be given more economic attention because they basically differed greatly from the USA and UK economic models.

These other nations have been in the GIR themselves for several decades, which succeeded and continued to do so with a different economic model. Northern EU, Japan, South Korea, and China are clear documented examples of a different economic model. For example, a key economic government program representing the GIR in the EU is the Feed-in-Tariff (FiT), which started in Germany during the early 1990s and was successfully taking route in Italy, Spain, and Canada as well as nations in the EU and Asia. While there are economic problems in Spain and Italy, Germany has decided to cut it back, the USA has not started a FiT in any significant, long-term planned policy programs on a national, let alone a state level. Some American communities and states have started very restrictive and modest FiT programs.

European Union Policies

Germany jumped out in the lead of the GIR in the EU with its FiT legislation in1990. Basically, the FiT is an incentive economic and financial structure to encourage the adoption of renewable energy through government legislation. The FiT policy obligates regional or national electricity utilities to buy renewable electricity at above-market rates. Successful models like that exist such as the EU tax on fuels and the California cigarette tax, both of which. The smoking tax cut smoking dramatically in California and the gas tax forced people to use mass transit and trains rather than drive their cars as much in the EU. But also provide incentives and metering mechanisms to sell excess power generated back to the power grid. Other EU nations, especially Spain, followed, and the policy is slowly being developed in Canada and some US states and cities. Chart 2.1 shows the economic impact of the FiTs. Over 250,000 "green" jobs created in Germany alone. The graphs in Chart 2.1 also show the growth in Germany of the solar and wind industries and how this expansion is becoming global.

Germany was the world's leading producer of solar systems until China took over in 2012 because it has more solar systems installed than any other nation based on the creation of world leading solar manufacturing companies, solar units sold and installed are measured by sales, amount of kilowatts per site and records keep

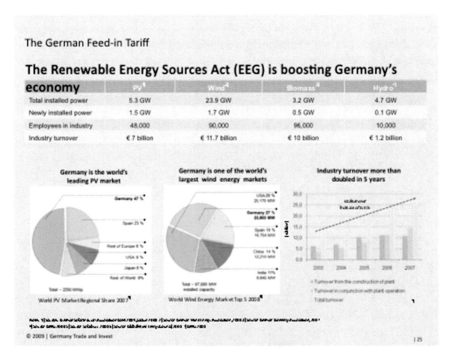

Chart 2.1 The Germany feed-in-tariff policy and results (1990–2010)

by the local and national governments (Gipe 2011). The extensive use of solar by Germany is despite the fact that the nation has many cloudy and rainy days along with significant snow in the winter is common to northern Europe. Japan implemented in 2010 a similar aggressive FiT system in order to stimulate its renewable energy sector and regain renewable energy technological (solar and system companies and installations) leadership that held in the early part of the *twenty-first century*. Technical and economic measurements were kept by the solar companies as well as local and national governments. Then MITI, the Japanese national research organization measures the use of renewable energy systems on a quarterly basis. However, the aftermath of the Japanese earthquake and destruction of the nuclear power plants in 2011 could actually expedite renewable energy growth and installation through a number of government programs and incentives that are being proposed.

Other European countries have similar GIR programs as well. Denmark, for example, will be generating 100% of its energy from renewable power sources by 2050. While trying to meet that goal, the country has created new industries, educational programs, and therefore careers. One good example of where the FiT policy has accomplished dramatic results is the city of Frederikshavn in the Northern Jutland region of Denmark. The city has 45% renewable energy power now, and by 2015, it will have 100% power from renewable energy sources (Lund 2009). In terms of corporate development in the renewable energy sector, for example, one Danish company Vestas is now the world's leading wind power turbine manufacturer with partner companies all over the world. Vestas was able to achieve that recognition for a number of reasons including FiT and its partnership and joint ventures in China since the early 1990s. Vestas continues to introduce improved third-generation turbines that are lighter, stronger, and more efficient and reliable. They also continue to design new systems, like those that can be installed offshore away from impacted urban areas.

Germany, Spain, Finland, France, UK, Luxembourg, Norway, Denmark, and Sweden are on track to achieve their renewable energy generation goals. Italy is fast approaching the same goals when in 2010, it took the distinction as having the most MW of solar installed from Germany. However, Denmark is one of the most aggressive countries due to its seeking 100% renewable energy power generation by 2050. Already Denmark has a goal of 50% renewable energy generation by 2015 (Clark 2009, 2010). Other EU countries are lagging behind, especially in Central and Eastern Europe. The EU has required all its member nations to implement programs like those in Western EU in order to be energy independent from getting oil and gas, especially now since most of these supplies come from North Africa, the Middle East, and Russia.

Various EU nations have widely different starting positions in terms of resource availability and energy policy stipulations. France, for example, is a stronger supporter of nuclear energy. Finland, recently, has installed a nuclear power plant due to its desire to be less dependent on natural gas from Russia. However, Sweden is shutting down its nuclear power plants. The UK and the Netherlands have offshore gas deposits, although with reduced output predictions. In Germany, lignite offers a competitive foundation for base-load power generation, although hard coal from

The German PV Market

The PV market in Germany boomed in 2004 following amendment of EEG

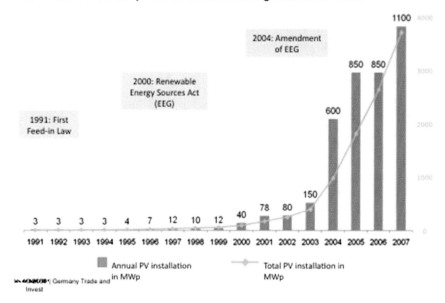

Chart 2.2 The Germany feed-in-tariff policy economic results (1990–2007)

German deposits is not internationally competitive. In Austria, hydropower is the dominating energy source for generating power, though expansion is limited.

Other EU directives toward energy efficiency improvement and greenhouse gas emission reductions also impact electricity generation demand. Many EU members have taken additional measures to limit GHG emissions at the national level. Since the EU-15 is likely to miss its pledged reduction target without the inclusion of additional tools, the European Parliament and the Council enacted a system for trading GHG emission allowances in the community under the terms of Directive 2003/87/EC dated 13 October 2003. CO_2 emissions trading started in January 2005 but have not produced the desired results due to the limitations of "cap and trade" economic measures and the use of auctions over credits given for climate reduction.

After being established for three years, by 2007, the results are not good, however, as the economics and "markets" are not performing as predicted. Basically the carbon exchanges have performed poorly and not as promised to either buyer or seller of carbon credits (or other exchange mechanisms). The initial issues are emission caps not tight enough with a lack of significant EU or local government oversight (EU 2009). By 2010, many of the exchanges have closed or combined with others. The problem is often cited as the lack of supporting governmental (EU or by nation) policies, but the real issue is that economics does not work as well as the control over carbon emissions. The trading and auction mechanisms furthermore do not provide direct and measurable solutions to the problem of emissions and its impact on climate change. A far more direct finance and economic mechanism as proposed by several EU nations and China would be to have a "carbon tax."

An important lesson from the FiT policies in Germany came from the two decades of the policies from 1990 through to 2007. As Chart 2.2 shows, Germany learned that a moderate or small FiT was not sufficient to push renewable energy systems like solar into the main stream of its economy. In short, a far more aggressive use of the FiT type of financing and/or direct carbon taxes need to be made.

On its own, the solar industry would not move fast enough into the GIR. In many ways, this is the lesson for other nations. In fact, the reality of the 2IR historically has been to have strong and continuous government incentives from the late nineteenth century to the present day. The definition and model of economics as a market remains critical in understanding how the USA can move into the GIR. Consider now how Japan and South Korea did just that: moved into the GIR with strong government leadership and financial support.

Japan and South Korea Are Leaders in the Green Industrial Revolution

While it took an extraordinary political transition to prompt Europe to open the door to the Green Industrial Revolution (GIR), Japan and South Korea in particular have taken a completely different path. And now China is moving aggressively ahead in to the GIR. Most of the information and data below will be focused on China (Clark 2009). For example, China led the USA and the other G-20 nations in 2009 for annual "clean energy investments and finance, according to a new study by The Pew Charitable Trusts" (Lillian 2010: 4):

> Living in a country with limited natural resources and high population density, the people of Japan had to work on sustainability throughout their history as a matter of necessity. With arable land scarce – some 70-80% of the land is mountainous or forested and thus unsuitable for agricultural or residential use – people clustered in the habitable areas, and farmers had to make each acre as productive as possible. The concept of "no waste" was developed early on; as a particularly telling, literal example, the lack of large livestock meant each bit of human waste in a village had to be recycled for use as fertilizer.
>
> Along with creating this general need for conservation, living in close proximity to others inspired a culture in which individuals take special care in the effect their actions have on both the surrounding people and environment. As such, a desire for harmony with others went hand in hand with a traditional desire for harmony with nature. Nature came to be thought of as sacred and to come into contact with nature was to experience the divine. Centuries-old customs of cherry blossom or moon viewing attest to the special place nature has traditionally held in the Japanese heart.

In April 2011, China became the world leader of financial investment in "clean tech" with $54 billion invested which was over $10 billion from second place Germany and almost double third place, USA (San Jose 2011:8). Wind was the favorite sector of renewable energy with $79 billion invested globally. This article noted in particular a comment by a senior partner in a venture capital firm, "a lot of the clean technologies are dependent on policy and government support to scale up. In some other parts of the world (not USA), you have more consistency in the way these types of funds are appropriated" (op cit.:8).

The Japanese have had a long cultural and business history in commercializing environmental technologies. The 2011 earthquake made Japan focus back on that historical tradition. The future has yet to become clear and will not be defined for some months and years ahead. However, in Japan, the environment took a backseat to industrial development during the drive toward modernization and economic development that began in the latter half of the nineteenth century. After nearly 300 years of self-imposed isolation from the world, Japan was determined to catch up to the industrialized West; in a fraction of that time, Europe and the USA made their transitions, eventually emerging Japan as a great power in the beginning of the twentieth century.

Economic development continued unabated until World War II, when its capacity was destroyed by American bombings. Economic growth restarted again in the postwar period at a rapid pace but with a distinctive orientation and concern for the limited nature resources of the island nation. By the 1970s, on the strength of its industry and manufacturing capabilities, Japan had attained its present status as an economic powerhouse. Companies like TOTO (concerned with bathroom water and waste conservation and technologies) along with the Japanese auto makers concerned with atmospheric pollution, emerged as global leaders. A large part of that success was the need for the government to invest in research and development organizations (e.g., METI) to support companies and business growth, what would now be called the GIR. For example, high-speed rail was started in Japan in the mid-1980s, and expanded. Such transportation systems were economically efficient along with being environmentally sound at reasonable rates.

While this incredibly successful period of development left many parts of the country wealthy, it also resulted in serious environmental problems. In addition, an oil crisis had hit Japan particularly hard because of its lack of natural resources, making it difficult for the industrial and manufacturing sectors to keep working at full capacity. To respond to the effects of pollution, municipalities began working in earnest on ways to reduce emissions and clean up the environment, while Japanese industry responded to the oil crisis by pushing for an increase in energy efficiency.

At the same time, Japan's economy was evolving more toward information processing and high technology, which held the promise of further increases in energy efficiency. Japan had created new innovative management "team" systems that were copied in the USA and the EU. Many manufacturing firms saw value in establishing plants in other developed countries in part to create a market for their products, employ local workers, and establish firm and solid roots. For example, Toyota and Honda established their Western Hemisphere Headquarters in Torrance, California. Other high-tech companies established large operations throughout the USA. In this way, Japanese government, industry, and academia have worked collaboratively with local and regional communities to reincorporate traditional Japanese values about conservation and respect for the environment in order to create sustainable lifestyles compatible with modern living.

Community-level government efforts in Japan, supported by national government initiatives, have led to unique advancements in energy efficiency and sustainable lifestyles, including novel ways of preventing and eliminating pollution. Japan is responsible for some 4% of global CO_2 emissions from fuel combustion, and through this is the lowest percentage among the major industrialized nations. Carbon is still what Japan intends to reduce, with a long-term goal of reducing emissions by 60–80% by 2050. With the majority of energy coming from coal, Japan also is attempting a major shift toward renewable energy.

As of November 2008, residential-use solar power generation systems have been put in place in about 380,000 homes in Japan. A close examination of the data on shipments domestically in Japan shows that 80–90% are intended for residential use. Thus shipments are likely to increase, as the government aims to have solar panel equipment installed in more than 70% of newly built houses by 2020 to meet its long-term goals for reductions in emissions. Current goals for solar power generation in Japan are to increase its use tenfold by 2020 and fortyfold by 2030. Furthermore large proposed subsidies for the installation of solar – 9 billion yen or $99.6 million total in the first quarter 2009 – along with tax breaks for consumers, will continue the acceleration of solar adoption by Japanese households.

In recent years, Europe, China, Southeast Asia, and Taiwan saw tremendous growth in energy generation almost entirely from solar power installations. However, these have mostly involved large-scale solar concentrated power facilities that do not fit for individual households. In Japan, however, as solar power generation systems for residential use become increasingly commonplace, they have become smaller thin film for creating sustainable communities through use on roofs of local homes and businesses.

The same is true with the LED light bulbs. Today, LED bulbs may cost a few pennies more, but they last far longer than a regular light bulb and can be recycled without issues of mercury and other waste contamination. The result is better lighting for homes and offices with significantly less costs in terms of the systems, demand and the environment. Some LED bulbs are guaranteed to last from 6 to 8 years (Nularis 2011). While energy demands in homes and offices continue to rise due to the internet, computers, and video systems, the installation of energy efficient and now cost-saving systems is very much in demand. Some states are even requiring by law to change from the less efficient light bulbs to the newer LED ones.

Distributed Renewable Energy Generation for Sustainable Communities

Adding more complications to the EU, Japan, and S. Korea's policy decisions are the reality of an aging grid and under capacity. The EU must crank up investment in this new generation. Estimates are coming that indicate to meet demand in the next 25 years, they will need to generate half as much electricity as they are now generating. According to the International Energy Outlook 2010, conducted by the U.S. Energy Information Administration (USEIA 2011), the world's total consumption of energy will increase by 49% from 2007 to 2035. This could result in a profound change in the EU's power generation portfolio, with options under consideration for new plants including nuclear energy, coal, natural gas and renewables.

Originally, when nations electrified their cities and built large-scale electrical grids, the systems were designed to transmit from a few large-scale power plants. However, these systems are inefficient for smaller scale distributed power from renewable sources (Clark 2006). Although some systems will allow for individual households to either buy power or sell power back to the grid, the redistribution of power from numerous small-scale sources are not yet managed well economically (Sullivan and Schellenberg 2011). As Isherwood et al. (1998 and then in 2000) document in the studies of remote villages, renewable energy for central power can meet and even exceed the entire demand for a village, hence making it energy independent and not needing to import any fossil or other kind of fuels. This model and program has worked in remote villages, but can also be applied to island nations and even larger urban communities or their smaller districts.

The grid of the future has to be "smart" and flexible and based on the principles of sustainable development (Clark 2009). As the Brundtland Report said in 1987 "as a minimum, sustainable development must not endanger the natural systems that support life on Earth: the atmosphere, the waters, the soils and the living beings" (Bruntland 1987: Introduction). With that definition in mind, a number of communities sought to become sustainable over the last three decades.

Integrated "agile" (flexible) strategies applied to infrastructures are needed for creating and implementing "on-site" power systems in all urban areas that often contain systems in common with small rural systems (Clark and Bradshaw 2004). The difference in scale and size of central power plants (the utility size for thousands of customers) with on-site or distributed power can be seen in the economic costs to produce and sell energy. Historically, the larger systems could produce power and sell it far less than the local power generated locally for buildings. Those economic factors have changed in the last decade (Xing and Clark 2009). Now on-site power particularly from renewable energy power (e.g., solar, wind, geothermal, and biomass) has become far more competitive and is often better for the environment. Large-scale wind farms and solar concentrated systems are costly and lose their efficiency due to transmission of power over long distances (Martinot and Droege 2011).

Developing World Leaders in Energy Development and Sustainable Technologies

Some of the major benefits of the Green Industrial Revolution are job creation, education, and new business ventures (Clark and Cooke 2011). Considerable evidence of these benefits (Next 10 2011) can be seen in the EU, especially Germany and Spain (Rifkin 2004). Many studies in the USA have documented how the shift to renewable energy requires basic labor skills and also a more educated workforce, but one that is also locally based and where businesses stay for the long term. This is a typical business model for almost any kind of business and is what has motivated EU universities to create "science parks" which take the intellectual capital from a local university and build new businesses nearby the campus (Clark 2003a, b).

Asia's shift to renewable energy will require extensive retraining. Consider the case of wind power generation in China. In the early 1990s, Vestas saw Asia and China as the new emerging big market. Vestas agreed to China's "social capitalist" business model (Clark and Li 2004; Clark and Jensen 2002), where the central government sets a national plan, provides financing, and gives companies direction for business projects over 5-year time frames, which are then repeated and updated. National plans like business plans are critical to any company, group or family, especially when set and followed by national and regional governments.

A major part of the Chinese economic model required that foreign businesses be co-located in China with at least a 51% Chinese ownership. This meant that in the late twentieth and early twenty-first centuries, the Chinese government owned companies or were the majority owners of the new spin-off government owned ventures, which established international companies or businesses started in China. Additionally, China required that the "profits" or money made by the new ventures be kept in China for reinvestments.

Additionally, the results, such as with the renewable energy companies like wind and solar industries, were that all the ancillary supporting businesses also needed to support the companies from mechanics, software, plumbing and electricity to installation, repair and maintenance, and other areas. Supporting industries were also needed such as law, economics, accounting, and planning, especially since the Chinese government began to create sustainable communities that required all these skill sets (Clark 2009, 2010). Hence, these businesses grew and became located in China.

However, the Chinese social capitalism model is not rigid with the government owning controlling percentage of a company. Many businesses were started by the Chinese government with its holding from 25–33% of shares, while the other firms were owned by the former government employees, until the companies went public (Li and Clark 2009). Yet in almost all cases, the companies are competitive globally and are performing remarkably well as demonstrated again in the renewable energy sector, where in early 2011, SunTech, a Chinese based publically traded company, became the world's largest manufacturer and seller of solar panels (Chan 2011). According to a press release by the company in February 2011, it has delivered more than 13 million PV panels to customers in more than 80 countries.

Today, China is the world leader in wind energy production and manufacturing with over 3,000 MW installed in China alone (Vestas 2011). The Chinese are now following a similar business model in the solar industry (Martinot et al. 2007–2010). As such, China and Inner Mongolia (IMAR) has contracted Vestas to install 50 MW for IMAR (op. cit. Vestas 2011), according to a report from the Asian Development Bank (Clark and Isherwood 2008 and 2010) which argues for targeted needs to:

- Create international collaborations between universities and industry.
- Conduct research and development of renewable energy technologies.

- Build and operate science parks to commercialize new technologies into businesses.
- Provide and promote international exchanges and partnerships in public education, government, and private sector businesses.

The end results for the EU are smart homes and communities. The Green Industrial Revolution starts in the home so that energy efficiency and conservation are a significant part of everyone's daily life. The home is the place to start. But it is also the place to start with the other elements of the GIR: renewable energy generation, storage devices, smart green grids for communities, and new fuel sources for homes and transportations.

Costs, Finances, and ROI

Government policy(s) and finance are critical for economic growth especially concerning the environment and climate change. The basis of the GIR in the EU, South Korea, and Japan can be seen in their articulation of a vision and financial programs. Most of these countries also had established government energy plans. China in fact has had national plans since the PRC was established in 1949. Having a plan is in fact the basic program and purpose of most business educational programs. Governments need to have plans, as most businesses do. Business plans are for themselves and their clients. Yet the USA continues without any national energy or environment plans. Most American states do not have them either, while an increasing number of cities and communities are developing them in order to plans for becoming sustainable.

This lack of planning has both long-term and short-term impacts. The finance of new energy technologies and systems (like any new technology) is often dependent on government leadership through programs in public policy and finance (Clark and Lund 2001). Fossil fuel energy systems in the 2IR have been funded and supported by the governments of western nations through tax reductions and rebates that continue today. For the GIR, it is only logical and equitable that such economic and financial support continues. That means the American national government should provide competitive long-term tax incentives, grants, and purchase orders for renewable energy sources rather than just fossil fuels.

Meanwhile, the EU, South Korea, and Japan took the leadership in planning, finance, and creation of renewable energy companies, while other nations including the USA did not (Li and Clark 2009). For example, because of the national policy on energy demand and use, Japan has one of the lowest energy consumption measurements in the developed world. This has been made possible by its continued investment in long-term energy conservation while developing renewable sources of energy and companies that make these products. Japan's per capita energy consumption is 172.2 million Btu versus 341.8 million Btu in U.S.A.

One critical of a long-term economic plan is the need for life-cycle analysis (LCA) versus cost-benefit analysis (CBA). While these two very different accounting processes are not discussed much in this chapter, elsewhere Clark and Sowell (2002) cover the topic in-depth as the systems apply to government spending. Each approach is critical in how businesses learn what their cash flow is and their return on investment (ROI). The CBA model only provides for 2–3-year ROI since that is what most companies (public or government) require for quarterly and annual reports. However, for new technologies (like renewable energy, but also even wireless and WIFI technologies), more than a few years are needed on the ROI. The same was true in the 2IR when oil and gas were first discovered and sold. Now in the GIR, economic and financial ROIs are needed.

LCA covers longer time periods, such as 3–6 years, and within renewable energy systems, some as long as 10–20 years, depending on the product and/or service. Furthermore, LCA includes externalities such as environment, health, and climate change factors, all of which have financial and economic information associated with them. The point is that cost-benefit analyses are limited. The basic concept is that the LCA consists of one long-term finance model in the USA today for solar systems; it is called a Power Purchase Agreement (PPA) that contracts with the solar installer or manufacturer for 20–30 years. PPA is a financial arrangement between the user "host customer" of solar energy and a third party developer, owner, and operator of the photovoltaic system (Clark 2010).

The customer purchases the solar energy generated by the contractor's system at or below the retail electric rate from the owner, who in turn along with the investor receives federal and state tax benefits for which the system is eligible on an annual basis. These LCA financial agreements can range from 6 months to 25 years and hence allow for a longer ROI. However, there are other ways to finance new technologies especially if they are installed on homes, office and apartment buildings. Today financial institutions and investors can see a ROI that is attractive when the solar system on a home, for example, is financed as a lease, part of tax on the home, or included in the mortgage itself like plumbing, lighting and air-conditioning are today.

What is interesting are some newer economic ideas on how to finance technologies that reduce "global climate change." One way to describe the GIR financial mechanisms is by looking at the analytical economic models that financed the 2IR. For example, the 2IR was based upon the theory of abundance. The misunderstand assumption was that the earth had abundant water and ability to treat waste, hence buildings, businesses, homes and shopping complexes all had plumbing for fresh water and drainage for waste. The same scenario occurred in electrical systems that took power from a central grid for use in the local community buildings. Locally and globally, people have found that systems work, but now especially with climate change there is the need to conserve resources and be more efficient.

When these economic considerations are factored into even the CBA rather than a LCA financial methodology, the numbers do not work (Sullivan and Schellenberg 2011). The financial consideration for energy transmission and then monitored by smart systems are needed, but costly. Long distances make them even more costly because the then impact of the climate (storms, tornadoes, floods, etc.) with required operation and maintenance is added today with security factors. The actual "smart" grid at the local level is where these and other uncontrolled costs can be eliminated and monitored.

The financing of water, waste, electrical, and other systems for buildings was over time incorporated into the basic mortgage for that building. In short, the 2IR infrastructure systems were no longer outside (e.g., the outhouse or water faucet) but inside the building. What this 2IR financial model does is set the stage for the GIR financial model. Much of the 2IR financing for fossil fuels and their technologies came about as leases or building mortgages. A variation of the 2IR model which is a bridge to the GIR is the PACE (Property Assessed Clean Energy) program started in 2008 in Berkeley, California, whereby home owners can install solar systems on their buildings, for example, and then pay for them from a long-term supplemental city tax that is on their property taxes. The financing is secured with a lien on the property taxes, which acquires a priority lien over existing mortgages. Thus program was put on hold in July 2010 when the Federal Housing Finance Agency (FHFA) expressed concerns about the regulatory challenge and risk posed by the priority lien established by PACE loans. Nevertheless, the US Department of Energy continues to support PACE.

The dramatic change to the GIR, however, moves past that financial barrier of a property tax. Mortgages are part of the long-term cost for owning a property. Therefore, in the GIR, the conservation and efficiency for the 2IR technologies in buildings can be enhanced with the renewable energy power, smart green grids, storage devices, and other technologies through mortgages that can be financed from one owner to another over decades (20–30 years or more). This sustainable finance mortgage model is long term or a LCA framework and provides for technologies and installation costs to the consumer that makes the GIR attainable with a short time. Changes, updated and new technologies, can easily be substituted and replaced the earlier ones. What needs to happen is that the banking and lending industries try this GIR finance model on selected areas. After some case studies, the financial model can be replicated or changed as needed.

Conclusions and Future Research Recommendations

The basic point of this chapter is to highlight the need for economics to be more scientific in its hypothesis and data collection. Furthermore, the economics of the 2IR and the GIR are very similar, if not parallel. That is, for example, the role of government since it must often take the first steps in directing, creating, and financing technologies. As the 2IR needed government to help drill for oil and gas as well as mine for coal, the government needed to build rail and road transportation systems to transport the fuels from one place to another.

The GIR is very much in the same economic situation. The evidence can be seen in Asia and the EU. And especially now in China, the central government plans for environment and related technologies help a nation move into the GIR. Moreover there is a strong need for financial support that is not tax breaks or incentives, but investments, grants and purchasing for GIR technologies, such as renewable energy. This can be seen in the USA today with the debate over smart grids. What are they?

And who pays for them? When the smart grid is defined as a utility, then the government must pay for them since they are part of the transmission of energy, for example, over long distances that must be secure and dependable.

But as the GIR moves much more into local on-site power, the costs of the smart grid are at homes, office buildings, schools and colleges, shopping malls, and entertainment centers. Local governments are also involved as they are often one of the largest consumers of energy in any region and hence emitters of carbon and pollution. Within any building, a smart grid must know when to regulate and control meters and measurement of power usage and conservation. The consumer needs the new advanced technologies, but the government must support these additional costs and their use of energy as they impact the local community and larger regions residential and business needs.

Economics has changed in the GIR. And yet, economics has a basis of success in the 2IR. Historically, 2IR economics was successful because the government was needed to support its technologies along with goods and services. The evolution into the neoclassical form of economics was far more a political strategy backed by companies who wanted control of infrastructure sectors. But the reality was that "greed" took over and has now forced a rethinking of economics as nations now move into the GIR.

References

Bruntland Commission Report (1987) Our common future. UN Commission General Assembly Resolution #38/161 for Process of preparation of the environmental perspective to the year 2000 and beyond, 1983. Oxford University Press

Chan S, President of SunTech (2011) Global solar industry prospects in 2011. In: SolarTech conference, Santa Clara

Chomsky N (2012) Plutonomy and the precariat: on the history of the US economy in decline. Tompatch.com, 8 May 2012, pp 1–5. http://truth-out.org/news/item/8986-plutonomy-and-the-precariat-on-the-history-of-the-us-economy-in-decline76

Clark WW II (2013) The economics of the green industrial revolution: energy independence through renewable energy and sustainable communities. Special issue, Contempy Econc Pol Jl (CEP), Western Economic Association International, Fall 2013

Clark WW II, Cooke G (2011) Global energy innovations. Praeger Press, New York, NY, Fall 2011

Clark WW II (2011) The third industrial revolution. In: Scott M, James H George B, Sustainable business practices: challenges, opportunities, and practices, ABC-CLIO, New York, NY, pp 263–278

Clark WW II (Author and Editor) (2010) Sustainable communities design handbook. Elsevier Press, New York

Clark WW II, Isherwood W (2010) Inner Mongolia Autonomous (IMAR) Region Report, Asian Development Bank, 2008 and reprinted and expanded on in Utility Pol J. Special issue. Environmental and Energy Sustainable Development, China

Clark WW II (Author and Editor) (2009) Sustainable communities. Springer Press, New York

Clark WW II, Fast M (2008) Qualitative economics: toward a science of economics. Coxmoor Press, London

Clark WW II, Lund H (Authors and Co-editors) (2008) Utility Pol J. Special issue. Sustainable energy and transportation

Clark WW II (2008) The green hydrogen paradigm shift: energy generation for stations to vehicles. Utility Pol J, Elsevier Press

Clark WW II (2006) Partnerships in creating agile sustainable development communities. J Cleaner Prod, Elsevier Press 15:294–302

Clark WW II, Rifkin J et al (2006) A green hydrogen economy. Special issue on hydrogen, energy policy, Elsevier 34(34):2630–2639

Clark WW, Bradshaw T (2004) Agile energy systems: global lessons from the California energy crisis. Elsevier Press, London

Clark WW II, Li X (2004) Social capitalism: transfer of technology for developing nations. Int J Technol Trans 3(1), Interscience, London

Clark WW II, Jensen JD (2002) Capitalization of environmental technologies in companies: economic schemes in a business perspective. Int J Energy Technol Pol 1(1/2). Interscience, London,

Clark WW II (2003a) Science parks (1): the theory. Int J Technol Trans Commercial 2(2):179–206. Interscience, London

Clark WW II (2003b) Science parks (2): the practice. Int J Technol Trans Commercial 2(2): 179–206. Interscience, London

Clark WW II, Sowell A (2002) Standard economic practices manual: life cycle analysis for project/program finance. Int J Rev Manage. Interscience Press, London

Clark WW II, Lund H (2001) Civic markets in the California energy crisis. Int J Glob Energy Issues 16(4):328–344. Interscience, UK

Economist (2012) The third industrial revolution. Special issue, 21 Apr 2012

Economist (2012) The rise of state capitalism: the emerging world's new model. Special issue with Lenin smoking a cigar on the cover. Two articles. 23 Jan 2012

Economist (2009) Modern economic theory: where it went wrong – and how the crisis is changing it. Special issue with Bible melting on cover. Editorial and three articles. 18 July 2009

EU, CCC (2009) Meeting carbon budgets: the need for a step change. Progress report to Parliament Committee on Climate Change. Presented to Parliament pursuant to section 36(1) of the Climate Change Act 2008. The Stationery Office (TSO). ISBN 9789999100076. http://www.official-documents.gov.uk/document/other/9789999100076/9789999100076.pdf. Retrieved 1 May 2010

Federal Housing Finance Agency. Retrieved from http://www.fhfa.gov/webfiles/15884/PACESTMT7610.pdf

Gipe P (2011) Feed-in-tariff monthly reports

Hawkins P, Lovins A, Lovins LH (1999) Natural capitalism: creating the next industrial revolution. Little, Brown and Company, Boston

Isherwood W, Smith JR, Aceves S, Berry G, Clark WW II, Johnson R, Das D, Goering D, Seifert R (2000) Remote power systems with advanced storage technologies for Alaskan Village. University of California, Lawrence Livermore National Laboratory, UCRL-ID-129289: Published in Energy Policy, vol 24, pp 1005–1020, Jan 1998

Knakmuhs H (2011) Smart transmission: making the pieces fit. RenGrid 2(3), www.renewgridmag. April 2011, p 1+

Li X, Clark WW (2009) Crises, opportunities and alternatives globalization and the next economy: a theoretical and critical review, Chap #4 In: Li X, Gorm W (eds) Globalization and transnational capitalism. Aalborg University Press, Denmark

Lillian J (2010) New and noteworthy. Sun Dial, Solar Industry, pp 3–4

Lund H (2009) Sustainable towns: the case of Frederikshavn. 100 percent renewable energy, Chap #10. Sustainable Communities, Springer Press, New York, NY

Martinot E, Droege P (2011) Renewable energy for cities: opportunities, policies, and visions. Series of reports from 2007–2010. http://www.martinot.info/Martinot_Otago_Apr01_cities_excerpt.pdf

Next 10 (2011) Many shades of green. Silicon valley and Sacramento, CA, 19 Jan 2011. www.next10.org

Nularis (2011) Data and information. www.nularis.com
Rifkin J (2012) The third industrial revolution. Penguin Putnam, New York
Rifkin J (2004) The European dream. Penguin Putnam, New York
San Jose Business Journal (2011) Clean energy financing jumps to record $243B. 1 April 2011, p 8
Sanders B (2011) US Senator, speech before California State democratic convention, Sacramento, CA, 30 April 2011
Sullivan M, Schellenberg J (2011) Smart grid economics: the cost-benefit analysis. RenGrid 2(3):12–13+. www.renewgridmag
Tickell J (Director and Producer) (2009) Fuel, independent documentary film, La Cinema Libra, Los Angeles
U.S. Energy Information Administration (US EIA). See annual reports at: http://www.eia.doe.gov/oiaf/ieo/world.html
Vestas (2011) www.vestas.com

Chapter 3
Market Solutions for Climate Change

Malcolm Dole, Jr.

Abstract This chapter is a nontechnical introduction to the basic economic issues surrounding the climate-change challenge. It discusses the design, pros and cons, and effectiveness of three market-based approaches: the free market (no-government approach), the carbon tax (price approach), and the cap-and-trade systems (quantity approach). None is a panacea. The free market fails to price the social cost of greenhouse gases; a carbon tax has to overcome the hostility to taxes and the actual reduction of greenhouse gases is uncertain; and cap-and-trade systems face the problems of cooperation among governments, the free-rider problem, negative manipulation of the system, and other concerns. Government implementation, including experience in Europe, Canada, and the United States of tax and trading systems, is provided to illustrate the effectiveness of those systems and to provide information for future design options and for the implementation of those systems.

The main challenge for a carbon tax, the choice of most economists, is to educate the public and policy makers that a carbon tax can reduce greenhouse gases (GHGs) while improving the efficiency of our current tax system and be revenue neutral. The challenges for policy makers in this area are formidable.

The author thanks Woodrow W. Clark II, CEO/Managing Director, Clark Strategic Partners, and Michael D. Intriligator, Professor of Economics, Political Science, and Public Policy at the University of California, Los Angeles, for their support in writing this chapter. The author also thanks Professor Janet Milne, Director of the Environmental Tax Policy Institute, and Professor of Law at Vermont Law School who made many good comments, Fereidun Feizollahi, manager of the Economic Studies Section of the California Air Resources Board for his support, and three anonymous referees for their comments of an earlier version of this chapter presented at the Western Economic Association International 2009, Pacific Rim Conference in Kyoto, Japan. Of course, any errors within are those of the author.

Malcolm Dole is an Independent Researcher who was with the California Air Resources Board for over three decades and can be contacted at E-mail: mdolearb@yahoo.com

M. Dole Jr., Ph.D. (✉)
California Air Resources Board, Sacramento, CA, USA
e-mail: mdolearb@yahoo.com

Abbreviations

ABEC	Americans for Balanced Energy Choices
ACCCE	American Coalition for Clean Coal Electricity
BLM	Bureau of Land Management
CAC	Command and Control
CAIR	Clean Air Interstate Rule
CAMR	Clean Air Mercury Rule
CAT	Cap-and-Trade
CC	Climate Change
CCS	Carbon Capture and Storage
CDM	Clean Development Mechanism
CEED	Center for Energy and Economic Development
CO_2	Carbon Dioxide
CT	Carbon Tax
EERS	Energy Efficiency Resource Standard
EPA	Environmental Protection Agency
EU	European Union
GDP	Gross Domestic Product GHGs: Green House Gases
GW	Global Warming
GWP	Global Warming Potential
NRDC	Natural Resources Defense Council
RECLAIM	REgional CLean Air Incentives Market
RGGI	Regional Greenhouse Gas Initiative
RTC	RECLAIM Trading Credits
SCAQMD	The South Coast Air Quality Management District
UN	United Nations
UNFCCC	United Nations Framework Convention on Climate Change
US	United States
USEPA	US Environmental Protection Agency
WCI	Western Climate Initiative

Introduction

Combating climate change (CC), caused by GHGs, is one of the major challenges of the twenty-first century. The costs and risks of CC have been estimated at least at 5% of global GDP (gross domestic product) per year (Stern 2006). In reducing the damage, society faces three challenges: developing a design for efficient solutions to this externality problem, the will to implement those solutions, and their successful implementation.

There are many ways to reduce GHGs: moral suasion by CC advocates; individual actions such as driving less, low-carbon diets, and buying Energy Star appliances; environmental protest movements such as Greenpeace; municipal movements such as the Cities for Climate Protection Campaign; state air pollution regulations and coordination such as the Regional Greenhouse Gas Initiative discussed below; and federal energy efficiency standards and federal grants for energy-efficient technologies. Economists' preferred solution is to adjust GHG prices to their total, private plus public, cost through market mechanisms.

Market systems are preferred as they have been shown to lower the cost of emission reduction compared to direct regulations,[1] produce prices that provide ongoing incentives for reduced energy use and improved emission-reduction technology, and provide estimates of the value/cost of marginal emission abatement. This survey reviews three market-based approaches for reducing GHGs and the case for and against them.

A practical design for the reduction of GHGs combines more than a single strategy to achieve the substantial reduction in GHGs desired. For example, California's plan, Global Warming Solutions Act (AB32 2006), includes direct regulations, alternative compliance mechanisms, monetary and nonmonetary incentives, voluntary actions, and market-based mechanisms such as a cap-and-trade (CAT) system. California's CAT system goes into effect in 2013 with the goal of reducing GHG emissions to 1990 levels by the year 2020 and an 80% reduction by 2050. See California Environmental Protection Agency web site, www.calepa.ca.gov/ and the California Air Resources Board web site, www.arb.ca.gov/.

Government experience with market approaches is included to provide insight into developing and implementing an efficient system to lower GHGs and reduce global warming (GW).

The No-Government Solution

Many people look to governments to deal with GW. Others advocate, letting the free market handle the problem with no-government intervention. Here, we look at three categories of advocates for the no-government approach.

One is people who think the unfettered market can handle the problem, two are skeptics who think GW is not a man-made problem, and three are free riders who benefit whether they pay or not or just don't want to pay the cost.

Market Advocates

The free market's relative price system efficiently allocates scare resources. Higher energy prices encourage energy conservation and research, leading to technological progress in energy efficiency. The price effects are pervasive and have increasing impact in the long run as demand is more elastic over time (Alchian 1983).

---

[1] The Congressional Budget Office in "Issues in the Design of a Cap-and-Trade Program for Carbon Emissions" November 25, 2003, stated, "In the case of sulfur dioxide, researchers estimate that the cap-and-trade program lowered the cost of meeting the emissions target by between 43% and 55% compared with the cost of requiring all regulated sources of sulfur dioxide to meet a uniform emission rate. Substantial cost savings would also be likely to occur under a cap-and-trade program for carbon dioxide."

Learning by doing compliments technological advancement. When output doubles, all costs (including administration, marketing, distribution, and manufacturing) decrease by a constant and predictable amount as measured by progress ratios and the cost of production which is generally in the 80% range, that is, a $100 cost would be $80 after output doubles (Dutton and Thomas 1984).

Higher oil prices mean consumers use less oil and switch to substitutes, for example, walking, bicycling, and using mass transit and solar power; entrepreneurs develop alternative energy sources such as wind, tidal, hydro, and solar power; and power companies use alternative sources of energy, for example, nuclear power and ethanol fuel. However, all substitutes are not environmentally friendly.

Nuclear Power: Support for nuclear power is increasing, despite the nuclear power plant explosion in Japan (Economist 2011), but there is the problem of storing the radioactive waste and of catastrophic events such as earthquakes and terrorist attacks. For a more complete list of problems, see Greenpeace, http://www.greenpeace.org.uk/nuclear/problems/.

Clean Coal: Coal is the world's main source of electricity and produces about half the electricity in the USA because it is inexpensive, abundant, and domestically available. A relative switch from foreign oil to domestic coal as oil prices rise reduces foreign oil dependence, creates jobs and income, and improves the US balance of payments but increases carbon emissions.

Pollution from coal plants produces dirty air, acid rain, and contaminated land and water. The Natural Resources Defense Council (NRDC) says, "The Health problems associated with coal pollution include childhood asthma, birth defects and respiratory diseases that take nearly 25,000 lives each year." There are also costs associated with the mining, transporting, burning, and disposing of coal's waste products. For example, on December 22, 2008, 1.1 billion gallons coal ash and water "burst through a dike … and blanketed several hundred acres of land, destroying nearby houses … and left much of the town uninhabitable" (Walsh 2009). "Coal is the single greatest threat to civilization and all life on our planet. – James Hansen, NASA's top climate scientist" (NRDC 2012).

Clean coal is a term that covers technologies that reduce GHGs released from coal burning, mainly carbon capture and storage (CCS). Clean coal is better than dirty coal, but, unfortunately, it is very expensive and "is nowhere near as 'clean' as a high-priced industry advertising campaign makes it out to be." [2, 3]

---

[2] The major lobbying organization for clean coal is the American Coalition for Clean Coal Electricity (ACCCE) which was formed in 2008 and combines the assets and missions of predecessor organizations – the Center for Energy and Economic Development (CEED) and Americans for Balanced Energy Choices (ABEC). One of those goals is "We're committed to ensuring that America's energy future is a clean one" (ACCCE web site).

[3] "ACCCE is made up of 47 companies in the coal business: disclosures that listed climate lobbying among their issues, have already topped $43 million in the first two quarters of this year [2008] … among the group's first acts was working to stop a bill … that sought to put the first nationwide limits on carbon emissions." (Lewis 2008).

On December 4, 2008, the Alliance for Climate Protection, League of Conservation Voters, National Wildlife Federation, Natural Resources Defense Council, and the Sierra Club announced the Reality Coalition, "a national grassroots and advertising effort to tell a simple truth: in reality, there is no such thing as 'clean coal.'" Natural Resources Defense Council (NRDC 2009). The better solution environmentalists say "is to repower America by investing in clean energy. Green technologies and renewable fuels will create millions of good-paying jobs, lift our poorest communities out of poverty, reduce dangerous pollution and help fight global warming" (NRDC 2012).

See also "Collapse of the Clean Coal Myth." New York Times Editorial, January 22 2009.

Controversial Sources of Oil: Rising oil prices increase pressure for offshore oil drilling and opening up Alaska's Arctic National Wildlife Refuge. (See H.R. 49, 112th Congress that repeals the prohibition against development of oil and gas production in the Arctic National Wildlife Refuge.) The resulting increase in supply would tend to lower oil prices and, if so, make "green" alternatives to oil less competitive and reduce industry's incentive to develop green alternatives, and "…oil drilling is a… destructive business that doesn't belong in environmentally sensitive areas…." (Pope 2009) See the web site for the Sierra Club, www.sierraclub.org/. For example, think of the BP oil spill in the Gulf of Mexico in 2010, which killed 11 people and spilled about 4.9 million barrels of crude oil, causing major damage to wildlife and the area's fishing and tourism industries.

Not to be overlooked is the fact that using up these domestic sources in the present makes the U.S., other things equal, even more dependent on foreign oil in the future when the cost of foreign oil will be higher.

Unconventional Sources of Oil: Higher fuel prices could bring unconventional sources (oil shale, tar sands, extra heavy oil, and coal liquids) of oil into play and could make the USA independent of foreign oil. For example, U.S. deposits of oil shale are about five times the proven oil reserves of Saudi Arabia (U.S. Dept of the Interior 2009); tar sands represent about 2/3 of the world's total petroleum reserves; and worldwide extra heavy oil is about six trillion barrels (National Petroleum Council 2007).

However, these sources are environmentally destructive. "Oil shale production is expected to emit four times more global warming pollution than … conventional gasoline…" (Mall 2008). Some good news is that "BLM is working to ensure that development of federal oil shale and tar sands resources will be economically sustainable and environmentally responsible" (U.S. Dept. of the Interior 2009, web site www.doi.gov/).

Natural Gas: Natural gas is a cleaner source of energy than clean coal, but there are problems with how it may be obtained, for example, hydraulic fracturing used to obtain shale gas can contaminate water supplies as well as cause other environmental damage. For more on this see the writings of *Robert Howarth*, Professor of Ecology and Environmental Biology, *Cornell* University. An indication of the seriousness of this is that France has banned hydraulic fracturing (Patel 2012).

Global Warming Skeptics

Some scientists say using scarce resources to reduce GHGs is wasteful because CC is caused by natural events, such as ocean currents, solar activity, or unknown natural causes,[4] and not significantly by humans. Some even assert that more carbon dioxide (CO_2) might benefit the world.[5]

Quite often these skeptics are wrong. For example, "You've probably heard about the accusations leveled against climate researchers – allegations of fabricated data, the supposedly damning e-mail messages of 'Climategate,' and so on. What you may not have heard, because it has received much less publicity, is that every one of these supposed scandals was eventually unmasked as a fraud concocted by opponents of climate action, then bought into by many in the news media." Why does this happen, "If you want to understand opposition to climate action, follow the money. The economy as a whole wouldn't be significantly hurt if we put a price on carbon, but certain industries – above all, the coal and oil industries – would. And those industries have mounted a huge disinformation campaign to protect their bottom lines." However, "Every piece of valid evidence – long-term temperature averages that smooth out year-to-year fluctuations, Arctic sea ice volume, melting of glaciers, the ratio of record highs to record lows – points to a continuing, and quite possibly accelerating, rise in global temperatures" (Krugman 2010, http://krugmanonline.com/).

Free Riders and Others

"Free riders" don't want to pay because they get the benefits whether they pay or not because air quality is a public good/bad. Some consumers support the no-government option as they just don't want to pay for reducing GHGs especially given the worldwide recession, high fuel prices, and unemployment. Some, especially senior citizens,

---

[4] These arguments provide talk show hosts, politicians, and businesses with arguments to oppose government policies to reduce GHGs. For example, a former chairman of the Environment and Public Works Committee proclaimed, "As I said on the Senate floor on July 28, 2003, 'much of the debate over global warming is predicated on fear, rather than science" (Inhofe 2008).

[5] Increased carbon may "be a good thing ... because carbon acts as an ideal fertilizer, promoting forest growth and crop yields. ...rising carbon may well be an ultimately benign occurrence" because "the warming...is not global but local, making cold places warmer rather than making hot places hotter." And "most of the evolution of life occurred on a planet substantially warmer than it is now... and substantially richer in carbon dioxide" (Dawidoff 2009).

believe they won't be around to reap the benefits. Many conservatives just oppose new taxes, more regulations, and bigger government.[6]

Many corporations, especially large coal and oil energy companies, oppose regulation because they would be hit with a significant amount of the bill,[7] and some politicians appear to be beholden to those large campaign contributors. "Senators and representatives feel in their bones (and campaign accounts) the interests of utilities and the coal and oil industries" (Wasserman 2010).

Finally, according to Paul Volcker, chairman of the Economic Recovery Advisory Board (since November 2008 the President's Council on Jobs and Competitiveness), "…any democracy has difficulty focusing on a problem which is not a crisis today but, with a high degree of probability, is going to be a major problem 10, 15, 20 or 30 years from now" (Volcker 2007).

Moral Suasion

Moral suasion by private citizens works alongside the free market to fight GW by persuading people by reason and logic and appealing to their sense of morality, civic duty, philanthropy, etc. Moral suasion includes letter writing, preaching, advertising, threats of regulation, public criticism, and mass protests. Well-organized groups like the NRDC, the Sierra Club, and Stop Climate Chaos, a coalition of 100 organizations with 11 million supporters, can gain media and public attention, leading to individual and political action.

---

[6] Michael Steele, past chairman of the Republican National Committee, wrote, "I ran for the job of RNC Chairman to lead our Party forward with its core principles…: shrinking the size of government…." and "If Obama and his liberal Democrat cohorts get their way, you and your family will be paying an additional $260 a month in energy taxes thanks to the Democrats' outrageous Cap and Trade legislation. That's $260 a month that you and your family should be allowed to spend, save or invest anyway you see fit" (Steele, emails February 11 2009 and May 18, 2009). The United States Environmental Protection Agency (U.S.EPA) estimated the average cost per household of "achieving the climate benefits" of the American Clean Energy and Security Act of 2009, H.R. 2454, at $80-111 per year (U.S.EPA 2009). The current RNC chairman wrote in the "Grassroots Quarterly," Winter 2011, "I believe…that 'government that governs least governs best.'"

[7] Companies, like Exxon Mobil, have "mounted a huge disinformation campaign to protect their bottom lines" and have "spent tens of millions of dollars promoting climate-change denial, or Koch Industries, which has been sponsoring anti-environmental organizations for two decades" (Krugman 2010).

Conclusion

To its advocates, this market solution means no new government legislation, taxes, regulations, international agreements, or bureaucracies, and the saved expenditures could be used better elsewhere, for example, lowering taxes. Society would automatically adjust to both the rising price of oil and the warmer weather leading to both green and nongreen actions. For example, some farmers would move into areas previously too cold to grow crops, and some would simply change to warmer weather crops.

However, market prices do not include the social costs[8] of CC and, thus, consumers and investors do not take into account the environmental costs when they make decisions about expenditures, that is, the market does not allocate resources efficiently with this market solution and, thus, does not solve the GW problem. More effective is to use carbon taxes (CT) and/or CAT systems to change prices to reflect total costs. As stated in the Washington Post, June 25, 2007, "Sooner or later, Congress will have to realize that slapping a price on carbon emissions and then getting out of the way to let the market decide how best to deal with it is the wisest course of action."

The Carbon Tax Solution

Introduction

Carbon is a source of energy coming from the burning of fossil fuels: coal, oil, and gas. A carbon tax (CT) is a price on CO_2 that translates to a tax on fossil fuels that produce GHGs. The CT acts like a market price that impacts the decisions of consumers, producers, and governments over a wide range of choices.

The economists' "theoretical solution" is to correct this market failure by equating the price of a GHG-producing activity to its private and social cost; that is, the price would equal the marginal damage at the optimum level of GHG emissions. The economists' two main solutions for achieving this are a CT and a CAT system.

---

[8] Higher temperatures and evaporation could lead to increased droughts and wildfires, heavier rainfall, and an increase in category 4 and 5 hurricanes. Hotter weather and heat waves increase the number of heat-related deaths, enable deadly mosquitoes to travel greater distances, and increase smog and intensify pollen allergies and asthma. Rising temperatures ravage coral reefs; speed the melting of glaciers and ice caps, melting the habitats of polar bears and penguins; and disrupt ecosystems, leading to extinction of species that cannot adapt (NRDC 2009).

Arguments for Carbon Taxes

Taxes have shown to be a powerful influence on what people do. For example, "higher cigarette taxes are a powerful tool to fight tobacco use.[9] Cigarette taxes encourage current smokers to quit, and they really work to prevent kids from starting to smoke. The larger the increase, the bigger the impact; every 10% increase in the price of cigarettes reduces the amount children smoke by about 7%" (American Lung Association 2012).

Positive Incentives: The CT puts a new, higher price signal on energy use equal to its private and social costs that encourage pollution-reducing decisions, for example, engineers developing more efficient carbon-reducing technologies; companies abandoning older, dirtier technologies for newer, cleaner ones; and businesses, consumers, and governments conserving on energy.

Simple Strategy: The CT has a relatively simple design, and enforcement mechanism, and administration with people deciding for themselves how best to economize on higher energy prices. This simplicity also protects a CT "Against the Perverse Incentives and Potential for Profiteering that Will Accompany Cap-And-Trade" (Carbon Tax Center web site, "pricing carbon efficiently and equitably").

Environmental Enhancement and Economic Efficiency: A CT provides a double dividend, environmental enhancement, and economic efficiency.

A Pigouvian CT corrects a market inefficiency by accounting for the environmental costs not priced by the market. The higher cost of polluting behavior reduces that behavior enhancing the environment.

The no-government solution does not take into account the externality. In a CAT system, while putting a price on polluting behavior, that price tends to be only on single energy sources such as power plants. A CT can more easily be on upstream fossil fuel U.S. production and foreign imports influencing and increasing the downstream market correction decisions of more people as they make more market correction decisions in more areas.

A CT, depending how the revenue is used, also increases the efficiency of taxation and production by taxing peoples' negative actions, polluting, rather than their positive actions, working.

Certainty: Predictable prices facilitate efficiency in long-run investment planning. A CT provides more cost certainty than CAT as tax rates are known in advance while under CAT prices can fluctuate.

Transportation: A CT provides incentives to switch to mass transit and energy-saving vehicles and to drive less which reduces traffic congestion, commute time

---

[9] For more information on the benefits of CTs, see the Carbon Tax Center web site, www.carbontax.org.

with fewer vehicles on the road, highway and vehicle maintenance, deaths, and injuries; and provides motorists with more time for more beneficial pursuits.[10]

Health benefits: A CT reduces GHGs and photochemical air pollution mitigating the negative health effects of those pollutants and the costs and suffering of related diseases such as asthma.

Equity: Because lower-income families spend a greater percentage of their income on energy than higher-income families, it is argued that a CT is a regressive tax that hurts the poor. On the other hand, higher-income families use more energy, paying more. Also the regressiveness of the CT can be reduced by rebates back to the lower-income families through dividends or tax shifting such as reducing the regressive sales or payroll taxes.

As an efficient CT taxes peoples' negative actions, polluting, rather than their positive actions, earning an income, it can be revenue neutral (the CT revenue used to reduce less efficient taxes) and improve tax equity, for example, tax credits for low-income households that spend a relatively high percentage of their income on energy consumption. A good example of revenue neutrality to ease the burden on lower-income taxpayers is British Columbia's Climate Action Plan and its Budget and Fiscal Plan 2012/13-2014/15.

The no-government solution does nothing to improve the equity of the tax system.

Because a CAT system tends to be on single energy sources, a CT can be more equitable by being on upstream fossil fuels, spreading the cost over more people, lowering the cost per person, and making the burden more equitable.

Finally, because CAT "relies on market participants to determine a fair price for carbon allowances on an ongoing basis, it could easily devolve into a self perpetuating province of lawyers, economists, lobbyists and other market participants bent on maximizing their profits on each cap-and-trade transaction" (Carbon Tax Center, "pricing carbon efficiently and equitably").

Cost-Effective: Compared to direct regulation, command and control (CAC), and CAT, the CT is less costly to implement as the implementation is more straightforward since it doesn't need a new bureaucracy, cooperation among governments, and the enforcement system is in place.[11]

Benefit Effective: The CT provides more pollution reduction (benefit) for the same cost than a CAT system because it affects more sources and produces more pollution-reducing decisions and less GW. For more information on the benefits of CTs, see the Carbon Tax Center web site, www.carbontax.org.

---

[10] "When gas prices shot up past $4 a gallon, average miles driven dropped significantly, as did energy consumption. Demand for fuel-efficient cars and overall energy efficiency skyrocketed." "Climate Change Solutions," Washington Post editorial, February 16, 2009.

[11] The costs of a CT or CAT would be about the same if administration costs were ignored, but "the net benefits of a GHG tax could be 30–34% higher than the net benefits of a cap" (Shrum 2007).

Tax Features

While a CT is relatively simple, there are still difficult decisions to be made. What should be taxed: just carbon or other GHGs such as methane, the content of the fuels, or the emissions produced? Who should be taxed: What consumers and producers? Where should the tax be levied, upstream or downstream: at the well head, mine mouth, port of entry, sale to the producer or consumer, or on the emissions? How high should the tax be to achieve the desired results and be politically viable? "Most estimates of near-term Pigouvian taxes…are in the order of about $5–25 per ton of CO_2. …Much higher Pigouvian tax estimates…are implied with a zero rate of time preference…. also…on the grounds of extreme catastrophic risks…"(Aldy et al. 2009).

How would the tax be adjusted if the original tax was too high or too low? Would it take legislation, might there be a special tax office that could adjust the tax more readily, or should a tax adjustment be built into the legislation, for example, indexed for inflation?[12]

Should there be exemptions, subsidies, and rebates? These are common. Richard Rosen says, "…a very high carbon tax would have to be combined with complementary carbon tax rebate programs… for two main reasons. First, the rebates would have to target the tax payments to exactly the kinds of new, more energy-efficient technologies that would be needed…. Secondly, the rebates would have to be allocated in ways so that the poor and middle class would not be thrown into poverty…" (Rosen 2006). How are exports and imports to be treated? What should be done with the revenue? What are the environmental, economic, revenue, and equity results? Is it politically realistic,[13] for example, who does it impact the most and how influential are they?

The Revenue

The revenue depends on the tax rate, the number of sources taxed, the amount of what is taxed, and the demand elasticity of the item taxed. Some uses of the revenue suggested by economists are the following: tax rebates to make the tax neutral, research grants to improve CC technology, climate adaptation projects, reduction of the national debt, lower capital income taxation, pay for Social Security and Medicare,

---

[12] Energy taxes in Sweden are indexed to the consumer price index. Nordhaus suggests, "The efficient tax would be equalized across space and growing over time at approximately the real carbon interest rate" (Nordhaus 2008).

[13] "Republicans won't enact a tax hike for any purpose" (Reich 2007). "POWERFUL anti-tax rhetoric has made legislators…afraid to talk publicly about a need to raise taxes" (Frank 2007).

and elimination of Third World poverty and disease. The leading use suggested by most economists seems to be making the tax revenue neutral by rebates or "dividends" back to the taxed or tax shifting, for example, reducing payroll taxes.

Arguments Against Carbon Taxes

Scientific Arguments: Researchers do not have definitive information on future emissions and social costs (think catastrophic damages) needed to set an optimal tax or to know the exact response to the tax or the impact on CC. What emissions and damages will be in the future depend on such unknown inputs as the cooperation of governments worldwide, population and economic growth, available emission-reduction technology and its use, energy-intensive firms moving to less regulated locations, temperature change (think GW and increased photochemical pollution), risk of catastrophic damages, and the appropriate discount rate.

Political Arguments: The tax will slow economic growth and decrease human welfare. It is not known how high the tax must be. It would be difficult to implement and adjust. Scientists and politicians should put their efforts into preparing for and adapting to GW. The tax may be used to increase the tax base rather than reducing other taxes alienating the public.

The biggest political problem is the aversion to new taxes. Presenting the tax as a means of making the current tax system more efficient should, however, reduce that aversion to new taxes.

The "It Doesn't Work" Argument: Some people have said where it has been tried it hasn't worked. "Denmark, Finland, Norway and Sweden have had carbon taxes in place since the 1990s, but the tax has not led to large declines in emissions in most of these countries.... An economist might say this is fine; as long as the cost of the environmental damage is being internalized, the tax is working – and emissions might have been even higher without the tax. But what environmentalist would be happy with a 43% increase in emissions?" (Prasad 2008)

Equity Argument: Because lower-income families spend a larger percentage of their income on energy than higher-income families, they would be disproportionately burdened. The money spent on CC would be better spent on alleviating poverty and providing better health care for the poor.

Inflation Argument: A tax on energy would be inflationary.

The Cap-and-Trade Solution

The third market solution is cap and trade (CAT) or cap and tax by its opponents. CAT "...is an environmental policy tool that delivers results with a mandatory cap on emissions while providing sources flexibility in how they comply. Successful

CAT programs reward innovation, efficiency, and early action and provide strict environmental accountability without inhibiting economic growth" (USEPA, Cap and Trade web site 2009, www.epa.gov/captrade/).

With a CAT system, permits to pollute are sold on a market giving the buyer property rights that allow the buyers to pollute and also to transfer those rights to other buyers.

The terms, permits, budgets, allowances, credits, and offsets all pertain to an allowable amount of emissions.

A CAT system could be designed for consumers such as rationing in World War II, but this has not been seriously suggested.

Elements of a Cap-and-Trade Program

First, the cap on emissions must be determined taking into account both environmental and other concerns such as cost, existing technology, political feasibility, and economy. The beginning cap for a polluting unit tends to be set at that unit's current emissions that provide the unit time to find the most efficient ways to reduce emissions. The complexity of this decision is increased because more than one political unit is normally involved and the cooperation of governments is a major challenge. Second, the units must be determined and defined. Third, the initial permits must be apportioned to the units and the operating permits approved. Permits may be given away, bought from the government, or purchased at auctions. Fourth, each source must set up a monitoring system and record, report, and assure the quality of its emission data. Fifth, fines must be set and collected for violations. Six, how fast the cap is ratcheted down must be determined, which depends on the costs and benefits of different levels of reduction and the time period allowed for compliance. Regulatory certainty is important so sources can plan for their reductions efficiently. This certainty is reduced with changing permit prices.

Permits: An application that identifies the budget unit and each source within that unit may be required to obtain permits. Permits may be given to the sources or sold in some fashion. The current consensus is to allocate permits by auctions.

Auctions: Auctions produce a market price indicating the value of the permits and raise revenue. An important element is auction design. For an idea of the complexity of designing an efficient auction, see Holt et al. (2007) and Milgrom (2004). Some criteria for successful/efficient auctions are the following: low administrative and transaction costs for bidders, understandable for bidders, perceived as fair by the parties involved, raising revenue for beneficial uses, coordination with federal agencies, public disclosure of the clearing price, identities of winning bidders and the quantity of allowances obtained by winning bidders, and agreement of all bidders with auction goals. Some potential problems are the following: collusion and hoarding of permits, corruption, price volatility, ties in bidding, nonpayment of bids, no bids at or above and setting the reserve price, and bids above the bidder's financial assurance level.

California will have its first quarterly auction of allowances (tradable permits) beginning in the summer of 2012. To assist in implementing this auction system successfully, a detailed instructional guidance document and training will be available for participants, and requests for proposals were sent out to select an auction and reserve sale operator and a company for auction and reserve sale financial services.

Revenue: Once permits are allocated, the revenue tends to go to emitters rather than the government as with the CT. In California, the revenue from the initial auction will go into a fund controlled by the legislature. The government receives revenue by initially charging for the permits[14] and collecting fines for emission violations. The revenue can be used for multiple purposes, for example, exemptions, subsidies, and rebates.

Efficiency: Permits can be traded among emitters allowing trades from low-cost emitters to high-cost emitters, reducing the total cost of meeting the cap.

Advantages of a Cap-and-Trade Program

While most economists advocate CTs, most governments and Robert N. Stavins, Director, Harvard Environmental Economics Program, favor a cap-and-trade system, at least in the short to medium term (Stavins 2007). In his arguments, "Besides providing certainty about emissions levels, cap-and-trade offers an easy means of compensating for the inevitably unequal burdens imposed by climate policy; it is straightforward to harmonize with other countries' climate policies; it avoids the current political aversion in the United States to taxes; and it has a history of successful adoption in this country." California's CAT program takes into consideration the potential impacts on disproportionately impacted communities.

Stavins proposes a free distribution of half of the allowances, primarily to those most burdened by the policy, to help limit potential inequities while bolstering political support. The other half would be auctioned to provide revenue for worthwhile purposes. Stavins suggests that in a perfect world a CT may be better because of its simplicity, but in the real world political reality tips the scales to a CAT system.

Global advantages include harmonizing with other countries' climate policies, providing for linkage with international emission-reduction credit arrangements, and providing for appropriate linkage with other actions taken abroad that maintain a level-playing field between imports and import-competing domestic products. California's CAT program includes working with British Columbia, Ontario, Quebec, and Manitoba (see WCI below) to develop harmonized CAT programs for efficient, cost-effective emission reduction.

---

[14] "Under the proposed federal cap-and-trade program, all GHG emission credits would be auctioned off, generating an estimated $78.7 billion in additional revenue in FY 2012, steadily increasing to *$83* billion by FY 2019" (US Department of Energy 2009).

Compared to standard government CAC regulations, a CAT program is more flexible, allowing for more efficient abatement of GHGs. Trading among low- and high-cost emitters lowers the cost of achieving a given amount of pollution and reduces the temptation to relax standards for sources facing economic or technical problems meeting the regulations.

The lower cost of emission reduction means more emissions can be reduced for the same cost or the same amount of emissions at less cost. This can reduce the opposition to the abatement program and increase the environmental goals of the program. Banking of allowances provides an incentive for earlier abatement of emissions, reducing the pollution earlier. Finally, a CAT system provides a stronger incentive for innovation in emission-reduction technology, further reducing the cost of reducing GW (Harrison 2007).

Offsets

An important part of the Kyoto Protocol (see United Nations Framework Convention on Climate Change UNFCCC web site) is tradable offsets that can be earned and then used to meet CAT requirements at a lower cost than if there were no offsets. High-cost emitters can meet their emission quota by investing in lower-cost projects that reduce or sequester GHGs in developing countries as long as they are real, additional, verifiable, enforceable, and permanent. If they meet these criteria, they obtain Clean Development Mechanism (CDM) approved Certified Emission Reductions which validate and certify the projects. This is a win-win situation if the projects would not have been done anyway. If they would have been done, there is no additional gain. There has been criticism of this happening.

Disadvantages of a Cap-and-Trade Program

Compared to a tax, the CAT's administration oversight of the system is significantly more complex. CAT needs a trading and enforcement mechanism, it has a more complex design, and it needs more coordination among governments. Permit price volatility brings forth the problem of uncertainty. This could be reduced through banking, borrowing, and flexibility of permit supply.

For more information on carbon trading, see the Cap and Trade Watch at http://www.carbontradewatch.org/issues/cap-and-trade.html and for a more technical, detailed explanation of the above issues, see Aldy et al. (2009) and Milne (2009).

Options Chosen

Each market solution has its supporters.

The No-Government Solution

Because of the weak economy, high unemployment, and high national debt, the Republican Party has chosen the no-government option over a CT or CAT. The Republican National Committee in this election year in the issues section of their web site wrote, in talking about President Obama, "…his global warming regulations will be a 'a nightmare scenario' for small businesses, killing real jobs"

The Cap-and-Trade Solution

Governments have chosen both the CAT and CT options in the past; however, their main focus today seems to be the CAT option; see the European Union below. The US federal government has relied mainly on CAC regulations; however, it has proposed CAT systems, but none have been passed, and given the current economic climate, it is very unlikely that any will pass soon.

State and local governments have implemented CAT systems; see RGGI, WCI, and SCAQM below.

The Carbon Tax Solution

While federal, state, and local governments have implemented environmental taxes, none have implemented a comprehensive CT.

The biggest supporter of the CT option is economists. Economists believe the no-government option is not a solution to the problem and a CT is more efficient than a CAT system for the following reasons:

1. A CT is much more comprehensive than a CAT system.
   A tax on the carbon content of fossil fuels burned can cover all CO_2 emissions, while a CAT system covers only those emissions covered by the particular system, for example, power plants. A CT can also be imposed early in the life cycle of fossil fuels to influence decision-makers' choices from carbon birth to death. A carbon tax can be thought of as a carbon disincentive.
2. A CT is simpler.
   With a CT the government establishes the tax rate and the IRS collects the revenue and enforces it with audits. No new bureaucracy or emission monitoring from smokestacks or tail pipes is needed. With a CAT system, the government needs to create a new legal entity, permits to emit CO_2 emissions, and determine the amount of permits to put on the market initially and over time. The government also needs to create a system to allocate those permits to emitters and rules

for transferring and tracking the permits when they are bought and sold in the market as well as ensuring that the emissions are authorized by the permits. The implementation of CAT is more complicated with the introduction of middlemen such as lawyers, lobbyists, and consultants.

3. Predictability.

   The predictability of future energy prices impacts the future energy decisions of producers and consumers. The prices of tradable permits are more volatile than tax rates because they are more impacted by economic and physical factors, for example, severe droughts. Thus, with a CT producers and consumers can make more informed decisions about their future purchases and investments, especially with regard to emission-reduction decisions.

4. Cost.

   A CAT system has higher costs than a CT. With a CT or CAT, the costs of actually reducing carbon emissions are about the same, but the costs surrounding that reduction are greater with a CAT system, for example, the costs of auctions; middlemen; market manipulation; setting up monitoring systems and the recording, reporting, and emission assurance that entails; enforcement; and other elements of a CAT system such as administering offsets.

5. Tax efficiency.

   An advantage of the CT as a tax is that it can be used to increase the efficiency of our tax system. Revenue neutral taxes can improve tax efficiency, for example, taxing people's negative actions, for example, polluting rather than their positive actions, working. Also, the revenues from a CT would go back to the people, while the revenues from a CAT system go to market participants and the increase in prices is hidden.

6. Faster implementation.

   Given the governments' resolve to reduce GHGs, a CT can be implemented faster as a CAT system is much more complicated to set up, for example, a new bureaucracy needs to be created and the cooperation of different government entities that have different ideas of how the CAT system should set up needs to be dealt with.

7. Special interests.

   A CT can be implemented with "far less opportunity for manipulation by special interests, while a cap-and-trade system's complexity opens it up to exploitation by special interest and perverse incentives that can undermine public confidence and undercut its effectiveness (Carbon Tax Center 2012)."

Government Experience

This section examines government experience with taxes and cap-and-trade systems that have been proposed or implemented.

Carbon Taxes

The US Government Experience

The USA has proposed and implemented environmental taxes. In the early 1970s, taxes were proposed on lead additives to gasoline and on SO_2 emissions, but not implemented.

The Energy Tax Act of 1978 placed a federal excise tax on the sale of new vehicles (except minivans, sport utility vehicles, and pick-up trucks) not meeting fuel economy standards.

The Comprehensive Environmental Response, Compensation, and Liability Act of 1980 led to taxes on petroleum, chemical feedstocks, and corporate income to finance the cleanup of hazardous waste sites. For example, the Woolfolk Chemical Works site was redeveloped into a new library, a tourist information center, a record storing facility, and a new building for Fort Valley University (U.S.EPA, Superfund).

In 1993 an energy tax was proposed by the Clinton Administration mainly as a deficit-reduction measure; however, the tax was also promoted as a way to reduce pollution and increase energy efficiency and independence. Renewable sources of energy such as solar, wind, and biomass were exempted. It was not passed, but a $0.043 excise tax on gasoline was.[15]

Seven market proposals were introduced in the 111th Congress. None of these were enacted.[16]

H.R. 594 puts an excise tax on fossil fuels of $10/t CO_2 to be raised $10 each year. This is an upstream approach taxing fuels as they enter the economy, including imports.

H.R. 1337 puts a CT on fossil fuels of $15/t CO_2 to be raised $10 each year. If emission targets (80% below 2005 emissions by 2050) are not met, the increase would be *$15*. The revenue would go for clean energy technology research, affected industry transition assistance, and lower payroll taxes and social security supplements.

H.R. 2380 is an upstream CT of $15/t CO_2 to increase to $100 in 30 years with the revenue going for payroll tax reduction split between employer and employee.

A review of the 112th Congress indicates no CAT and only one CT bill.

H.R. 3242 would amend the Internal Revenue Code of 1986 to reduce emissions of carbon dioxide by imposing a tax on primary fossil fuels based on their carbon content. This bill was referred to committee but has little to none chance of passing.

---

[15] For political lessons to be learned from this, visit www.vermontlaw.edu/envirotax.

[16] To track federal government legislation, see The Library of Congress, http://thomas.loc.gov/cgi-bin/thomas or GovTrackus, www.govtrack.us/congress/ and for more on this see the Carbon Tax Center web site, www.carbontax.org.

The tenor of the 112th Congress is indicated by the environmental bills with the most cosponsors: *H.R. 49* the American Energy Independence and Price Reduction Act repeals the prohibition against development of oil and gas production in the Arctic National Wildlife Refuge, *H.R. 910* the Energy Tax Prevention Act of 2011 amends the Clean Air Act to prohibit the U.S.EPA from promulgating any regulation taking into consideration the emission of GHGs, and *H.R. 872* the Reducing Regulatory Burdens Act of 2011.

The European Experience

European governments have used environmental taxes to provide incentives for reducing air pollutants including CTs.

The positive impacts of environmental taxes would be increased if all competing emitters faced similar taxes across nations.

Because of concerns about competitiveness with other countries, a common feature of environmental taxes is exemptions and rebates. The taxes have had some success. To have a bigger impact, the taxes could be higher, more comprehensive, with less exemptions and rebates, and supplemented with other policies. Some of the revenue has been used to reduce personal and corporate income taxes and to implement energy-saving projects.

Finland: Finland, the first country to institute a CT (in 1990), has a plethora of environmental taxes, charges, fines, and fees on, for example, exceeding GHG emission limits, hazardous waste, waste from ships, oil release, air traffic noise, fishing and hunting, soil extraction, forest management, mining, water, oil release, auto idling, nuclear waste, and beverage containers.

Finland's CT goal is to reduce GHGs 80% by 2050 from their 1990 level. The government says the tax has stimulated investment in renewable energy technology, and CO_2 emissions are 5% lower despite exemptions and rebates (Finland 2008).

Denmark: In Denmark CTs led to per capita CO_2 emissions about 15% lower in 2005 than 1990 levels despite "…posting a remarkably strong economic record and without relying on nuclear power. What did Denmark do right? … if we want to reduce carbon emissions, … tax the industrial emission of carbon and return the revenue to industry through subsidies for research and investment in alternative energy sources, cleaner-burning fuel, carbon-capture technologies and other environmental innovations (Prasad 2008)."

Sweden: Sweden enacted a CT in 1991 of $100/ton; discounts to lessen the impacts reduced the effective tax to about $25. Exemptions for renewable sources of fuel such as ethanol, methane, biofuels, and peat are given which resulted in a major expansion of the use of biomass for heating. "The Swedish Ministry of Environment projected that the tax policy lowered carbon dioxide emissions in 2000 by 20–25% from 1990 levels…." (Shrum 2007) From 1990 to 2006, carbon emissions fell 9%, while economic growth increased 44%. The Swedish environment minister, Andreas

Carlgren, said without the CT, emissions would have been 20% higher and "a carbon tax is the most cost-effective way to make carbon cuts and it does not prevent strong economic growth" (Fouche 2008).

Norway: A carbon tax was implemented in 1991. "Data for the development in CO_2 emissions … provide a unique opportunity to evaluate carbon taxes as a policy tool. To reveal the driving forces behind the changes in the three most important climate gases, CO_2, methane and N_2O in the period 1990–1999, we decompose the actually observed emission changes, and use an applied general equilibrium simulation to look into the specific effect of carbon taxes.… we find a significant reduction in emissions per unit of Gross Domestic Product (GDP) over the period due to reduced energy intensity, changes in the energy mix, and reduced process emissions despite extensive tax exemptions and relatively inelastic demand in the sectors in which the tax is actually implemented" (Bruvoll and Larsen 2002).

The European Union (EU): The EU consists of 27 countries and has high environment standards. "Today the main priorities are combating climate change, preserving biodiversity, reducing health problems from pollution and using natural resources more responsibly. While aimed at protecting the environment, these goals can contribute to economic growth by fostering innovation and enterprise" (Europa 2010).

The EU's trading system was introduced in 2005 and covers about 12,000 energy-intensive industries like power companies and steel and cement makers that account for about half of the EU's CO_2 emissions. In the future, more industries like petrochemical companies will participate. Companies can buy spare permits from more efficient businesses if they need to, and they will be able to buy offsets for projects in non-EU countries.

The CC goal is to reduce GHGs by at least 20% by 2020 from 1990 levels, raise renewable energy's share of the market to 20%, and cut energy consumption by 20%.

In October, 2011, the European Commission released its annual report on meeting its Kyoto Protocol target for reducing GHGs. The EU reduced emissions by 15.5% since 1990 while the economy grew by 41%. Emissions have fallen six consecutive years up to and including 2009. "The EU-15 remains firmly on course to meet its 8% emission reduction target under the Kyoto Protocol and is most likely to overachieve it" (Europa 2012).

The Canadian Experience

Quebec: Quebec implemented Canada's first CT in 2007. They expect to raise $200 million/year to pay for energy-saving initiatives, for example, improvements to public transit. Quebec's CT will be 0.8 cents/l on gas and 0.938 cents on diesel. Expected revenues from oil companies: $69 million gasoline, $43 million heating oil, and $36 million diesel fuel. A major concern was whether or not the tax will be passed on to consumers and how much (data submitted by the Group of Eight to the UN's Climate Change Secretariat).

British Columbia: In 2008 British Columbia instituted the Western Hemisphere's first major CT. The tax will be phased in over time, protects lower-income consumers, and is designed to be revenue neutral through a package of tax cuts (on personal and business income taxes) and credits. The tax applies to gasoline, diesel, natural gas, coal, propane, and home heating fuel. The tax was raised in 2009 from $10 to $15 (Canadian) per metric ton of CO_2 and is now at $30/t. In 2009 Premier Gordon Campbell, who instituted the CT, was reelected for the fourth time. British Columbia has shown the rest of Canada, a country with high carbon emissions per head, that a CT can achieve multiple benefits at minimal cost (The Economist 2011).

The US Local Governments Experience Boulder, Colorado

In 2002 the Boulder City Council committed to reducing GHG emissions to 7% below 1990 levels by 2012. In 2006 the city voted to initiate the first carbon tax in the US.

Residents pay $0.0022 per kWh, commercial customers $0.0004, and industrial customers $0.0002 which comes to approximately $12-13 per ton of CO2 (Brouillard and Van Pelt 2007).

"In 2008, our collective efforts helped keep roughly 81,000 metric tons of carbon dioxide from entering the atmosphere. We're not to our goal yet, but we're heading in the right direction" (City of Boulder 2009).

The CT is augmented by programs to retrofit existing buildings and replace appliances to improve energy efficiency, to promote energy-conserving behavior, *to* maximize opportunities for energy efficiency in new buildings, to promote use of renewable energy sources for individual buildings and sites, to increase renewable sources in their regional energy supply, to reduce and eventually eliminate the amount of waste going to landfills, and to plant more trees and protect the existing urban forest.

Bay Area Air Quality Management District (San Francisco, California Area)

In 2008 the BAAQMD prepared a draft rule for a CT of 4.2 cents per metric ton of CO_2 applying to all district facilities emitting GHGs which received much publicity. It appears in 2012 this is Regulation 3, Schedule T: Greenhouse gas fees. The GHGs include CO_2, methane, nitrous oxide, and other GHGs, but not biogenic CO_2. The fees for a permitted facility are based on the sum of the CO_2 equivalent ($0.048/metric ton) of all the GHG emissions, which are determined by multiplying the annual emissions, by a global warming potential (GWP) value. The "GWPs compare the integrated radiative forcing over a specified period (i.e., 100 years) from a unit mass pulse emission to compare the potential climate change associated with emissions of different GHGs."

The audited revenue and transfers from GHG fees in 2010 were $1,240,070.

Information

For more information on tax rates, annual revenue, and revenue distribution for the above countries and Boulder, see the Carbon Tax Center and Sumner et al. (2009).

Cap-and-Trade Programs

US Congress: With a history of standards and CAC regulation, the USA has begun to introduce bills for market-driven incentives[17] to reduce GHGs.

Several CAT bills have been introduced over the years, but none has passed.[18]

The Waxman-Markey American Clean Energy and Security Act H.R. 2454 has been the most successful passing the House in June 2009, 219–212 but now is dead.[19] Its goals were to create clean energy jobs, achieve energy independence, reduce global warming, and be a transition to a clean energy economy. This bill provided 15% of the auction revenues for low-income assistance.

Congress has approved no CT or CAT systems. In the 112th Congress, it seems the no-government action "solution" is the market choice of the majority.

U.S. Environmental Protection Agency (U.S.EPA): While Congress has not implemented a CAT program, the U.S.EPA has. Their successful acid rain trading program is the major forerunner of air quality, market-based approaches in the US.[20]

The Clean Air Interstate Rule (CAIR): CAIR was issued on March 10, 2005 covering 28 eastern states and the District of Columbia. "This rule provides states with a solution to the problem of power plant pollution that drifts from one state to another." The rule gives the states two options: first, requiring power plants to participate in an EPA-administered interstate CAT system and second, to "meet an individual state emissions budget through measures of the state's choosing." When this CAT program is fully implemented, the U.S.EPA expects SO_2 and NO to decline by 70% from 2003 levels.

By the year 2015, CAIR is expected to result in the following: nearly $100 billion in annual health benefits annually preventing 17,000 premature deaths, millions of

---

[17] The Obama Administration has proposed a CAT program. Emission credits would be auctioned off and the "proposed budget directs $15 billion per year of the funds toward clean energy technologies, while directing the remaining funds toward a tax cut" (US Department of Energy 2009).

[18] To follow the current status of congressional bills, see www.gov.track.us.

[19] "…Republican leaders called the legislation a national energy tax and predicted that those who voted for the measure would pay a heavy price at the polls next year" (Broder 2009).

[20] The goal is to achieve SO_2 and NO_2 reductions at the lowest cost while encouraging pollution prevention and energy efficiency by allowing trading of authorized allowances in the open market. This program reduced annual SO_2 emissions by 56% compared to 1980 levels and 52% compared to 1990 levels. Sources emitted 1.9 million tons below the emission cap of 9.5 million tons which is even below the 2010 cap of 8.95 million tons (U.S.EPA 2008)

lost work and school days, and tens of thousands of nonfatal heart attacks and hospital stays,; nearly $2 billion in annual visibility benefits in southeastern national parks; and significant regional reductions in sulfur and nitrogen deposition, reducing the number of acidic lakes and streams in the eastern USA.

However, electric power producers challenged this ruling in court on the grounds that EPA usurped its authority by requiring greater pollution reductions than called for in the 1990 Clean Air Act and on July 11, 2008 the U.S. Court of Appeals for the District of Columbia Circuit, unanimously struck it down. However, a December 23, 2008 ruling leaves CAIR in place until EPA issues a new rule to replace it (U.S.EPA, CAIR).

The Clean Air Mercury Rule (CAMR)

CAMR was the first nationwide CAT control rule. Its goal was to reduce coal-fired power plant emissions of mercury from 48 t/year to 15 t/year. This involved a moving cap from 38 t in 2005 to 15 t in 2018. As with CAIR this went to court and the U.S.EPA lost. Environmental organizations and Native American tribes filed the suit because the program removed power plants from the Clean Air Act's list of toxic sources and replaced it with a CAT system.

The USEPA has proposed standards to limit mercury, acid gases, and other toxic pollution from power plants to replace the court-vacated Clean Air Mercury Rule (U.S.EPA 2005).

State and Local Governments: Not all CAT systems spring from the federal government. Three non-federal government programs are discussed below: the Regional Greenhouse Gas Initiative (RGGI), the Western Climate Initiative (WCI), and the South Coast Air Quality Management District (SCAQMD) (Los Angeles area).

Regional Greenhouse Gas Initiative (RGGI):[21] RGGI, initiated in 2003, consists of nine states[22] and is the first mandatory, market-based program for the reduction of GHGs in the USA. Its goal is to reduce CO_2 emissions from power plants 10% by 2018. Emission allowances are sold through sealed bid, uniform price auctions, and the proceeds used to benefit consumers by providing clean energy technologies such as energy efficiency and renewable energy.[23]

The RGGI program highlights some of the issues in setting up a CAT system. The framework for RGGI's CAT program, based on EPA's NO_x Budget Trading

---

[21] The information here is mainly from the RGGI web site at www.rggi.org/.

[22] Connecticut, Delaware, Maine, Maryland, Massachusetts, New Hampshire, New York, Rhode Island, and Vermont. New Jersey is no longer a member of RGGI.

[23] "The fourth quarterly auction of carbon allowances raised more than $104 million for 10 Northeastern states to invest in energy efficiency and renewable energy programs. Officials announced Friday that all 30.8 million allowances offered on June 17 were sold for *$3.23* each" (RGGI 2009).

Program for State Implementation Plans, provides guidance, consistency, and flexibility for the participating states in preparing their trading programs. The states were given some leeway in formulating their plans with regard to applicability and source exemptions, allowance applications, and set-asides. Consistency between states is important "to provide for participation in a regional allowance trading program." Only fossil fuel-fired electric-generating units serving a generator of 25 MW or above are included which emit about 95% of the CO_2 from power plants. Emissions from eligible biomass (sustainable harvested woody and herbaceous fuel sources available on a renewable or recurring basis) may be deductible.

Banking of emissions is allowed. Each source must install a monitoring system and record, report, and quality assure all the data that is to be kept for 10 years. Each allowance is for one ton of CO_2 and each ton in excess of the budgeted amount is a separate violation.

The 2012 Auction Schedule is available at http://www.rggi.org/market/co2_auctions/upcoming_auctions.

Allowances are tradable making for mutually beneficial exchanges.

The initial cap has emissions near the level at the beginning of the program and would stay at that level 2009–2014. From 2015 to 2018, there would be a 2.5% emission reduction. Sources that are subject to the Acid Rain Program may submit a statement that they already met the requirements. The system provides offset allowances for projects outside the capped area if they reduce or sequester GHGs.

These offsets if they are real, additional, verifiable, enforceable, and permanent may be used to satisfy a limited amount of the sources' reduction requirements. Offsets are limited to landfill methane capture and destruction; reduction of sulfur hexafluoride; sequestration of carbon due to afforestation; avoided methane from manure management operations; and reduction or avoidance of CO_2 emissions from natural gas, oil, or propane end-use combustion due to end-use energy efficiency in the building sector. The CAT also includes programs for early reduction of emissions and voluntary ratepayer purchases of qualified renewable energy. The early reduction allowance program grants allowances for qualifying reductions made before the program starts. The voluntary ratepayer purchases allow the cap to be lowered if the purchases are voluntary, eligible, and renewable energy on behalf of retail customers.

This program with its auctions provides "market signals and regulatory certainty so that electricity generators begin planning for, and investing in, lower-carbon alternatives…without creating dramatic wholesale electricity price impacts and attendant retail electricity rate impacts."

Western Climate Initiative (WCI)[24]: The WCI was founded in 2007 when Arizona, California, New Mexico, Oregon, and Washington agreed to develop a regional

---

[24] Information here is from the WCI web site.

target for reducing GHGs, to participate in a multi-state registry to track and manage GHG emissions, and to develop a market-based program to reach the target. In 2007 and 2008 British Columbia, Manitoba, Ontario, Quebec, Montana, and Utah joined the original five states to deal with CC on a regional level and produced a *Design for the WCI Regional Program* released in July 2010. British Columbia, California, Ontario, Quebec, and Manitoba continue to work together through the WCI to develop and harmonize their emission trading program policies. They also work with Western, Midwestern, and Northeast states on climate and clean energy strategies through the North America 2050 Initiative.

They have forged a comprehensive strategy to mitigate climate change that will spur investment in clean energy technologies, create green jobs, and reduce dependence on import oil. When fully implemented, the plan will reduce GHG emissions to 15% below 2005 levels by 2020 (WCI 2012).

California is working closely with British Columbia, Ontario, Quebec, and Manitoba through the *Western Climate Initiative* to develop harmonized CAT programs that will deliver cost-effective emission reductions. The WCI jurisdictions have formed a nonprofit corporation, WCI, Inc. to provide coordinated and cost-effective administrative and technical support for its participating jurisdictions' emission trading programs. Just as with other voluntary agreements that ARB establishes with local air districts, states, federal government, and contractors, ARB's agreement with WCI, Inc. does not confer any decision-making authority; decisions concerning the ARB's CAT regulation are made by ARB at the direction of the ARB Board. More details on the organization and operation of WCI, Inc., can be found at: http://www.wci-inc.org/.

The South Coast Air Quality Management District (SCAQMD): The SCAQMD designed a CAT system for SOx and NOx that was adopted in 1993 called RECLAIM for REgional CLean Air Incentives Market. There were problems initially with emissions being higher than allowed. However, the U.S.EPA evaluated the program and concluded, "SCAQMD has been effective managing RECLAIM and modifying the program to adopt quickly to changing conditions." They also identified issues related to the successful operation of CAT programs. First, significant planning, preparation, and management are needed for development and for the life of the program. Second, consistency in the market and policies is needed for confidence and trust in the program. Third, the design of the program must enable the program to react quickly and effectively to unforeseen developments. Forth, "Periodic evaluation, revisiting of program design assumptions, and contingency strategies are crucial to keeping programs on track" (U.S.EPA, Region 9).

Another facet of the RECLAIM program is how they handled "hot spots." Hot spots result when several sources in a given area buy RECLAIM trading credits (RTC) in a similar time frame, resulting in excess pollution in that area. SCAQMD handled this by dividing the region into two zones: a coastal zone and an inland zone. Sources could then buy RTCs only from sources in their own zones.

Conclusion

This chapter provides a nontechnical introduction into the major economic issues involved in combating climate change (CC). It is hoped that this overview allows the readers to quickly come up to speed as to the broad view of economic issues facing the problem of global warming. For decision makers involved in the more technical and complicated aspects of designing and implementing efficient solutions to the problem, for example, how to design efficient auctions under a cap-and-trade (CAT) system, references are included for further, more detailed, and technical sources.

This chapter was written with policy makers in mind who have decided something must be done and provided information to help make complicated decisions a bit easier. First, there is the choice between direct regulation and market mechanism. Second, because direct regulation has been shown to be less efficient than a market solution, the rational choice comes down to choosing among the three market solutions discussed in this chapter. Because the no-government market solution does not take into account the externality problem from the burning of fossil fuels, the final serious choice comes down to choosing between a carbon tax (CT) or a CAT system.

While both options have their positive attributes, most economists are advocating the simpler CT over a CAT system; however, the CT would have to be implemented by a government where both political parties are averse to raising taxes in this political climate, especially republicans who control the House of Representatives.

The answer to this dilemma lies in education and understanding, which is the main purpose of this chapter. This education is in two parts: First, exposing disinformation for what it is, for example, the past chairman of the RNC wrote that the additional cost of the government CAT plan would be $260/month per family, while the USEPA estimated the same cost at $80-111/year and see also the Krugman, 2010 reference. Second, educating people that while a CT is a tax, if implemented properly, it would not raise taxes, rather it would be a more efficient distribution of tax revenue and incentives, for example, taxing people's negative actions, polluting, while lowering taxes on their positive actions, earning a living.

Given the state of the economy today and the current political climate implementing either an efficient CT or a CAT system[25] will be extremely difficult which increases the importance of educating the public and policy makers on the benefits of these two market solutions.

---

[25] For a comprehensive review and critique of policy designs and the costs of reducing GHGs, see Congress of the United States 2008 and "The Costs of Reducing Greenhouse Gas Emissions", November 23, 2009.

References

AB32 (2006) Assembly Bill 32, California Global Warming Solutions Act of 2006

Alchian A (1983) Exchange and production: competition, coordination and control. Wadsworth Publishing, Belmont

Aldy JE, Krupnick AJ, Newell RG, Parry WH, Pizer WA (2009) National Bureau of Economic Research working paper no. 15022

American Lung Association web site (2012) State of tobacco control, "State cigarette excise tax"

Broder J (2009) House passes bill to address threat of climate change. New York Times, 26 June 2009

Brouillard C, Van Pelt S (2007) Community takes charge: boulder's carbon tax

Bruvoll A, Larsen B (2002) Greenhouse gas emissions in Norway do carbon taxes work? Discussion papers337, Research Department of Statistics Norway

California Environmental Protection Agency, Air Resources Board. AB32 Scoping Plan

Carbon Tax Center. www.carbontax.org

City of Boulder (2009) Community guide to boulders' climate action plan

Congress of the United States, Congressional Budget Office (2008) Policy options for reducing CO2 emissions

Dawidoff N (2009) Respected scientist takes on climate 'ideology. The Global edition of the New York Times, 28–29 Mar 2009. The scientist quoted is Freeman Dyson of the Institute for Advanced Study, Princeton

Dutton J, Thomas A (1984) Treating progress functions as a managerial opportunity. Acad Manage Rev 9(2):235–247

Economist (2011) We have a winner: British Columbia's carbon tax woos sceptics, 21 July 2011

Europa (2010 and 2012) The EU's web site, www.europa.eu/.

Finland (2008) Finnish ministry of the environment web site, 14 Jan 2008, www.environment.fi

Fouche G (2008) guardian.co.uk, 29 April2008

Frank R (2007) Reshaping the debate on raising taxes. New York Times, 9 Dec 2007

Harrison D (2007) Per Klevnas and Daniel Radov. Worldwide: emissions trading for air quality and climate change in the United States and Europe – part 1, 20 Sept 2007

Holt C et al (2007) Auction design for selling CO_2 emission allowances under the regional greenhouse gas initiative. Final report, 26 Oct 2007, investigators: Holt C, Shobe W, University of Virginia. Dallas Burtraw, Karen Palmer, Resources for the future. Jacob Goeree, California Institute of Technology

Inhofe J Sen. R-Okla (2008) Senate speech, 1 Apr 2005, from his web site www.inhofe.senate.gov/

Krugman P (2010) Who cooked the planet? New York Times, 26 July 2010, p 23

Lewis M (2008) Big coal's big-time lobby. I watch news web site, 26 Aug 2008

Mall A (2008) Energy policy meltdown: bush administration and oil shale. NRDC press release 22 July 2008

Milgrom P (2004) *Putting auction theory to work*. Cambridge University Press, New York. ISBN: 0521536723

Milne JE (2009) Carbon taxes versus cap-and-trade: the relative burdens and risks of market-based administration. In: Lye L-H et al (eds) Critical issues in environmental taxation: international and comparative perspectives, vol VII. Oxford University Press, Oxford, England

National Petroleum Council (NPC) (2007) Heavy oil. Topic paper 422, working document of the NPC global oil and gas study made available, 18 July 2007 NPC web site www.npc.org/

New York Times Editorial (2009) Collapse of the clean coal myth, 22 Jan 2009

Nordhaus W (2008) After Kyoto: alternative mechanisms to control global warming, 27 Mar 2008

NRDC (2009) The consequences of global warming. The NRDC web site, 22 July 2009

NRDC (2012) Quotes from the Natural Resources Defense Council web page, www.nrdc.org/

Patel T (2012) France vote outlaws 'fracking' shale for natural gas, oil extraction, Bloomberg web site, www.bloomberg.com/. 3 Aug 2012

Pope C, Sierra Club Executive Director (2009) Bush administration's answer to BP oils spills and shutdown: more drilling in Alaska, 23 Aug 2006 from the Sierra Club web site www.sierraclub.org/ 22 July 2009

Prasad M (2008) On carbon, tax and don't spend. York Times, 25 Mar 2008

RGGI (2009) RGGI web site, www.rggi.org/rggi

Reich R (2007) Inherit the windfall: tax oil company profits to pay for alternative energy initiatives. American Prospect, 7 Feb 2007

Rosen RA (2006) The threat to the planet: an exchange by Richard A. Rosen, Ruth F. Weiner. The New York Review of Books, 21 Sept 2006

Shrum TA (2007) Greenhouse gas emissions: policy and economics. Report prepared for the Kansas Energy Council Goals Committee, 15 June 2007

South Coast Air Quality Management District. REgional CLean Air Incentives Market (RECLAIM) web site, www.aqmd.gov/

Stavins R (2007) A U.S. Cap-and-trade system to address global climate change. Hamilton Project discussion paper, Oct 2007. On line document available at Brookings, search cap and trade, www.hks.harvard.edu/fs/rstavins/cvweb.html

Steele M (2009) Chairman of the Republican National Committee. 11 Feb 2009 and 18 May 2009 emails

Stern N (2006) The economics of climate change: the stern review. Cambridge University Press, New York

Sumner J, Lori B, Smith H (2009) Carbon taxes: a review of experience and policy design considerations. U.S. Dept. of Energy/National Renewable Energy Laboratory, Golden, Colorado

U.S. EPA. Superfund web site, http://www.epa.gov/superftind/

U.S. Dept of the Interior, Bureau of land management (2009) Oil shale and tar sands. Web site, www.doi.gov. July 2009

U.S. Dept. of Energy (2009) President's budget draws clean energy funds from climate measure. EERE, 4 Mar 2009

U.S.EPA (2005) Web site news releases by date, EPA announces first-ever rule to reduce mercury emissions from power plants, release date: 15 Mar 2005, www.epa.gov/camr

U.S.EPA (2008) Web site Clean Air Markets, 2008 emission, compliance, and market analyses web site, www.epa.gov/airmarkets. 23 Sept 2009

U.S. EPA (2009) Web site cap and trade, 15 Sept 2009

U.S. EPA (2009) EPA analysis of the American clean energy and security act of 2009 H.R. 2454 in the 111 *Congress,* 23 June 2009

U.S. EPA (2009) Web site clean air markets, clean air interstate rule (CAIR) web site www.epa.gov/cair/. Accessed 23 Sept 2009

U.S. EPA Region 9, EPA's evaluation of the RECLAIM program in the south coast air quality management district

UNFCCC, United Nations framework convention on climate change, home page web site, unfccc.int/

Volker P (2007) From an internet news story, Associated Press, 6 Feb 2007

Walsh B (2009) Exposing the myth of clean coal power. Time, 10 Jan 2009

Wasserman L (2010) Four ways to kill a climate bill. New York Times, 25 July 2010

WCI (2012) Online information can be found at www.westernclimateinitiative.org

Chapter 4
Qualitative Economics: The Science Needed in Economics

Michael Fast and Woodrow W. Clark II

Abstract This chapter is about science from a book that on Qualitative Economics (Clark and Fast 2008), specifically building a science of economics, grounded in understanding of organizations and what is beneath the surface of structures and activities. Economics should be, as a science, concerned with its assumptions and how to develop and formulate theories of ideas and reality that produce descriptions of how to understand phenomenon that create experiences, hypotheses generation, and replicable data for prediction, which need to be connected to everyday business life. Economics has to start with a discussion involving the philosophy of science.

There is a "disconnection" between economics which focuses on statistical structures and universal laws from those that are in contrast with the everyday of life of business activity, which are processual and dynamic. This discussion is the central issue in the chapter and is discussed from the perspective of interactionism (Blumer 1969). It is a perspective developed from the lifeworld philosophical traditions, such as symbolic interactionism and phenomenology, seeking to develop the thinking of economics through the use of linguistics (Clark and Fast 1968).

The argument is that economics first of all is about two things; it is about interaction and it is about construction. If we are not able to understand and describe how people interact and construct, we cannot develop any theory of economics or understand human dynamics that is scientific.

M. Fast (✉)
Department of Business and Management, Aalborg University, Aalborg, Denmark
e-mail: fast@business.aau.dk

W.W. Clark II, Ph.D.
Clark Strategic Partners, Beverly Hills, CA, USA
e-mail: wwclark13@gmail.com

Introduction

Economics is based upon human interaction and construction in everyday of life. So to develop economics into a science that can describe and understand human dynamics, the focus must be on the demands for such a science in relation to its ontology and epistemology. Thus language and the theories from linguistics play a critical role in making economics a science.

The dominant and traditional view on economics is that it is a matter of construct theories that can explain the laws invisible to the eye and under the surface. This is the tradition developed during the nineteenth and twentieth centuries and has its roots in positivism (Comte 1991 and, Durkheim 1991) and rationalism (Descartes) and later on in system theory (Bertalanffy 1971). The epistemological question here is if the factors and laws are connected, not in relation to reality but to the models and the constructed theoretical universe. There are no empirical arguments for if and in which way reality is constructed as a system or as a mathematical reality and if it is possible that reality can be explained strictly on numbers or if there are universal laws which are only assumed by the tradition.

An alternative to those concepts of science comes from the philosopher in connection with the development of a subjectivistic approach, Immanuel Kant (1724–1804). He is one of those philosophers giving inspiration to an alternative philosophic tradition and scientific conception. Kant thought that the inner activities of man as conceptualized in the minds of human beings must be brought into focus. Our thoughts are not turned toward the objects, as they are represented or defined in themselves, independent of human intersubjectivity. Science has only to understand the world in so far as we have shaped it ourselves by forming ideas of it. If therefore the sciences shall have at least an element of truth in their analyses, pronouncements, and validity, they must build on the relative necessity,[1] which is maintained by the intersubjective everyday life reality experienced by man. Sciences do not constitute a reference system standing above, abstracted and removed from the world to justify the validity of everyday life. The scientific conceptualization rests on preconditions, which mankind places into science itself, by being a participant in the experiential world of everyday life. It is not necessary that the single scientist knows everything about the organizing of an experience. Therefore, he does not necessarily see the viewpoint presupposed by science or the basis of which he works himself. Kant's view of the relation between science and everyday life throws light on science as a human endeavor in which we are responsible ourselves for its outcomes. Schutz (1973b: 22) underlines that from a phenomenological perspective with the observation that social scientists' facts, events, and data are of a totally different structure than in the objective approach. The social world is not structureless in its nature. The world has a special meaning and structure of relevance to those people that live, think, and act in it. Human beings have pre-chosen and pre-interpreted this world through a set of common-sense constructions of everyday of life reality. Such

---

[1] That is the general understanding of man.

a construct of the world outlines those topics of thoughts that determine individual's actions, defines the aim for their actions, the means to achieve them, and that are accessible to reach them. This perspective helps people to orientate themselves in their natural and sociocultural milieu and to become comfortable with in it. The topics of thoughts that are constructed by the social scientist refer to and are founded upon the topic of thoughts that are constructed by an individual's common-sense thinking as they live their everyday lives among other people. The constructions therefore that the scientist use are thereby constructions of a second-order, namely, constructions of the constructions that are performed by the actors on the social scene.

If we are looking for what is meaningful in understanding reality, we must have concepts of what that reality is. This is the area of ontology and in relation to economics, we have to connect the discussion of economic figures, relations, forces, etc., to where they arise and in which way they are meaningful. The only way to do this is to take the departure in the subject and the subject relation to the phenomenon: both the economic actor and the researcher who is trying to understand the subject. We need a moving picture of what the economic actor is and what his realities are, and we need a focus upon how knowledge of this is produced.

In order to develop such a picture of everyday economic interactions, we have to focus upon what will be described as "qualitative economics," as a perspective and understanding of economics. Qualitative is seen in the complex construction by the actors of the economic organizing. The roots in this are in the traditions of *"lifeworld"* and interactionism. Lifeworld comes from the German *die Lebenswelt*, with its roots in the eighteenth century philosophy of Kant and later on Husserl, Heidegger, Schutz and Gadamer and can also be seen in the tradition of American philosophers' Mead and Blumer from the early to mid-twentieth century. The theoretical development from this philosophical tradition is seen in different schools of contemporary social science thought ranging from phenomenology, hermeneutic, ethnomethodology, linguistics, and symbolic interactionism. The lifeworld tradition and its interactionistic theoretical development is an approach to theorizing, describing, understanding, and explaining everyday life and is therefore creating the science of qualitative economics.

The aim of this chapter is therefore – through the everyday life tradition, culture, language and their interactionism– to discuss the central issues and basic concepts in order to understand and develop a qualitative economic perspective.

The Logic of Qualitative Economics: The Object of Thought

The *reality* of economics has been investigated and explained in many ways. But the discussion of how to understand the business research and how the research is done along with the (ontological and epistemological) assumptions lying behind the research and its reality in everyday life are rarely discussed. Discussions of philosophy in science and methodology are not only important but the departure for understanding reality and theorizing on its applications in everyday life. It is precisely these connections among philosophy of science that theorizing and methodologies arise to capture the reality, which must be in the center of any scientific discussion. Furthermore, openness and a specific discussion of an alternative philosophical approach to the established traditional way of seeing science and reality are necessary. Thinking and reflection are critical in the scientific investigation of reality together with and related to the basic philosophical assumptions. It is only in this connection that we can talk about something being true (e.g., correct) or false.

We will discuss how to understand the very concept of organizations and how organizations are constructed and developed. We need to have an understanding of what people are and what they bring to the organizational economic context by interacting with one another and in groups.

When the functionalistic economic theory fails to understand business life, the root to the problem is in the lack of a conceptual discussion on the very understanding and meaning of business activities within the firm. This section focuses on interaction and the firm as a social construction and upon understanding the process of change and development of the firm. The purpose is to discuss a conceptual understanding of the firm as a subjective, interactionistic, and processual phenomenon. The discussion focuses upon the way in which actors in their everyday of life create an understanding of business reality and through their actions and interactions construct and change the firm.

The Constitution of the "Firm"

All business and economic activities are conducted by individuals communicating, where the relations consist of concrete meetings. The words *"organization"* or *"firm"* are (only) a concept, which we use to describe a phenomenon. It is a conceptualization of what we believe and do and what we orient our actions toward. Organization is a concept in the same way as the concepts of family, class in school, a football team, a union, etc. In other words, organization is a phenomenon that we experience when and where we see more than one person involved in activities over time.

Thus, organization becomes a collective arrangement where people try to give the situation and the activities meanings. In line with Blumer (1969), organizations consist of *the fitting together of lines of activity – the interlinking of lines of action*. Actors mixing, sharing, competing, and cooperating are parts of the interactive process that define groups and organization. And that is why most organizations, by definition, change and move dynamically in space and time. By fitting together the lines of action and interaction as logically prior in organization, we are discouraged from mistakenly regarding organizations as "things" or simply "solid entities" such as a building or structure. Organizations are not concrete, immutable, or even lifelike objects that, somehow independent of our conscious intentions or unconscious motives, shape and determine what we do. The technical term for this kind of cognitive error is "reification," an unconscious tendency to forget or be obvious to the role of human agency in creating, sustaining, and transforming social relations (Hummel 1990: 12). We actively construct our social reality through our language, through a process of symbolization by forming words and sentence to describe our experiences as well as our wants and desires. We create our organizational existence and live within it.

The language we share and use from our culture and traditions constitutes our relationships (White 1990: 82). An organization should therefore be understood through the actors who by their actions and knowledge create the firm in their everyday pursuit of life. In this the relation between action and knowledge is the central issue of interaction.

The actions exist in a context created by the actor through his/her actions. The action is related to the actor's interpretation and understanding of the situation in the context of meanings imparted in the interaction of the phenomenon (Blumer 1969; Schutz 1972; Mead 1962; Brown 1978; Jehenson 1978 among others). The actor has motives and definitions of the situation that makes the social world into an inner logic, which have rules and lines of action derived from the situation itself. Actions also happen in connection with expectations. When the actors are involved in the society, they expect suitable actions from themselves and from others: They are capable of understanding meanings of action by others and make their own point of view on themselves based on the response of other actors. They associate meanings to situations and to other actor's actions and act in relation to their interpretations of these meanings. This can be understood in relation to typifications, formed by the earlier experiences of the actor, which define his/her *"thinking in future"* of others' possible reaction to his/her actions.

The typifications that the actor uses in a situation are dependent on his/her knowledge in everyday life that is *"the stock of knowledge"* and *"the generalized other"* as Blumer (1976a, b, c, d) described the phenomenon. These typifications give the individual a frame of reference that the actor can use to create actions and make sense of others' actions. See Blumer's (ibid) notion of "reflections," for example. Typifications are thereby expectations to others actions containing symbols in relation to community and collective interpretations.

This social reality is predefined in the language by which we are socialized. The language gives us categories that both define and emphasize our experiences. The language spoken and dialogue among actors within an organization can be seen as communication of meanings and actions. But such language usage is also a means to create a new understanding, changes in meanings, and a new worldview. Language is the base line from which we understand and can interpret knowledge. Thus, knowledge, as expressed in language usage, can thereby be understood as moving pictures of reality: experiences and information are produced through actions and transformed (by interpretation and retrospection) to the knowledge that the actor's experiences are useful and relevant.

The world with which the actor is confronted is composed of experiences which the process of consciousness will develop or simplify toward different paths (or structures) and then become transformed into actions (again). The actor uses and develops a scheme for interpretation to connect episodes of social action in a sensible way. A "scheme" should be understood as active information seeking pictures that accept information and orient actions continuously (Weick 1979; Bartunek 1984). The action-knowledge process gives an understanding of the way in which people think, act, reflect, and interact. Simultaneously, it shows that the actors are engaged in their environment by means of interpretation and orientation with one another.

The focus on the understanding of the organization is the way organizational members interpret their organizational world, which is nothing else than a special sphere of the individual's lifeworld. Lifeworld refers to the fact that in any real-life experience, there is something that is given in advance or something that exits in advance and, thus, taken for granted. This taken-for-granted world includes our everyday life and whatever prejudices and typical interpretations we may derive from it. Acting as a member of an organization, therefore, does not differ essentially from acting as an individual, for "whether we happen to act alone or, cooperating with others, engage in common pursuits, the things and objects with which we are confronted as well as our plans and designs, finally the world as a whole, appears to us in the light of beliefs, opinions, conceptions, certainties, etc., that prevail in the community to which we belong" (Gurwitsch, in Jehenson 1978: 220).

The important characteristic of this experience in any organization becomes the typical form of everyday life or as described by Schutz (1990a: 7): "The individuals common-sense knowledge of the world is a system of constructs of its typicality." In social interaction, the role of typification is important and can be expected to vary according to the nature of the relationship.

Environment

The actors in their "environment" construct reality and knowledge. It is precisely because knowledge is related to and has an orientation toward the "environment" through interactions that the environment itself can be defined as *the experiential space* and as *the interpretation space*.

The experiential space is what is close and concrete, where, for example, the actors move around and interact. This can be seen in the consciousness of human beings in "the natural attitude" first of all being interested in that part of the actor's everyday of lifeworld that is in his reach and that in time and space are centered around him/her (Schutz 1973b: 73). The place where the body occupies the world, the actual here, is the point from which one orientates oneself in the space. In relation to this place, one organizes elements in the environment. Similarly, the actual now is the origin of all the time perspectives under which one organizes events in the world as before and after, and so on. This experiential space is experienced by the actor as the core of reality, as the world within their reach. It is the reality in which all humankind are engaged.

The interpretation space can be seen as the reality beyond the actor's knowledge (e.g., through stories and tales) where something which the actor relates to, but which is not centered around his or her everyday of life, for example, not in time. In relation to this, we can see the distinction that Weick (1999: 2) talks about when he says that humans live in two worlds – the world of events and things (or the territory) and the world of *words* about events and things (or the map). In this, the process of abstraction is the process that enables people to symbolize (Blumer 1969) and is described as "the continuous activity of selecting, omitting, and organizing the details of reality so that we experience the world as patterned and coherent." This process becomes necessary but inherently is inaccurate because the world changes continuously and no two events are the same. The world becomes stable only as people ignore differences and attend to similarities. In a social constructed world, the map creates the territory. Labels of the territory prefigure self-confirming perspectives and action.

This perspective also means that the development of knowledge has its start in the actor's existing knowledge or as Weick (1999: 5) put it: "it takes a map to make a map because one points out differences that are mapped into the other one. To find a difference, one needs a comparison and it is map like artifacts which provide such comparisons."

The development can be seen in relation to the actor's everyday experiences with the attempt to orient him/herself and to solve problems. When the actors act in their experiential space, they thus widen their understanding of reality by interpreting and relating themselves to the result of the actions. Development of knowledge involves interpretation and retrospection whereby the actors create their experiential space: Reality is what one sees; hence, it changes every time the actor constructs a new concept or a picture of connections. Development of knowledge thus demands that the actor reflects and relates to an understanding of the situation and the experiential space.

The essence is in the idea that we all develop knowledge through actions and that actions are the means by which we engage ourselves in the reality; our actions construct and keep us in touch with the world (Garfinkel 1967; Morgan and Ramirez 1984). The action-knowledge discussion is built upon the assumption that we only have a reality in force of that we are engaged in it: reality is socially constructed. This does not imply that people are in full control over the process of constructing the reality or that they have possibilities to change it basically because they do not act alone and because it is an ongoing process.

It is necessary to take this discussion of actors, actions, and knowledge and develop an understanding of the way in which people are orientated toward each other and in which way the organizational reality actually becomes a reality.

Interaction and Knowledge

Interaction is symbolic in the sense that actors respond to the actions of others, not for some inherent quality in them, but for the significance and meanings imputed to them by the actors. Meanings shared in this way, in an intersubjective way, form the basis for human social organization (Singelmann 1972: 415). People learn symbols through communication (interaction) with other people, and therefore many symbols can be thought of as common or shared meanings and values (Rose 1962: 5). This mutually shared character of the meanings gives them an intersubjectivity and

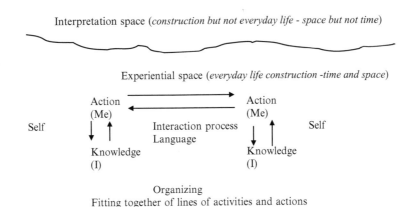

Fig. 4.1 Knowledge and interaction

stresses that it is interaction and intersubjectivity that constitute the firm as a reality for the actors. Interaction in this relation should be understood as a complete sequence of interaction, as a process of interaction (see also Mangham 1978).

The central point in this is the time perspective and the dependency of the context and the acts: It is the actions by the actor and the process of interaction that give and make the firm over time. The "firm" therefore both has a past (the experiences of the actors) and a present (the actors' interpretations and pictures) and a future in relation to the actors' fantasies of the future and orientations. The processes related to interaction are presented in the figure below.

Figure 4.1 outlines interaction between the actors in the firm. It is a process of knowledge development, which occurs through the process of interaction in an experiential space. It is intersubjective and can be seen as a moving picture that defines what the actors experience as important and real. Thus, knowledge has an impact on future actions and is central for an understanding of the actors' orientation and the organizational actions. The actors act in relation to the picture and definition they have of the experiential space and the situation. Each action means possibilities for experiences and information and for strengths or weaknesses in interpretation of connections in the situation. In every situation there is the possibility of several different interpretations. This means that changes in the experiential space create ambiguity, and the actors are tempted to use previous successful actions and interpretations – the existing picture of reality.

Organizing: Fitting Together of Lines of Activities and Actions

Through the processes of interaction, the actors construct some results: the interaction means organizing and creation of the firm, and the actors create a moving picture of and a relation to the experiential space. The actors create intersubjective moving pictures of the reality, which is an organizational paradigm.

The actors create over time something we define a "firm." The processes that occur can be understood as organizing, which not only focuses upon action and interaction but also on moving pictures of reality and intersubjectivity. Essentially, the firm can be understood as overlapping interactions. The actors create the firm through interactions, but "it" has also an influence upon them through their interpretation of "it." This dialectical perspective appears from the view that the firm only exists through the interactions between the actors and thus is viewed as a corollary of these interactions. Simultaneously, the organization is historically to the individual member: The individual enters into an already existing organizational everyday of life, which sets the institutional parameters for his self-development. Self and organization thus develop together and because of each other in a dialectical process of mutual transformation (Singelmann 1972: 415; Mead 1962; Berger and Luckmann 1966; Benson 1977; Arbnor and Bjerke 1997).

The actors have to live with and exist with uncertainty and ambiguity. In other words, the way in which the actors handle themselves is in itself uncertain and exposed to many different interpretations and understandings. To reach security, the actors attempt to organize their activities. Organizing means assembling the actions and should be seen in relation to interpretation and understanding by the actors. The actors form their actions so as obtain information and experiences that give meanings to the organizational world. This is organized by the actors in an attempt to construct an understanding. In organizing, the dependent actions are oriented toward removing contradictions and uncertainty: the actors seek to define and make sense in their situation, and thus they both create the firm and the experiential space. Organizing is to be seen as a social, meaning-making process where order and disorder are in constant tension with one another and where unpredictability is shaped and "managed." The raw materials of organizing are people, their beliefs, actions, and their shared meanings that are in constant motion (Sims et al. 1993: 9).

There is a similarity between the phenomenological meanings of the practical activity of organizing and theorizing – the act of sensemaking is in fact the central feature of both. Theorizing is most fundamentally an activity of making systematic as well as simplified sense of complex phenomena that often defy understanding by everyday, common-sense means. Theorizing might also be seen as a means by which people in organizations make their own and other's actions intelligible by reflective observations of organizing processes; through these processes, novel meanings are created and possibilities for action are revealed. Theorizing becomes an act of organizing, first, when it is a cooperative activity shared in by several or even all of the actors in an organizational setting and, second, when its purpose is to reveal hidden or novel possibilities for acting cooperatively. Organizing is cooperative theorizing and vice versa (Hummel 1990: 11). In short, the firm is a social construction and a collective phenomenon.

Interaction between actors in a situation allows for many different interpretations whereby the actors are facing multiple realities. The interaction between different opinions means that new conceptions may arise. The reality is seen differently which produces changes. Brown states that the organizational change could be seen as an analogy with scientific change (see also Imershein 1977):

> ...most of what goes on in organizations, involves practical as well as formal knowledge. That is, the relevant knowledge is often a matter of application, such as how to employ the official procedures and when to invoke the formal description of those procedures, rather than abstract knowledge of the formal procedures themselves. Paradigms, in other words, may be understood not only as formal rules of thought, but also as rhetoric and practices in use. (Brown 1978:373)

Bartunek (1984: 355) talks about an organizational paradigm as interpretive schemes, which describes the cognitive schemata that map our experience of the world through identifying both its relevant aspects and how we are to understand them. Interpretive schemes operate as shared, fundamental (though often implicit) assumptions about why events happen as they do and how people are to act in different situations.

The structures of meaning arises in and is institutionalized through the action of human beings, our own and those of our fellow men, and those of our contemporaries and our predecessors. All objects of culture (tools, symbols, language systems, social institutions, etc.) point back, through their origin and meaning, to the activities of human subjects. Intersubjectivity, therefore, can be seen as a common subjective state or as a dimension of consciousness that is common to a certain social group who mutually affects each other. The social connections are rendered possible through the intersubjectivity such as through a mutual understanding of common rules that are, however, experienced subjectively. Intersubjectivity refers to the fact that different groups may interpret and experience the world in the same way that is necessary at a certain level and in some contexts out of regard for collective tasks.

Human behavior is part of a social relationship, when people connect a meaning to the behavior, and other people apprehend it as meaningful. Subjective meanings are essential to the interaction, both to the acting person who has a purpose with his action and to others who shall interpret that action and react in correspondence with the interpretation (Blumer 1969; Ritzer 1977: 120). The basis for intersubjectivity is the social origin of knowledge or the social inheritance in which the acting persons are socialized to collectively typify repeated social events as external, objective events (which shall be seen in relation to structures of meaning). However, in consciousness such a typification is experienced as subjective reality.

Essence of all this is that the meaning people create in their everyday reality gives the understanding of why people are like they are which can be seen in their interaction and intersubjectivity, including their common interpretations, expectations, and typifications. As long as organizational actors act as typical members, they tend to take the official system of typification for granted as well as the accompanying set of recipes that help them define their situation in an organizationally approved way. The emergence of other, non-organizationally defined typifying schemes results from the breaking down of the taken-for-granted world when the actors enter into face-to-face relationships.

Connections of Everyday of Business Life: The Process of Thinking

Kant thought that the problem with all classical objective metaphysics was that it forgot to investigate the meaning and cognitive reach of its own concepts (cf. Wind 1976: 17). Kant was of the opinion that all cognition starts with the experience and that knowledge was a synthesis of experiences and concepts: without sensing we cannot be aware of any objects (the empirical cognition); without understanding we cannot form an opinion of the object (the a priori cognition).

> There can be no doubt that all our knowledge begins with experience. For how should our faculty of knowledge be awakened into action did not objects affecting our senses partly of themselves produce representations, partly arouse the activity of our understanding to compare these representations, and, by combining or separating them, work up the raw material of the sensible impressions into that knowledge of objects which is entitled experience? In the order of time, therefore, we have no knowledge antecedent to experience, and with experience all our knowledge begins. (Kant 1929/1787:41)

However, there are limits to knowledge. Kant distinguishes between the phenomena (the world of phenomena) and reality (the noumenal world): We cannot apprehend the mysterious substance of the thing, what he called *"das Ding an Sich"* (the-thing-in- itself). If we try to go outside the world of phenomena, that is, if we wish to use the concepts outside the limits of the comprehensible world, it will lead to paradoxes, fallacies, and pure self-contradictions. Kant argued that the traditional metaphysical arguments about the soul, immortality, God, and the free will all exceed the limits of reason. Reason can only be used legitimately in the practical sphere, that is, if we try to acquire knowledge of the world. If we cannot reach das Ding an Sich, then we must be satisfied with *"das Ding für Uns"* (the things as they present themselves to us).[2]

This is the question that we have to raise when we are studying the field of economics: What are the things, who are the actors, and in which way do the I (the economist) understand?

The primary goal of the social sciences is to obtain organized knowledge of social reality. Schutz understands social reality as the sum total of objects and occurrences within the social cultural world *as experienced by the "common-sense" thinking of men* living their daily lives among their fellow men, connected with them in manifold relations of interaction (Schutz 1970: 5). It is a world of cultural objects and social institutions in which we are born, in which we have to find our bearings and to come to terms with. Seen from outside, we experience the world we live in as a world which is both nature and of culture, not as a private world, but as an *intersubjective world*. This means that it is a world common to all of us, either actually given or potentially accessible to everyone, and this involves intercommunication and languages. It is in this intersubjective world that action shall be understood.

---

[2] Note also Husserl (1962) concept of intentionality.

In this everyday lifeworld, the actors use "common-sense knowledge" as kind of knowledge held by all socialized people. The concept refers to the knowledge on the social reality held by the actors in consequence of the fact that they live in and are part of this reality. The reality experienced by the actors as a "given" reality; that is, it is experienced as an organized reality "out there." It has an independent existence, taking place independently of the individual. However, at the same time, this reality has to be interpreted and made meaningful by each individual through his experiences – we experience reality through our common-sense knowledge, and this knowledge is a practical knowledge of how we conduct our everyday lives.

All our knowledge about the world involves constructions, that is, a set of abstractions, generalizations, formalisms, and idealizations which are specific for the organizational level of thoughts in question (Schutz 1973b: 21). Such things as pure and simple facts do not exist. According to Schutz (1973b: 47), social science must deal with the behavior of man and common-sense interpretation in the social reality, based on an analysis of the entire system of projects and motives, of relevances, and structures. Such an analysis refers necessarily to the subjective viewpoint, that is, to interpretation of the action and its surroundings from the viewpoint of the actor. Any social science that wishes to understand "social reality" must adopt this principle. This means that you always can and for certain purposes must refer to the activities of the subjects in the social world and their interpretation through the actors in project systems, available means, motives, relevances, etc.

To be able to understand the social reality and handle the subjective views, science must construct its own objects of thought, which replace the objects of common-sense thinking. This approach allows for an understanding of research work on models of parts of the social world, where typical and classified events are dealt within the specific field in which the research worker is interested. The model consists of viewing the typical interactions between human beings and to analyze this typical pattern of interaction as regards its meaning to the character types of the actors who presumably created them. The social research worker must develop methodological procedures to acquire objective and verifiable knowledge about a subjective structure of meaning.

In the sphere of theoretical thinking, the research worker "puts in brackets" his physical existence and thus also his body and its system of orientation, of which his body is the center and the source (Schutz 1973b: 96). The research worker is interested in problems and solutions, which in themselves are valid to anybody, everywhere, at anytime, anywhere, and whenever certain conditions, from which he starts, are present. The "jump" in theoretical thinking involves the decision of the individual to suspend his subjective viewpoint. And this very fact shows that it is not the undivided self, but only a partial self, a role player, and a "Me," that is, the theorist, who acts in scientific thinking. The features of the epoché, which is special for the scientific attitude, can be summarized through the following. In this epoché the following is put in brackets: (1) the thinking subjectivity as man among fellow men, including his bodily existence as psychophysical human being in the world, (2) the system of orientation through which the everyday Lifeworld is grouped in zones within actual, restorable, achievable reach, etc., and (3) the fundamental anxiety and the system of practical relevances, which originate from it (ibid.: 97).

The system of relevances, reigning within the province of scientific contemplation, arises in the random act of the research worker, when he chooses the object of his further exploration, that is, through the formulation of the existing problem. Thus, the more or less anticipated solution to this problem becomes the summit of the scientific activity. On the other hand, the mere formulations of the problem, the sections, or the elements of the world, which are topical or may be connected to it as relevant concerning the present case, are determined at once. After that this limitation of the relevant field will pilot the investigation.

The difference between common-sense structures and scientific structures of patterns of interaction is small. Common-sense structures are created on the basis of a *"Here"* in the world. The wide-awake human being in the natural attitude is first of all interested in the sector of his everyday lifeworld, which is within his reach and which in time and space is centered around him. The place that my body occupies in the world, my topical Here, is the basis from which I orient in the space. In a similar way, my topical *"Now"* is the origin of all the time perspectives under which I organize events in the world, like before and after, past and future, and presence and order (ibid.: 73). I always have a Here and a Now from which I orient and which determines the reciprocity of the assumed perspectives and which takes a stock of socially derived and socially recognized knowledge for granted. The participant in the pattern of interaction, led by the idealization of the reciprocity of the motives, assumes that his own motives are joined with those of his partner, while only the manifest fragments of the actions of the actors are available to the observer. But both of them, the participant and the observer, create their common-sense structures in relation to their biographic situation.

The research worker has no Here in the social world which he is interested in investigating. He therefore does not organize this world around himself as a center. He can never participate as one of the acting actors in a pattern of interaction with one of the actors at the social stage without, at least for some time, to leave his scientific attitude. His contact is determined by his system of relevance, which serves as schemes for his selection and interpretation of the scientific attitude which is temporarily given up to be resumed later. The research worker observes, assuming the scientific attitude, the pattern of interaction of human beings or their results, in so far as they are available to become observations and open to his interpretation. But he must interpret these patterns of interaction in their own subjective structure of meaning, unless he gives up any hope of understanding "social reality" on its own merits and within its own situational context.

The problematic that Schutz brings up here and the understanding that one may reach of the subjective knowledge of another person can be expressed in the following way. The whole stock of my experience of another from within the natural attitude consists of my own lived experiences of his body, of his behavior, of the course of his action, and of the artifacts he has produced. The life experiences of another's acts consist in my perceptions of his body in motion. However, as I am always interpreting these perceptions as "body of another," I am always interpreting them as something having an implicit reference to "consciousness of another." Thus, the bodily movements are perceived not only as physical events but also as a sign that the other person is having certain lived experiences, which he is expressing through those movements. My intentional gaze is directed right through my perceptions of his bodily movements to his lived experiences lying behind them and signified by them. The signitive relation is essential to this mode of apprehending another's lived experiences. Of course, he himself may be aware of these experiences, single them out, and give them his own intended meaning. His observed bodily movements become then for me not only a sign of his lived experiences as such but of those to which he attaches an intended meaning. The signitive experience of the world, like all other experience in the Here and Now, is coherently organized and is thus "ready at hand" (Schutz 1972: 100).

The point is how two "streams of consciousness" get in touch with each other and how they understand each other. Schutz expresses it quite simply, when he talks about the connection, as the phenomenon to "grow old together" and to understand the inner time (*durée*) of each other. In fact, we can each understand all others by imagining the intentional acts of the other when they happen. For example, when someone talks to me, I am aware – not only of the words but also of the voice. I interpret these acts of communication in the same way as I always interpret my own lived experiences. But my eyes go directly through external symptoms to the internal man of the person talking. No matter which context of meaning I throw light on, when I experience these exterior indications, its validity is linked with a corresponding context of meaning in the mind of the other person. The last context must be where his present, lived experiences are constructed steps by step (Schutz 1972: 104).

The simultaneousness of our two streams of consciousness does not necessarily mean that we understand the same experiences in identical ways. My lived experiences of you are, like the surroundings that I describe to you, marked by my own subjective Here and Now and not by yours. But I assume that we both refer to the same object that thus transcends the subjective experiences of both of us. But at the same time, not all your lived experiences are open to me. Your stream of lived experiences is also a continuum but where I can catch detached segments of it. If I could become aware of all your experiences, you and I would be the same person. Hence, the very nature of human beings is that they do not have exactly the same interpretation of experiences and therefore are different. It is precisely this human diversity that distinguishes humans from other life forms yet creates conflict and turmoil within societies and between them.

We also differ in other ways, how much of the lived experiences of the other we are aware of and that I, when I become aware of the lived experiences of the other, arrange that which I see within my own meaning context. And in the meantime the other has arranged them in his way. But one thing is clear: This is that everything I know about your conscious life is really based on my knowledge of my own lived experiences. My lived experiences of you are constituted in simultaneity or quasi-simultaneity with your lived experiences, to which they are intentionally related. It is only because of this that, when I look backward, I am able to synchronize my past experiences of you with your past experiences (ibid.: 106). My own stream of consciousness is given to me continuously and in all its perfection, but that of the other person is given to me in discontinuous segments and never in its perfection and exclusively in "interpreted perspectives." This also means that our knowledge about the consciousness of other persons can always be exposed to doubt, while our own knowledge about our own consciousness, based as it is on immanent acts, is in principle always indubitable. In the natural attitude, we understand the world by interpreting our own lived experiences of it. The concept of understanding the other is therefore the concept: "Our interpretation of our lived experiences of our fellow human beings as such." The fact that the you confront me as a fellow human being and not a shadow on a screen – in other words that the others duration and consciousness – is something that I discover through interpretation of my own lived experiences of him. In this way, the very cognition of a "you" also means that we enter into the field of intersubjectivity and that the world is experienced by the individual as a social world.

So in this discussion of how to understand phenomena and meaning, we have to focus on the central dimension: language.

The Use of Linguistics as Science for Economics

Connected to symbolic interactionism and phenomenology is Chomsky (1975) theory of languages such that natural language is common "to discover 'the semantic and syntactic rules or conventions (that determine) the meanings of the sentences of a language' (Swanson 1970), and more important, to discover the principles of universal grammar (UG) that lie beyond particular rules or conventions" (Chomsky 1975: 78). Chomsky's "primary purpose is to give some idea of the kinds of principles and the degree of complexity of structure that it seems plausible to assign to the language faculty as a species-specific, genetically determined property" (ibid.: 79). He does this by distinguishing between "surface" and "deep" structures.

Chomsky describes the surface structure as the basic everyday words and sentences we use to communicate. On the surface, we understand each other or think that we do and proceed to communicate and behave based on those sets of assumptions. At the surface level, we can form "various components of the base interact to general initial phrase markers, and the transformational component converts an initial phrase marker, step by step, into a phonologically represented sentence with its phrase marker" (ibid.: 81). In short, we can take everyday discussions and mark the sentences into a theoretical form for further detail and analysis. This process leads to the transformational derivation which is "The sequence of phrase markers generated in this way…" to form sentences (ibid.). From this process, we have the syntax of a language.

The basic terms are structure and deep structure which refer "to non-superficial aspects of surface structure, the rules that generate surface structures, the abstract level of initial phrase markers, the principles that govern the organization of grammar and that relate surface structure to semantic representations, and so on" (ibid.: 86). The deep structures are the semantics that give meanings to the sentence and words of the surface structures. Figure 4.2 illustrates the relationship between surface and deep structures. Transformational relations or rules connect the two structures.

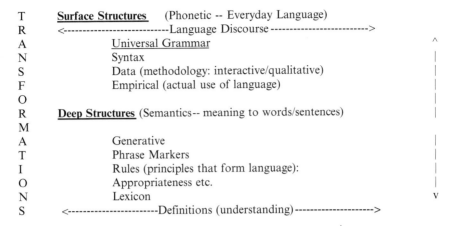

Fig. 4.2 Linguistic transformation theory (Chomsky 1975)

"We use language against a background of shared beliefs about things and within the framework of a system of social institutions" (Chomsky 1980: 247). Transformations are rules (shows the occurrence of a word corresponding to a yes-no question) which "map phrase markers into (other) phrase markers" (Chomsky 1975: 80). Transformation component is "One component of the syntax of a language consists of such transformations with whatever structure (say, ordering) is imposed on this set" (ibid.). For the transformation component to function in generating sentence structures, must have some class of "initial phrase markers" (ibid. 1975).

The concept of universal grammar indicates that all languages contain the components in Fig. 4.2. In other words, the transformational theory can apply to all languages. "The study of language use must be concerned with the place of language in a system of cognitive structures embodying pragmatic competence, as well as structures that relate to matters of fact and belief" (Chomsky 1980: 247).

A number of useful concepts can be borrowed from linguistic theory for the understanding of economics. The basic premise of linguistic theory is that language has its own order. The use of grammar to connect ideas requires the definition and meanings of words, phrases, and sentences to be understood. To that requires the scientific method which consists of hypotheses, observation, data collection, and analyses with the ability to replicate experiments (in this case language) in order to validate the hypotheses. Linguistic theory does this through the examples of deep and surface structures which need to be understood through the interactions of transformational rules (Chomsky 1988). The application of linguistic theory and science to economics can be done with a focus in four areas.

First, as noted, language distinguishes human beings from all other forms of life. Humans do have complicated language and therefore communication systems that allow them to send messages, symbolize, create, and build on a body of knowledge. Human language is composed of complicated sets of symbols that when used interactively allow messages to be transmitted. Second, linguistic theory argues that language is divided into two components: surface and deep structures. The surface structures are those symbols that people use in their everyday life to speak and write. The surface structures are the part of the grammar that cultures devise in order to record their history, communicate, and transact business. The deep structures are an entirely different phenomenon. Language has meaning attached to words and combinations of words (sentences) that are not expressed in the communication act itself. Furthermore, many of the deep structures are not defined in dictionaries or other guides to the language. In short, deep structures constitute the real core and understanding of any language and therefore of any culture and people's actions. Third, individuals learn surface structures (speaking and dialogue of a language) throughout their lives. Some of the aspects of language can be taught. However, empirical studies show that people understand or learn the deep structures (grammar and syntax) at an early age.

The qualitative perspective focuses on understanding of the meaning and definitions behind the interactive dynamics of human change within society. Qualitative methods and language therefore become crucial for describing, understanding, and perhaps predicting the human condition. Quantitative methods on the other hand do not provide an adequate framework or even set of tools to understand the creativity of innovation and its adaptation in everyday business life. Moustakas (1994: 21), in discussing qualitative methods, talks about the common qualities and bonds of human science research as being (1) recognizing the value of qualitative designs and methodologies, studies of human experiences that are not approachable through quantitative approaches, (2) focusing on the wholeness of experience rather than solely on its objects or parts, (3) searching for meanings and essences of experience rather than measurements and explanations, (4) obtaining descriptions of experience through first-person accounts in informal and formal conversations and interviews, (5) regarding the data of experience as imperative in understanding human behavior and as evidence for scientific investigations, (6) formulating questions and problems that reflect the interest, involvement, and personal commitment of the researcher, and (7) viewing experience and behavior as an integrated and inseparable relationship of subject and object and of parts and whole.

The qualitative perspective is strongly humanistic, with focus upon the understanding of the human being, the human condition, and of science. An empirical science has to respect the nature of the empirical world that is its objects of study, and the empirical world is understood as the natural world created by group life and conduct. To study it is to involve and interact with the actual group of actors to understand how they carry on in their lives – social life appears in their natural environment – in their everyday of life. In seeing the organization as an organization of actions, interactionism seeks to understand the way in which the actors define, interpret, and meet the situations at their respective Here and Now. The linking together of this knowledge of the concatenated actions yields a picture of the organized complex.

In a qualitative perspective, some general demands to scientific constructions are needed. The discussion of science and its demands on the structure of models for the understanding of the social or business reality can be categorized in four principles (Schutz 1973b: 56 and 126):

1. *The Demand for Logical Consistency.* The system of typical structures drawn up by the research worker must be established with the largest extent of clearness and precision in the frame of concepts implicated and must be fully compatible with the principles of formal logic. The fulfillment of this demand guarantees the objective validity of the objects of thought constructed by the research worker, and their strictly logical character is one of the most essential features with which scientific objects of thought differ from the objects of thought constructed by common-sense thinking in everyday life which they are to replace. In other words, a logically connected system implies that the means-goal relations together with the system of constant motives and the system of life plans must be constructed in such a way that (a) it is and remains accepted by the principles of formal logic, (b) all its elements are drafted in full clearness and precision, and (c) it only contains scientifically verifiable assumptions which must be totally accepted by all our scientific knowledge.
2. *The Demand for Subjective Interpretation.* The researcher must, to explain human action, ask which model can be constructed by an individual consciousness and which typical content must be ascribed to it in order to explain the observed facts as a result of such an activity of consciousness in an understandable relation. The acceptance of this demand guarantees the possibility of referring all kinds of human action or its result to the subjective meaning that such an action or its result has to the actor.
3. *The Demand for Adequacy.* Any expression in a scientific model referring to human action must be constructed in such a way that a human act carried out in the lifeworld by an individual actor in the way which is indicated by the typical structure is rational and understandable to the actor himself as well as to his fellow men in the common-sense interpretation of everyday life. The demand for adequacy is of the greatest importance to social scientific methodology. Adequacy makes it possible for social science to refer to events in the lifeworld at all. The interpretation of the researcher of any human act and situation could be the same as that of the actor or his partner. Accordance with this principle therefore guarantees the consistency of the data of the researcher with data in the common-sense experience of everyday business reality.
4. *The Demand for Ethics.* Ethics must be applied to research in everyday business life. Because the interaction between the researcher and the subjects is intense and often revealing, it is important that the results of the work reflect the concerns and well-being of those who provided the data. Dire consequences could come to people if certain business secrets (as in the case presented in Chaps. 9 and 10 below regarding intellectual property of commercialized inventions) or strategies are revealed. Everyday business life has numerous hazards attached to it; the work of the researcher should not be one of them. In the end, the researcher should be able to contribute and enhance the well-being of the everyday business

activity under study. And this is precisely the purpose of action research: to contribute to the business situation through interaction.

Summary and Conclusions

The business actions of people, groups, and their networks and organizations are about people interacting in everyday of life, trying to construct the future and making sense of the present. In the science of economics, we have to focus upon that, but the dimension in this is to create theories that make a difference.

Weick (1999) talks about that and end up with some qualities as possible properties of such moving theories: (1) Analysis is focused on what people do. (2) Context of action is preserved, and context-free depiction of elements is minimized. (3) Holistic awareness is attributed to the actor. (4) Emotions are seen to structure and restructure activity. (5) Interruptions are described in detail with careful attention to what people were doing before the interruption, what became salient during the interruption, and what happen during resumption of activity. (6) Activity is treated as the context within which reflection occurs, and reflection is not separate from, behind, and before action. (7) Artifacts and entities are portrayed in terms of their use, meaning, situated character, and embedding in tasks rather than in terms of their measurable properties. (8) Knowledge is seen to originate from practical activity rather than from detached deductive theorizing or detached inductive empiricism. (9) Time urgency rather than indifference to time is treated as part of the context. (10) The imagery of fusion is commonplace, reflecting that activity takes place prior to conceptualizing and theorizing. (11) Detachment from a problem and resort to general abstract tools to solve it is viewed as a last resort and a derivative means of coping rather than as the first and primary means of coping (whatever else people may be, they are not lay social scientists).

In Weick's discussion of theorizing and understanding, he points to important issues in science and theorizing: What is interesting science in terms of saying something meaningful about reality, and what is not? What is important to people in their search for understanding of their reality and to organize their everyday of life, and what is not important?

In the discussion of the "firm" and its constant economical and organizational changes, it is important to have an understanding of both organizing and time and space as a subjective and intersubjective phenomenon. The process of organizational activities and actions comes from interpretation and understanding of the situation by those actors involved in the actions. It is thereby a discussion of interaction processes and the way in which the actors interpret the processes and how the interpretations effect changes in the organizational development of the firm.

The development of the firm is a complex phenomena but also an everyday of life reality for people and thus very simple on another level of understanding. It is not something one experiences as abstract. Individuals are engaged in and related to the firm and are thinking about it in very concrete ways. Firms are unique phenomena, simply by the reason that people are unique. To understand a firm – an organization – we have to treat it as subjective and qualitative phenomena. In this, the central issue in understanding the firm is an understanding of the actors' subjectivity and intersubjectivity with their motives and intentions in their everyday business life. People understand themselves retrospectively and act accordingly, but additionally they are thinking-in-future: What are the projects they are thinking upon? In which way do they try to realize them? How do the projects change through the process of action and interaction? People construct their organizational reality through actions in everyday life and they build paradigms in order to orient themselves to their own reality. We have to relate ourselves to this discussion in economics if it is the empirical reality and not the theoretical "reality" in which we are interested. In other words, understanding of the social construction of people's organizational life and activities is the context of their everyday business life within the firm. NOTE: where do get back to the issue of economics (use of language) as a science?

References

Arbnor I, Bjerke B (1997) Methodology for creating business knowledge. Sage, London
Bartunek JM (1984) Changing interpretive schemes and organizational restructuring: the example of a religious order. Adm Sci Q 29:355–372
Bengtsson J (1993) Sammanflätningar – Husserls och Merleau-Pontys fenomenologi. Daidalos, Uddevalla
Benson JK (1977) Organizations: a dialectical view. Adm Sci Q, Mars 22:1–21
Berger PL, Luckmann T (1966) The social construction of reality: a treatise in the sociology of knowledge. Doubleday and Company, New York
Bjurwill C (1995) Fenomenologi. Studenlitteratur, Lund
Blumer H (1969) Symbolic interaction: perspective and method. Prentice-Hall, Englewood Cliffs
Blumer H (1976a) Qualitative methods: lectures. University of California/Institute for Qualitative Research, Berkeley
Blumer H (1976b) Social interaction: lectures. University of California/Institute for Qualitative Research, Berkeley
Blumer H (1976c) A critique of the conventional scientific paradigm: social action: lectures. University of California/Institute for Qualitative Research, Berkeley
Blumer H (1976d) The self and social action: lectures. University of Calif./Institute for Qualitative Research, Berkeley
Brown RH (1978) Bureaucracy as praxis: towards a political phenomenology of formal organizations. Adm Sci Q 23(3):365–382
Chomsky N (1975) Reflections on language. Pantheon Books, New York
Chomsky N (1980) Rules and representations. Columbia University Press, New York
Chomsky N (1988) Language and problems of knowledge. MIT Press, Boston
Clark WW II, Fast M (2008) Qualitative economics: toward a science of economics. Coxmoor Publishing, Oxford
Comte A (1991) Om Positivismen (Discours préliminaire sur l'esprit positif, 1844). Korpen, Göteborg
Durkheim (1991) Sociologins metodregler (Les règles de la méthode sociologique, 1895). Korpen, Göteborg
Fast M (1992) The internationalization subjectivity and intersubjectivity: a development of a hermeneutic understanding of companies internationalization as an alternative to mainstream theory. Institute of Development and Planning, International Business Studies, Aalborg University

Fast M (1993) Internationalization as a social construction. In: Proceedings of the 9th IMP-conference, Bath, UK
Gadamer H-G (1975) Truth and method. Sheed and Ward, London, 1993
Gadamer H-G (1986) Reason in the age of science. The MIT Press, Cambridge, MA/London
Garfinkel H (1967) Studies in ethnomethodology. Prentice-Hall, Englewood Cliffs
Heidegger M (1927) Being and time. Blackwell, Oxford, 1992
Hummel RP (1990) Applied phenomenology and organization. Adm Sci Q 14(1):9–17
Husserl E (1962) Ideas. Macmillan, New York
"Imagination and Perception" in *Experience and Theory*. In: Foster L and J. W. Swanson (eds) Amherst: University of Massachusetts Press, 1970
Imershein AW (1977) Organizational change as a paradigm shift. In: Benson JK (ed) Organizational analysis: critique and innovation. Beverly Hills, London
Jehenson R (1978) A phenomenological approach to the study of the formal organization. In: Psathas G (ed) Phenomenological sociology: issues and applications. Wiley, New York
Kant I (1929) Critique of pure reason (*"Kritiken der reinen Vernunft" (1781/1787)*). Macmillan, Hong Kong
Katz D, Kahn RL (1966) The social psychology of organizations. Wiley, New York
Mangham I (1978) Interactions and interventions in organizations. Wiley, Bath
Mead GH (1962) Mind, self, & society – from the standpoint of a social behaviorist. The University of Chicago Press, Chicago, 1934
Merleau-Ponty M (1994) The phenomenology of perception. Routledge/Kegan Paul, London (1962)
Morgan G, Ramirez R (1984) Action learning: a holographic metaphor for guiding social change. Hum Relation 37(1)
Moustakas C (1994) Phenomenological research methods. Sage, California
Polkinghorne D (1983) Methodology for the human sciences: system of inquiry. State University of New York Press, Albany
Ritzer G (1977) Fundamentale perspektiver i sociologien. Fremad, Odense
Rose AM (1962) A systematic summery of symbolic interaction theory. In: Rose A (ed) Human behavior and social processes: an interactionist approach. Routledge/Kegan, Paul, London
Schutz A (1970) Reflections on the problem of relevance. Yale University Press, New Haven/London
Schutz A (1972) The phenomenology of the social world. Heinemann Educational Books, London
Schutz A (1973a) Some leading concepts of phenomenology. In: Collected papers I: the problem of social reality. Matinus Nijhoff, Haag
Schutz A (1973b) Hverdagslivets sociologi. Hans Reitzel, København
Schutz A (1978) Phenomenology and the social sciences. In: Luckmann T (ed) Phenomenology and sociology. Penguin, London
Schutz A (1982) Life forms and meaning structure. Routledge/Kegan Paul, London
Schutz A (1990a) Collected papers I: the problem of social reality. Kluwer, The Netherlands
Schutz A (1990b) Collected papers II: studies in social theory. Kluwer, The Netherlands
Schutz A (1990c) Collected papers III: studies in phenomenological philosophy. Kluwer, The Netherlands
Schutz A (1990d) Collected papers IV. Kluwer, The Netherlands
Schutz A, Luckmann T (1974) The structure of the life world. Heinemann, London
Silverman D (1983) The theory of organisations. Heinemann, London
Sims D, Fineman S, Gabriel Y (1993) Organizing "Organisations". Sage, Great Britain
Singelmann P (1972) Exchange as symbolic interaction: convergences between two theoretical perspectives. Am Sociol Rev 37:414–424
von Bertalanffy L (1971) General system theory. Allen Lane/The Penguin Press, London
Weick KE (1979a) The social psychology of organizing. Addison-Wesley, Reading, (1969)
Weick KE (1999) That's moving – theories that matter. J Manage Inq 8:134–142
White JD (1990) Phenomenology and organizational development. Adm Sci Q 28:331–496
Wind HC (1976) Filosofisk hermeneutik. Berlingske leksikon bibliotek, København

Chapter 5
Energy Planning for Regional and National Needs: A Case Study – The California Forecast (2005–2050)

Gary C. Matteson

Abstract Achievement of a sustainable balance between energy consumption and energy resources has become a critical component for energy planning at the regional, national, and international levels. For regional planners to estimate energy requirements, they must define the population growth, per capita consumption, and applicable energy conservation. They must also determine the technical capacity for energy supply from the respective energy resources. An energy consumption projection and energy resource plan has been developed for the State of California covering the period of 2005–2050.

Introduction

In 2006, the US Department of Energy estimated the world's consumption of energy to be at 407 quads BTU in 2003, with a predicted total usage of 721.6 quads BTU in 2030 (USDOE, EIA 2006b). This almost doubling of energy consumption dictates that all governments and their respective industrial partners must plan carefully for their future energy usage. Concern for global warming and climate change now makes the sustainable-energy approach a critical component for energy planning at regional and national levels. For planners to estimate their demands for energy, they must estimate population growth, per capita consumption, and technologies for energy efficiencies that could be applied. Next, they must determine the technical capacity for the supply of the respective energy sources, for example, natural gas (heating), gas/diesel, hydro-generation, coal, electricity, nuclear, biomass, solar,

G.C. Matteson, M.S., M.B.A. (✉)
Department of Biological and Agricultural Engineering, University of California,
One Shields Avenue, Davis Campus, Davis, CA 95616, USA
e-mail: gcmatteson@ucdavis.edu

wind, small hydro, geothermal, and "prospective" generation. This assessment is based on the technical capacity of the respective fuels to be available to meet the demand over a planning period. Technical capacity identifies the amount of energy available for use in a given time period, while consideration of the types of systems will be used to harvest and handle the energy feedstock. Environmental, social, and political constraints may affect technical capacity.

An analysis, based on the assessment of energy consumption and energy resource projections, has been developed for the period of 2005–2050. This analysis has been applied to data obtained for the State of California. The results of this analysis demonstrate how difficult it is to wean a nation, state, or region off fossil fuels, especially foreign oil. The analysis also shows that governments and their business partners need to formulate a sustainable-energy strategy through the development of their own consumption and energy resources parameters, set policy, make the necessary energy procurements, and allocate human resources and capital funds to develop the required technical capacity of energy resources and energy distribution systems.

This chapter was originally written and presented, in early 2007, at the Bren School's Western Forum on Energy and Water Sustainability located on the Santa Barbara campus of the University of California. Since that presentation, some of the references have been updated, but the original data for the year 2005 and projections for each of the subsequent 5-year periods, extended to 2050, have been retained and verified over the last five years. Presently, the actual results for the year 2010 are not yet available, but the patterns of generation and consumption are emerging from data coming forth for the years 2006 through 2009. These recent trends will be discussed at the end of this chapter.

The Long-Term Vision: A Sustainable-Energy Strategy

In April 1987, the World Commission on Environment and Development published a report. Chairwoman Gro Harlem Brundtland observed in the "foreword" to that report: "the 'environment' is where we all live; and 'development' is what we all do in attempting to improve our lot within that abode. The two are inseparable" (Brundtland 1987a). Later, in the report's "overview" section, sustainable development is defined: "Humanity has the ability to make development sustainable – to ensure that it meets the needs of the present without compromising the ability of future generations to meet their own needs" (Brundtland 1987b). Clark and Yin addressed the concept of sustainability of biomass energy in the Pacific Northwest. They concluded that there are four dimensions of sustainable development: economic, social, environmental, and institutional (Clark, and Yin 2007).

The BP oil spill in late April 2010 in the Gulf of Mexico further alerted regions, governments, and businesses to the negative environmental impact of fossil fuels, should a problem or accident occur. Along with the growing evidence of "peak oil and gas" (Hicks and Nelder 2007; Sorrel et al. 2010) and the impact of global warming, there is an urgency to plan and implement energy strategies using one or more of the following: (1) explore and produce fossil fuels while installing methods of CO_2 sequestration, (2) expand nuclear plants, (3) develop sustainable renewable-energy resources, and (4) apply existing and new energy-efficiency technology. The last two options appear to be the best choices to meet the definition of sustainable energy, but the other options remain. The case analysis for California that will be presented indicates total reliance on renewable-energy resources and energy conservation over the period of 2005–2050, but it will not meet the expected total demand for energy. This model would, therefore, be significant for the Unites States and certainly throughout the world, especially to develop the skill sets and experts in the areas of sustainable communities (Clark 2010).

Sourcing Energy 2005–2050

The sourcing of energy for a nation or region requires energy planners to establish a long-term vision. An analysis for energy consumption and resources for the years 2005–2050 is contained in this chapter. The analysis projects to 2050 because some of the possible strategies (hydrogen fuel and fuel cells) do not become significant factors until after 2030. This analysis is displayed as a spreadsheet in Appendix 1. This spreadsheet format permits an energy planner to display a region's goals and strategies in terms of energy consumption and specific quantities of energy supplies.

As a nation's or region's energy planners construct an energy consumption and resource spreadsheet for their area of operation, they first must estimate their energy requirements. This requires the determination of population growth, expected standard of living (per capita consumption), level and types of industry, as well as an estimate of technologies for energy efficiencies that could be applied to the per capita consumption data for the period projected to 2050. Next, planners must conduct an assessment of technical capacity of the respective energy sources, for example, natural gas (heating), gas/diesel, hydro-generation, coal, electricity, nuclear, biomass, solar, wind, small hydro, geothermal, and "prospective" energy resources. As noted above, technical capacity identifies the amount of energy resources available to meet the expected demand for energy while considering a number of sustainable production factors. A number of authors have developed models for evaluating the sustainability of energy streams. Wing has proposed an equilibrium model of the US economy, which shows electric power's technological margins of adjustment when impacted by carbon taxes (Wing 2005).

There is an increasing demand to conduct biomass to energy conversions using a sustainable model, based on environmental, social, and economic principles (Forest Stewardship Council 2008, and Roundtable 2008). These practices are based on principles of greenhouse gas balance, carbon sinks, existing food supplies, biodiversity, land availability, water availability, air quality, applicable laws, local economic development, social well-being of employees, and transparency to the public. These same principles can be applied to determine the sustainability of the other sources of energy.

The planners will need to differentiate their sources of energy by indentifying the energy imports from outside a nation, state, or region and the sources of energy from within their borders. The planners will then use their local sustainable-energy sources, for example, biomass, solar, wind, small hydro, and geothermal, to the fullest capacity when they become available. The planners will next define the costs of various strategies employed to construct and maintain a sustainable-energy system. Finally, the planners will determine if their energy mix is readily available and affordable. Some examples of sustainable communities are documented from around the world in Clark (2009).

As noted earlier, the US Department of Energy estimates an almost doubling of the world's consumption of energy by 2030 (USDOE, IEO 2006a). This dramatic increase in energy consumption dictates that every nation's or region's government and their business partners should be planning today for future energy needs. Nations and regions are confronted with geopolitical realities of economics and security. These factors point to the need for energy independence.

Countries with established government energy planning offices should be working with business partners and their utilities to achieve a sustainable-energy system. The consequence of a "do nothing" approach will be readily apparent as the planners develop the spreadsheet contained in this chapter's case analysis. The consequences of an aggressive sustainable-energy program should show reduction of fossil fuels consumption, increasing use of renewable energy, reductions in the rate of CO_2 releases to the atmosphere, improvements in efficiencies of consumption, and achievement of security by having required and affordable energy available when needed.

Energy Consumption and Resource Analysis for a Region

An energy consumption and resource analysis for a region during the period 2005–2050 could be developed to show the optimal level of sustainable energy that can be achieved in a defined time period and at a calculated cost. The analysis would show different mixes of energy resources for the region and its respective geographic areas. Solar energy, for example, may be more available and commercially accessible in one region and less in another. The same is true for wind, biomass, geothermal, and ocean or wave power.

To perform the analysis for a region, the base case numbers must be determined for the respective fuels. These numbers have to be qualified as to which amounts of the fuels are available from within the region and what amounts are currently being imported. Next, a resource assessment must be made for each fuel. Once these numbers are obtained, then the assumptions must be entered for energy consumption per capita, population growth, and the protocol for future energy transmission (electricity, hydrogen, or methanol).

Next, projections for both local and on-site power, energy use per capita, and energy conservation and efficiency must be developed. Models that track global warming also should be examined and integrated into the region's sustainable-energy strategy.

The technical capacities for renewable-energy sources must be determined for all renewable-energy systems that feed energy to the region. The determination of a technical capacity for a given energy resource is based on environmental, as well as social and economic principles, to assure that it is secure, renewable, accessible locally, affordable, and obtained with minimal impacts on the environment.

Another strategy for securing sustainable energy in a region is the use of hydrogen in stationary and mobile systems (Clark 2006a). Hydrogen not only serves as a method for delivering energy to the region's transportation fleet, but also is a reservoir of energy for use on demand. Duane Myers of Directed Technologies has also come to the conclusion that future energy transmission in a region should be hydrogen (Myers et al. 2003). His team determined that hydrogen should become a major transporter of energy in the years after 2030. This conclusion is based on cost and availability. His team calculated that the cost of maintaining the existing gasoline infrastructure per vehicle supported is up to two times more expensive than the estimated costs of maintaining either a methanol or a hydrogen fuel infrastructure. Second, the cost of transmission of hydrogen from distances over 540 miles is less than the cost of transmitting the equal amount of energy in the form of electricity. Third, the team determined that it is possible for a region to generate sufficient hydrogen annually from renewable-energy resources in the 2030–2050 years time frame. They selected wind and biomass as the primary producers of hydrogen. They found these forms of renewable energy to be technically available and at a comparatively low cost for energy conversion (when compared to solar and geothermal).

One must define the technical capacity for all of the fuel streams, including the geographical location (intra-country or foreign imports) for these streams. Once these data elements are defined and reflected in a spreadsheet similar to the one shown in Appendix 1, various scenarios can be run to find the optimal path. Mark Jacobson and Mark Delucchi have been researching this area on a global basis (Jacobson and Delucchi 2010).

Case Analysis: State of California

This case analysis focuses on the State of California because data is available for both consumption patterns for the period of 2005–2050 and estimates of in-state energy resources and the imports of energy resources for the same period. All energy data is presented in quads (one quad equals 10^{15} BTUs or 10^{18} J). It is important to note that the technical capacities shown for the various energy resources, as provided by the referenced authors and agencies, do not follow all of the

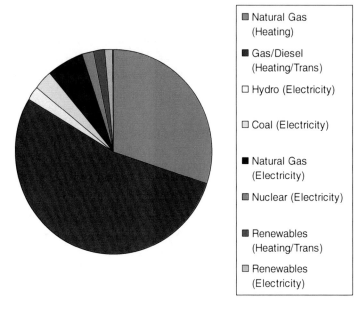

Chart 5.1 Percentage of energy consumed 2005 by source

sustainability criteria (environmental, social, and economic principles) provided in Sect. Sourcing Energy 2005–2050. The referenced authors and agencies did consider social and political constraints that may affect their access and use of selected energy resources. The inclusion of the environment (air, water, and land impacts) is observed in some reports and not in others. As authors and agencies update their forecasts of the energy resources, all of these parameters need to be incorporated in their projections of technical capacity.

The present sources of California's energy streams are largely unsustainable. The following review describes California's governor's orders, laws, and regulations and actions by the corporate sector and municipal sectors to formulate an energy strategy for the state. The California Energy Commission annually updates an Integrated Energy Policy Report with projections on available energy resources projected to 2020. An example is the California Energy Commission's Oil Supply Sources to California's Refineries (CEC Oil Supply to CA Refineries 2005). To date, there is no oil consumption and oil resource projection for the period of 2020–2050. Examples of the unsustainable sourcing and consumption of energy are greenhouse gases released from combustion of fossil fuels including natural gas, oil, and coal; the operation of nuclear reactors using nonrenewable feedstock; and the need to store the waste from these reactors for very long periods in secured pools of water or vaults. Appendix 1 defines energy consumption and energy resource projections for the period 2005–2050. Charts 5.1 and 5.2 are derived from Appendix 1 and show how various forms of renewable energy can move the state toward a predominately sustainable-energy supply in 2050.

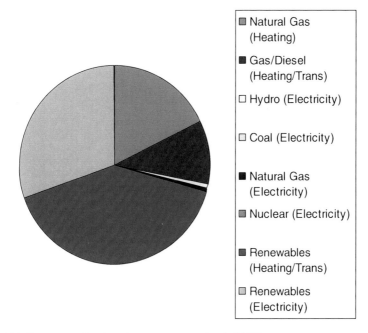

Chart 5.2 Percentage of projected energy consumption in 2050 by source

Conditions and Assumptions Used in the Case Analysis

The case analysis of sustainable-energy strategy for California assumes the main driver on the demand side will be the growth of population and its related growth in consumption of energy. Population growth is addressed in many economics treaties, and it is the focus for many "think tanks," such as Optimum Population Trust (Optimum Population Trust 2002). Desvaux observes the impact of population growth on the environment (through exploit of resources). He notes the impacts are directly proportional to population size, affluence, and technology. He defines biocapacity as the world's total resources generated in a year to support human populations on the planet. He calculates that the world's present population of 6.3 billion has exceeded the available biocapacity. He calculates that there is only enough biocapacity to support 5.1 billion people. He also notes that biocapacity can be seriously impacted by global warming (Desvaux 2007).

California's population was just over 36 million in 2005 and in 2012 exceeded 39 million, and it is forecasted to grow to more than 64 million by 2050. This data is drawn from the Public Policy Institute of California's "What Kind of California Do You Want" (Baldassare and Han 2005). The percentage increase from the "present migration" column, as shown in Graph 5.1 (below), is used in Appendix 1.

Desvaux and Weizsacker observe that the increases in demand for energy are determined largely by growth of the population's wealth (Desvaux 2007 and

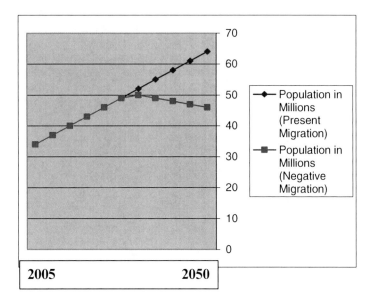

Graph 5.1 Projection of population growth for California

Von Weizsacker 1998). There are two forces at work on a per capita basis. The first force is the higher energy consumption per capita, as represented by the increase in miles driven per capita (Highway Statistics 2004), and the purchase of "McMansions" (McGhee 2006). The second force, and a counter to the first force, is a combination of economic and technological factors: (1) the application of existing and new technologies in energy efficiency, (2) the higher cost of energy, and (3) the application of decentralized energy generation. Clark addresses this last factor in his text on the evolution of sustainable communities by providing on-site power generation (Clark 2006b).

Haas has examined energy services per capita, energy conversion efficiency, increases in renewable energy, energy prices, and regulatory policies to identify necessary changes to achieve a more sustainable-energy system. He observes that "every advance in end-use efficiency enhances the demand for energy services" (Hass et al. 2008).

The State of California GDP per capita was $37,848 in the year 2000, and it grew to $42,376 in "chained dollars" by 2007 (US Census Bureau 2005, and US Bureau of Economic Analysis 2007), but by the end of 2011, due to the global economic crisis had gone down by 25%. If it continues at this rate of 1.6% GDP per year, which it will not, but for the sake of data modeling, to the year 2050, and with the population growing at the rate described by the state's Department of Finance, this would amount to a 2.2-fold increase (CA Department of Finance 2011). This level of population growth and GDP per capita growth breaches our plan for sourcing sustainable energy after 2050. If the state assumed a slower population growth rate as is the reality from the economic crisis, say the "Negative Migration" line in Graph 5.1, the 1.6% GDP per capita could still prevail. There would be fewer people consuming energy, but the individual wealth could remain the same.

Appendix 1 has end points for consumption. California's present overall consumption is about 6.7 quads. This consumption should grow, in line with growth of the population and its related consumption, to about 11.5 quads in 2050. Appendix 1 also shows energy consumption for transportation, based on population growth and its related per capita consumption growth. The transportation group starts out at 3.73 quads in 2005 and grows to 6.4 quads in 2050. These end points are derived from data shown in "California Climate Change Center 2005 Report to the California Energy Commission" (Murtishaw et al. 2005).

Data was defined for the Appendix 1 spreadsheet cells by three methods. The first approach was to take the 2005 data and then apply the present population growth line, as described in Graph 5.1, coupled with increases of 1.6% annual growth in overall consumption per capita (it assumed that growth in energy consumption parallels overall growth in consumption). This approach was, for example, used in the state's small hydro analysis. The second method was to use target data developed at the national level and apply these growth figures to the California 2005 data. This approach, for example, was used for the hydrogen fuel derived from geothermal and wind.

The third method was to use arbitrary decisions and hypotheses based on listed (below) legislation that was enacted and perceived strategic goals for the State of California. The coal and nuclear data in the analysis reflect these strategic goals:

- Exceed Kyoto Protocols for elimination of global warming (CA Assembly Bill 32 2006)
- Reduce greenhouse gases from combustion of fossil fuels, including coal (CA Assembly Bill 32 2006)
- Eliminate the need to refine and consume imported foreign oil (Schwarzenegger, Executive Order 2004)
- Create a hydrogen highway (Schwarzenegger, Executive Order 2005)
- Reduce the consumption of nonrenewable resources such as natural gas (CA Senate Bill 1078 2002)
- Reduce the need to run nuclear reactors and store spent uranium from those reactors (McCarthy 2005; Schaffer 2011)

Findings from the Case Analysis

The renewable-energy flows in California are dependent upon development of new technologies and the rates of capital investment in these systems. There may not be a significant impact until the year 2030. At this point in time, the spreadsheet shows the renewables flowing in the California energy mix could jump from 3% (in year 2005) to 33% (in year 2030). Concentrated solar power systems and flat-panel solar PV could be employed with sufficient size and capacity to cause a significant shift in California's energy mix. By 2030, solar could be contributing 83% toward California's renewable energy with the other renewable-energy sources – wind, biomass, and geothermal – contributing together the other 17% (note: large hydro is excluded from California's Renewable Portfolios Standard Keese 2003). After 45 years, the amount of renewable energy in California's energy mix could reach a limit of 62%. This observation has been recently affirmed by Jane Long of Lawrence Livermore National Laboratory and Miriam John of Sandia National Laboratory (Long and John 2011).

Given global warming and legislative initiatives from elected officials, the deployment of renewable energy is desirable, but certainly not the total solution to meeting California's energy requirements. The population and its desire for additional energy consumption, on a per capita level, start to push the energy consumption past the available sustainable-energy mix after 2050. The renewable-energy portion of the sustainable-energy mix reaches the limit of its technical capacity in the 2040–2050 time period. While Californians certainly do not want to return to the days of dependence on unsustainable-energy sources, the issue of falling back into that trap or conserving more (on a per capita basis) is clearly before the state. A more efficient transportation system could lower the energy demand. Jacobson and Delucchi suggest a scenario in which all "wind, water, and sunlight" renewable powered transportation "would require approximately 30% less end use power than… a conventional fossil fuel scenario" (Jacobson and Delucchi 2010). There remains the possibility of the development of new sources of energy, such as ocean wave/tide power systems, and fusion. A change in migration patterns (as suggested in Graph 5.1 "negative migration pattern") could also have a big impact. These consumption patterns assume no changes in relative prices. If prices on fossil fuels were to rise dramatically, due to changes in supply and demand or the initiation of a Pigouvian tax, then the shifts to renewable and conservation would occur quicker, and there would be a bit more time before the technological limits of the renewable-energy resources would be reached. Knittel has discussed the impacts of a Pigouvian tax on petroleum product (Knittel 2012).

Appendix 1 demonstrates that California can reduce greenhouse gases from combustion of fossil fuels in the transportation sector, eliminate the need to refine and consume imported foreign oil, and maintain present generation levels for nuclear reactors and storage of spent uranium from those reactors. The ability of the citizens of California to bear the costs of moving to sustainable-energy resources must be addressed. Before the issue of cost is discussed, a better understanding of present and proposed use of fossil fuels and nonfossil fuels is presented.

Specific Findings from the Case Analysis: Fossil Fuels

California's consumption of natural gas totaled more than 2.4 quads in 2005 (Marks 2005). Of this amount, about .4 quad was used in the generation of electricity (with the remainder going to heating).

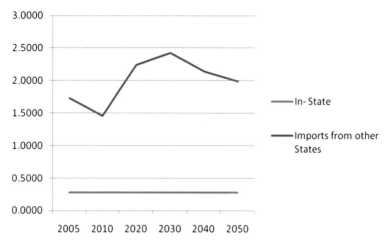

Graph 5.2 Natural gas consumed for heating in quads

Natural Gas for Heating

Graph 5.2 shows the data points in Appendix 1 for natural gas used for heating. This graph shows the early affect that renewable-energy sources have on natural gas consumption in the 2005–2010 time frame. During the period of 2005–2010, the ramping up of renewable-energy systems outpaces the increases in consumption related to growth of population and its per capita wealth. This allows for a drop in the amount of imported natural gas to .2 quad. The most recent 2010 data from the California Energy Commission shows the 2010 forecast to be only a .1-quad drop (Byron 2009). The smaller reduction in use of natural gas was due to the state's inability to ramp up its renewable-energy program at the earlier projected rate (CA Senate Bill 1078 2002).

During the 2010–2020 period, natural gas consumption will rise, and this demand will be met by importing natural gas from other states. The rise in consumption is due to growth in population and its per capita wealth. The increase in natural gas for heating is also due to the renewable-energy sources that were used for heating being diverted to the transportation sector. Production of in-state produced natural gas heating is held constant.

In the 2020–2030 time period, the ramping up of renewable-energy systems will be less than population growth, thus pushing the consumption of the United States-sourced natural gas to 2.4 quads. The ramping up of renewable-energy systems in the 2040–2050 time frame holds the need for natural gas drawn from other states to 1.9 quads. Since natural gas is the cleanest form (in the energy conversion to heat) of the fossil fuels, it is the preferred form of fossil fuels to use when using fossil fuels for heating.

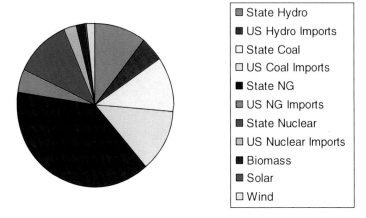

Chart 5.3 Percentage of electricity generation consumed by fuel source in 2005 by source

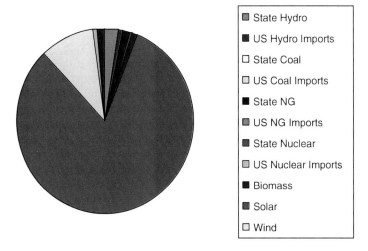

Chart 5.4 Percentage of electricity generation consumed by fuel source projected for 2050

Natural Gas for Generating Electricity

Charts 5.3 and 5.4 (below) show that while the quantity of natural gas consumed in electrical generation remains flat during the period 2005–2050, the percentage of natural gas' contribution drops dramatically. The state consumption of natural gas for generation electricity was .4 quad in 2005 (Marks 2005). Appendix 1 holds the base year (2005) levels of consumption of natural gas for electricity generation level out to 2050. The ramping up of electricity generation from renewable sources per the state's Renewable Portfolio Standards (CA Senate Bill 1078 2002) is the primary factor holding the gas-fired electricity generation level. Concentrated solar and photovoltaic solar are the renewable-energy sources that replace natural gas.

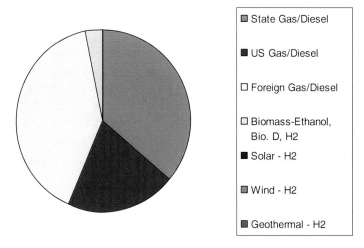

Chart 5.5 Percentage of heating and transportation energy consumed in 2005 by source

Natural Gas

The data in Appendix 1 indicates California relies on some form of fossil fuels throughout the 2005–2050 survey. The data also shows that significant reductions in oil can be achieved, but natural gas remains a vital source of energy. A recent report by MIT researchers indicates that "natural gas will assume an increasing share of the United States' energy mix over the next several decades" (Moniz et al. 2010a). The base year of this study was 2005, and during that period, the issue for natural gas was scarcity of on-land natural gas drilling production. In the next few years, the industry moved to shale production through the use of horizontal drilling and hydraulic fracturing. Now there is a concern about water management with respect to the hydraulic fracturing required to gain access to natural gas in shale. The MIT report calls for the US Department of Energy to sponsor additional research and development to ensure that the "shale resource is exploited in the optimum manner" (Moniz et al. 2010b).

The US energy and climate change policy will play a significant role in the future use of natural gas. Under the scenario with 50% reduction of CO_2 to 2050, "the principle effects of the associated CO_2 emission price are to lower energy demand and displace coal with natural gas in the electric sector" (Moniz et al. 2010c).

Gas/Diesel

Charts 5.5 and 5.6 show how components of the gas and diesel consumption diminish over the projection period, in favor of renewable-energy sources. Concentrated solar production of hydrogen takes a prominent role.

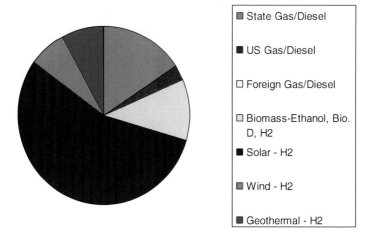

Chart 5.6 Percentage of projected heating and transportation energy consumption for 2050 by source

Appendix 1 shows a decrease in the dependence on crude oil, both foreign and domestic, as renewable-energy streams become established. California's gasoline and diesel consumption in 2005 totaled in excess of 3.6 quads (CEC Oil Supply Sources 2005). This consumption increases to 3.9 quads by 2010 because the rate of population and GDP growth (and its related increase in consumption per capita) increase exceeds the rate of deployment of renewable-energy alternative fuels.

Graph 5.3 shows the forecasted conversion of a sizable portion of the transportation fleet from internal combustion to electrical drives drawing energy from batteries or hydrogen fuel-cell systems. The low point of oil consumption is in the 2040–2050 time period with foreign imports at zero. Foreign oil imports could rise after 2050. The requirement for foreign oil is correlated with inability of the renewable-energy source to keep up with consumption of the increasing population.

Coal

The state's electrical energy generated from coal is about .2 quad in 2004 (CEC Gross System Power 2004) and has been at this level for some time. The state's consumption in 2005 was assumed to be at the same level. The imported electrical energy from coal was just over .1 quad. Most of this generation was produced in Utah or Arizona and delivered to the Los Angeles Department of Water and Power under long-term contracts (Fisher 2011). The California-based coal-to-electricity generation is just under .1 quad. Coal is carbon intensive, and its emissions of CO_2 and mercury, as a by-product, have negative impacts on the environment. Economic solutions for carbon sequestrations are being sought (Al-Juaied and Whitmore 2009; Vliet et al. 2011). Given these limitations, there is no reason to increase the use of coal in the energy mix for the State of California. With the ramping up of

5 Energy Planning for Regional and National Needs...

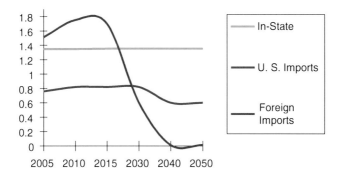

Graph 5.3 Gas/diesel consumption

renewable-energy production, the use of coal is allowed to remain at the current level through 2050.

Specific Findings from Case Analysis: Nonfossil Fuels

Each of the nonfossil fuel groups in the California analysis, Appendix 1, will now be analyzed. The data presented for California is drawn from various cited studies. The biomass, geothermal, and wind energy date agree with the corresponding nationwide forecast for consumption of these renewables in the time period 2020–2030 (Dutzik 2006).

Hydroelectric

The state's large hydroelectric production in 2004 was just a bit more than .1 quad (CEC Gross System Power 2004). In California, large hydroelectric power generation (over 30 MW) is not considered renewable energy, due to environmental impact reasons. In-state hydro-generation is about .1 quad, and out-of-state is .05 quad. The production of the state hydroelectric is held constant through the time span of this analysis. There are no additional dams on the planning horizon. In fact, some of the existing dams are being evaluated for demolition to address environmental concerns. The imported hydroelectric from other states in the United States is also not expected to change. It is held at the present levels because hydroelectricity currently flows, depending on the time of day and the season, as exports and imports of energy to California. Neighboring states depend on this service. It is observed that hydroelectric/pumping facilities may be constructed in the future. These facilities utilize electricity generated by wind and solar systems to pump water to higher elevations in the Sierra. The water is then allowed to flow down through hydroelectric dams when there is high electrical demand on cloudy and windless days. These hydro storage facilities would not change our production numbers for this category.

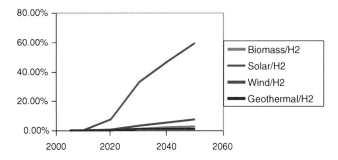

Graph 5.4 Renewable-energy consumption as a percentage of the total heating/transportation energy consumption (Data obtain from Appendix 1)

Nuclear Energy

The state's consumption of nuclear energy production 2004 was slightly more than .1 of a quad (CEC Gross System Power 2004). Nuclear energy production is relatively constant with changes only occurring when there is equipment failure.

Nuclear energy supplies about 20% of the electricity in the United States but contributes only 2% to California's energy needs. It has been 30 years since any new nuclear power plants have been installed. With the development of a very high-temperature reactor (VHTR), a nuclear rebirth has been anticipated in the industry. This new technology promises better utilization of the feedstock and less waste. These reactors are expected to enter service in the second half of the twenty-first century and outside the time horizon of this study (Boer 2009). Kessides has researched the economic risks of the nuclear option. It is his position that "volatile fuel prices, concerns about the security of energy supplies, and global climate change are coinciding to strengthen the case for building new nuclear power generation capacity." He then goes on to state, "Finally, even in a carbon-constrained world, nuclear power may be less economically attractive than a host of decentralized energy-efficiency and distributed generation technologies" (Kessides 2010). With the ramping up of renewable-energy production, this source of energy, in our California case analysis, is held at present levels through 2050.

Renewable Energy

Graphs 5.4 and 5.5 show how renewable energy can, over the next 45 years, meet a significant portion of California's heating, transportation, and electrical energy needs. The data points in these graphs are drawn from the Appendix 1. For an explanation as to why there is a dip in the solar generation line in Graph 5.5, see sections "Solar" and "Hurdles to be Overcome in California Case Analysis." These sections describe the routing of solar generation into hydrogen production for California's transportation fleet. During the period of 2015–2040, the solar production is shown in Graphs 5.4 and 5.5.

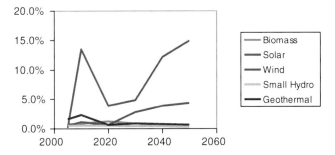

Graph 5.5 Renewable-energy consumption as percentage of the total electrical energy consumption (Data obtained from Appendix 1)

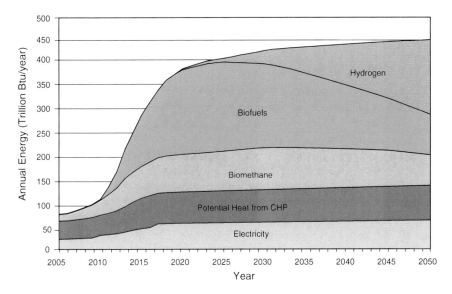

Graph 5.6 Biomass consumption

Biomass

The data shown in Graph 5.6 for biomass is the "technical capacity" information from the California Biomass Collaborative's Roadmap (Jenkins 2006a). This graph shows the initial use of biomass for electricity, heat, and transportation. Imports of biomass to the state are not included in this table. Biofuels and biomethane are the intermediate forms of the feedstock. The graph starts out with .07 quad of energy in 2005 and goes to .45 quad in 2050.

The production of methane/ethanol/biodiesel/heating-combined heat and power constitutes 80% of the energy production during the years 2020–2030, and then hydrogen production increases and eventually takes up 36% of the entire production by 2050. Note that the capacity to produce energy and energy carriers is reached in 2030. From that point on, it is just a resorting of the energy and energy carriers. This time frame for the production of hydrogen as a transporter of energy is also seen in solar, wind, and geothermal energy systems.

The imported feedstock that is converted to ethanol and the ethanol produced in other states but shipped into California is shown separately in Appendix 1. The data points were developed by Perez and Platts (Perez and Platts 2005). In 2005, this energy feedstock amounted to about .07 quad and is expected to grow to .4 quad in 2050 (BTAC 2002). After 2030, this feedstock is directed, in part, to the production of hydrogen.

Solar

The performance of solar is shown as a percentage of the total energy mix in two graphs – Graph 5.4 for heating and transportation and Graph 5.5 for electricity production. The percentage jumps around significantly in Graph 5.5. Graph 5.5 shows the solar energy percentage to be initially quite strong. This reflects the contribution of solar energy coming from flat-panel PVs, but the percentage diminishes as the demand from electric cars comes online while the solar generation cannot keep the pace. Solar energy from concentrated solar energy farms eventually enters into the mix, and the percentage again rises. As noted above, some of the energy is directed to hydrogen fuel and used in transportation. Graph 5.4 shows the steady growth of the solar contribution to the transportation sector.

The 2004 production of electricity from flat-panel solar PV was .024 quad (CEC Gross System Power 2004). The production for 2005 was assumed to be at the same level because of the existing subsidies held for at least another year. The future for solar PV could be concentrated solar electrical production systems and flat-panel solar collectors. The sustainable-energy strategy for California assumes that solar power employs hydrogen as the energy transporter and storage system when infrastructure becomes installed in 2020. (Note: 1 quad of hydrogen can be generated from 1.32 quads of electricity, but quad of hydrogen displaces 2.2 quads of gasoline/diesel.) This strategy also assumes there is a role for solar energy in the operation of electric automobiles, but the electric demand from transportation will be far greater than the growth in solar production.

Stoddard, from the National Renewable Energy Laboratory (NREL) in Golden, CO, provided the electrical production estimates from concentrated solar parabolic mirror systems for California (Stoddard et al. 2006a). This chapter defines a clear opportunity for meeting about half of California's energy needs over the next 45 years. Appendix 2 is an analysis of the amount of land required and the potential production of electricity from concentrated solar photovoltaic energy systems. Stoddard has proposed a concentrated solar parabolic mirror system that started production after 2010. The sustainable-energy strategy assumes that "technical capacity" will be approached with Stoddard's "high deployment scenario" in the years out to 2020 and "technical capacity" achieved before the year 2050.

Chaudhari and her team from Navigant describe the flat-panel PV contribution for the United States (Chaudhari et al. 2004). The data for California in the period of 2010–2025 has been extracted for Chaudhari's analysis and is shown in Appendix 3. The subsequent data points for our Appendix 1 analysis are based on a linear build-out of the flat-panel PV production to its "technical capacity" in 2050. In 2050, the expected production of electricity or hydrogen is 5.25 quads. This level of generation is about one-half the total energy required by California in 2050.

The energy presently captured by solar hot water heating systems in California has not been quantified and included in Appendix 1.

Wind

The wind levels of production were .0145 quad for 2005. The data from the California Energy Commission's strategic value analysis/economics of wind energy was used (Yen 2005). The build-out of wind generation follows the California Energy Commission's deployment schedule out to 2016 and then runs the deployment out to "the technical capacity" by 2050. The "technical capacity" is 1.033 quads or about 10% of California's total energy needs. A portion of the wind is diverted to hydrogen production after the year 2030.

Geothermal

The geothermal levels of production were 0.048 for 2005. The data from the California Energy Commission's strategic value analysis/economics of geothermal energy was used (Sison-Librilla 2005). The build-out of geothermal generation follows the California Energy Commission's deployment schedule from 2005 out to 2017 and runs the deployment out to the "technical capacity" by 2050. Again, a portion of the geothermal is diverted to hydrogen production after the year 2030.

Small Hydroelectric

The 2004 production of electricity from small hydro or "run of the river" is .016 quad (CEC Gross System Power 2004). The California Energy Commission (CEC) stopped compiling gross system power for small hydro in 2004 and now compiles only the data on small hydro purchases made by the utilities for the Renewable Portfolio Standard measurement. There is no growth in small hydroelectric production seen in the recent CEC reports (Nyberg 2009). The small hydro systems in California are primarily located on irrigation canals. The sustainable-energy strategy, as reflected in Appendix 1, assumes that growth in small hydro tracks population and consumption increases.

Prospective Energy Sources

This category was included in Appendix 1 as a placeholder for future energy production from sources not shown in the other categories. Fusion has been suggested as a possible candidate for this category, but it is still under active development and demonstration in the United States, European Union, and Japan. Costs prevent it from being ready for the market in the next 25–30 years (Rosenberg 2007). California also has its own laser fusion energy project under way at the Lawrence Livermore National Laboratory's National Ignition Facility. The facility's Director Mike Hatcher is expecting the giant laser system to generate fusion with energy gain by the end of 2012 (Hatcher 2012; Long and John 2011).

Hurdles to Be Overcome in California
Case Analysis

Concentrated solar power has significant barriers to overcome before it can be applied at the levels shown in Appendixes 1 and 2. Certainly the identification and purchase of the 5,900 square miles is an obstacle. George Simons of the California Energy Commission has identified the best locations for installing concentrating solar power systems. He found that the locations in the southern part of the state (Riverside, Imperial, and San Diego counties) and northern part of the state (Modoc and Lassen counties) have the greatest potential (Simmons 2005a).

Once the locations are identified, permits must be obtained from the state and federal governments. BrightSource has been working on a utility-scale 370-MW thermal power plant for the State of California. This $1.37 billion project has won federal loan guarantees for its Ivanpah project but is into the 30th month of the permit process (Baker 2010). The solar thermal power industry is meeting some resistance as "conservation, labor and American Indian groups are challenging the projects on environmental grounds" (Woody 2011).

Once land is identified and permits obtained, the cost of energy is a significant barrier. Simons has examined the levelized cost of energy of power towers and parabolic trough systems. Stoddard has also examined the levelized costs of energy for trough systems. Depending on their respective assumptions, the cost for trough systems ranges from $.073/kWh (Simmons 2005b) to $.173/kWh (Stoddard et al. 2006b). Tower system's cost of energy (for a 2004 system) was $.16/kWh (Simmons 2005b). Operating a parabolic mirror system also requires the ability to clean and maintain these systems on a routine basis.

The California sustainable-energy strategy uses hydrogen from solar, wind, and geothermal generation to balance out the consumption patterns of the transportation sector and the need for gas-fired combined cycle electrical generation. Only a portion of these renewable-energy sources are diverted to hydrogen production. This elective is dependent upon how much and how fast hydrogen can be used to replace fossil fuels in the transportation sector. If the transportation sector moves more toward electric batteries, then less hydrogen would be required. If transportation moves toward fuel cells, then more hydrogen would be needed. The production of hydrogen is also dependent upon the economics and efficiency of the electrolysis or biological processes to make hydrogen out of water. Knittel has recently completed a thorough analysis of hydrogen vehicles in his paper titled "Reducing Petroleum Consumption from Transportation" (Knittel 2012).

Appendix 1 shows hydrogen production initially coming from concentrated solar power systems and ramping up in 2020, as described by Stoddard et al. (2006b). Authors Myers (Myers et al. 2003) and Lipman (Lipman 2004) show established hydrogen production occurring in 2030 and 2040, respectfully. Political leadership in the European Union under Ramon Prodi (Pordi 2003) demonstrated the more immediate need for hydrogen. Governor Schwarzenegger made the same aggressive public policy in 2004 with an executive order for a California Hydrogen Highway (Schwarzenegger 2005). Clark argued for far more aggressive "green hydrogen economy" with the creation of renewable hydrogen energy stations as in Germany and Japan (Clark 2006a). Private industrial companies have been demonstrating hydrogen fuel cells throughout California and eastern United States (New York, in particular).

The linkage between solar generation and hydrogen as a fuel storage medium is a critical factor in meeting California's transportation requirements. Hydrogen not only serves as a method for delivering energy to the state's transportation fleet, but also it is a reservoir of energy for use on demand. Thus, the intermittent generation of renewables (solar and wind) can be resolved, and these energy flows applied almost immediately for our new hybrid vehicles (Clark and Morris 2002). The potential abundance for concentrated solar power in 2020 and the observation that concentrated solar power generation will be localized geographically to a few sites in California (minimizing transmission logistics) are the factors supporting this thesis.

The "State Hydrogen Highway" was to address the transportation and storage issues for hydrogen. With state government leadership and seed money, the highway was to start with connections to stationary energy conversion and micro-turbine power plants and then move into the vehicle transportation market (Clark 2006b). California's Hydrogen Highway has been refocused on particular cities (clusters). The intent is to have retail-like stations in more targeted neighborhoods/communities with demographics matching potential fuel-cell customers. By 2017, the California Air Resources Board "estimated 50–100 retail hydrogen stations – roughly 10 stations per year – will be needed to satisfy the demand created by the vehicle and bus deployments" (Achtelik 2011).

Meanwhile, there have been a series of articles chronicling the development of hydrogen in home power systems (Pyle 1998). Honda is also developing its FCX Clarity hydrogen fuel-cell car to operate from a home refueling station. Carpenter has asked Honda, "How quickly will home hydrogen re-fueler follow, and how much it will costs? Honda won't say" (Carpenter 2010).

Economic Feasibility for California Case

The data contained in Appendix 1 drives the timeline for capital investment shown in Appendix 4. The capital investment is made of renewal of existing but exhausted facilities, construction of nonrenewable-energy systems, construction of new renewable-energy generation, and construction of transmission lines. There will be trade-offs between capital investment in upgrading existing facilities and/or construction of nonrenewable-energy facilities and construction of new renewable energy. These trade-offs will be based on the adopted sustainable-energy strategy. In addition, considerable amounts of new monies will be needed for energy transmission infrastructure.

The level of renewable-energy sources identified in Appendix 1 will need incentives, both legislative and financial. The current transfers of wealth from the fossil fuels component will need to be carefully evaluated in terms of total internal and external costs.

Appendix 1 indicates a strong trend toward the increased use of renewable energy for electricity generation or for the production of hydrogen to fuel the transportation system. Renewable-energy power plants need to be built in sufficient numbers to take the load off existing and future electrical generation plants running on coal, nuclear, and natural gas. The cost of the capital assets for the needed renewable-energy power plants could well exceed a trillion dollars, if one assumes that all the technical capacities identified in California are placed in service by 2050. To place this projected cost in perspective, it is noted that the Energy Information Association's projection for total US energy expenditures for the single year, 2025, is $1.3 trillion (USDOE, EIA 2006b).

The following capital cost estimates are summarized below from the detailed calculations in Appendix 4. They amount to a capital investment of more than $44 billion per year over the next 45 years. All calculations are in 2006 US dollars. These calculations are not adjusted for the changing value of the dollar nor are the expected reductions in cost per unit reflected.

- Biomass (Jenkins 2006b) $20 billion
- Concentrated solar power (Stoddard 2006a) $1.5 trillion
- Solar PV (Chaudhari 2004) $360 billion
- Wind (Yen 2005) $62 billion
- Geothermal (Sison-Librilla 2005) $10.6 billion

The above numbers can be placed in the context of current spending for power systems and petroleum/coal refining. The United States invested in new power systems and oil/coal refining at the level of $38.6 billion per year in 2004 and $37.7 billion in 2005 (Department of Commerce 2004). If one looks at just the generation side of this investment (about 40% per Department of Commerce), then the number is about $15 billion per year. The investment in generation facilities for California, in 2005, was roughly 10% of this number. It becomes clear that if California wants to achieve the mix of renewable-energy generation resources identified in Appendix 1, it must invest in renewable resources at a pace greater than the 2005 investment for United States' power systems and petroleum/coal refining. At this level of spending, the generation will be covered, but the transmission of that energy is another matter and additional cost.

The natural gas combined cycle plant is often used as the bench mark for cost of electricity (COE) generation. Based on Market Gas Referent (MPR 2005), prices for 2007, $6.40/MMBtu, and assuming a 500-MW combined cycle plant operating at a 40% capacity factor, the COE is $104 per MWh. A 100-MW centralize solar power system operating with 6 h storage, 30% investment tax credit, and operating at 40% capacity factor is $157 per MWh (Stoddard et al. 2006c). Future improvements in the efficiency of centralized solar power plants could narrow the COE. The incorporation of externalities is not included in the above cost figures. An energy strategy for California that is sustainable requires construction of centralized solar power plants to meet the renewable-energy requirements as opposed to natural gas combined cycle plants to meet the fossil fuel energy requirements.

There is an urgency that is clearly demonstrated by data in the 2010 and 2020 columns of the Appendixes 1 and 4. These numbers will never materialize unless long-term energy contracts are established, lines of credit established with guarantees, and the people of California and their respective energy suppliers embrace the concept that energy costs include not only the feedstock, refining, and distribution but also the environmental impacts. Borenstein discusses the market and nonmarket valuation of electricity generation from renewable energy, as well as the costs of the subsidies that are available. He finds, "On a direct-cost basis, renewable are expensive. But, the simple calculations fail to account for many additional costs and benefits of renewables." He proposes levelized cost of electricity estimates be "thoughtfully adjusted for the market value of the power generated and for the associated externalities" (Borenstein 2011).

Technical Feasibilities in the California Case

Some of the technologies described in this chapter are yet to be tested at the commercial scale in California. The thermal concentrated solar power system noted in section "Solar" falls into this category. Some of the technologies need to be researched and developed to perform to the economic standards and efficiency standards suggested in this chapter. The conversion of electricity to hydrogen certainly fits in this category. The management of feedstocks must be greatly improved for biomass to meet its targets.

One of the largest hurdles facing the renewable-energy industry is the transmission of energy from the point of generation to the point of use. Given the location requirements for biomass, wind, and concentrated solar power, this generation will be quite a distance from the urban load. The existing grid could be augmented with new overhead lines, but many would oppose these efforts. Chauncey Starr, the former president of Energy Power Research Institute, and Paul Grant, who was involved in the discovery of high-temperature superconductivity, have come up with a system that could transport 5–10 gigawatts of power. This system uses underground superconducting direct current cables that use hydrogen for cooling. This grid could transmit both electricity and hydrogen (Overbye 2006). No cost data is available for this system.

Finally, the California Independent System Operator (Cal ISO) observes, "As renewable sources come in the system, procurement of fossil-fuel generation will decline just when it is needed to offset the intermittency of variable resources." The Cal ISO finds "the fossil-fuel generation will remain needed for reliability until other technologies such as storage or demand response mature" (Beberich 2012).

The California Renewable Portfolio Standard

California established its Renewable Portfolio Standard (RPS) in 2002 (California Senate Bill 1078 2002). This RPS requires the state's investor-owned utilities (IOUs) to include 20% of the electricity in their sales portfolios to be generated from renewable energy by 2017. California's past governor, Arnold Schwarzenegger, has recommended that this goal be increased to 20% by 2010 and 33% by 2020 (Schwarzenegger 2004). While Appendix 1 indicates that the State of California had a good chance of achieving the original RPS, in fact the state's IOUs fell short of the mark (IOUs were only able to purchase 17%). They may achieve the 20% level by 2017. The State of California's RPS is based on California's IOU purchases of renewable energy, and it is a lower percentage than the numbers in Appendix 1 would generate. The state's RPS omits residential and commercial rooftop solar and on-site power generation. The projected RPS (including residential commercial rooftop solar and on-site power generation) for Appendix 1 is 2010 – 10%; 2020 – 13%; 2030 – 33%; 2040 – 50%; and 2050 – 62%.

Current Status

As noted earlier, this case analysis focuses on the State of California because data was available for both consumption patterns and estimates of in-state energy resources and the imports of energy resources for the period of 2005–2050. The period of 2005 to the present is now providing real numbers, which can be compared to the model's projections. The US Energy Information Administration provides periodic reports in its State Energy Data System (SEDS). On June 30, 2011, they released SEDS data for 2009. The data flowing from these reports reveal that the slowing economy almost eliminated the projected growth in natural gas for heating and electric consumption. Gas generation did increase by the rate of 4.2% per year, but this was primarily due to a drop in hydroelectricity production, related to low rainfall in the recent years. The most disturbing trend coming out of the recent data is the slow pace of growth for the renewable energy. Wind only grew at 6% per year (the projection was for 25% per year). Solar was projected to have tremendous growth, due to solar farms arising in the deserts of California. This projected growth was slowed, due to environmental impact report concerns. The data shows only a 9.6% per year growth, and this was largely attributable to solar rooftop installations. Ethanol was projected to grow at the rate of 5% per year, but the data shows only 1.2% per year rate of growth. The economy also held down the growth of petroleum consumption for transportation. The projected growth was 1.6% per year, but the actual growth, due to the poor economy, was −1.3% per year.

The population was projected to grow by the rate of 1.6% per year. The data from the California Department of Finance shows the population growth of only 36.1 million to 36.9 million or about .07% per year during this time period (CA Department of Finance 2011). The energy consumed per capita was projected to remain at .0% per year. The total energy consumed per capita actually grew 1.1% per year.

In summary, the data presented in Appendix 1 for the base year 2005 hardly changed in the years leading up to 2010. In terms of a sustainable-energy generation and consumption, this is both positive and negative. The positive side is the consumption did not increase. The negative side is that renewable energy failed to achieve the exponential growth required to meet the levels of renewable energy in Appendix 1's 2050 projection

Conclusion

The California case analysis shows how regions and nations can define their energy needs and energy sources. It also demonstrates the critical need to identify future demands for energy and technical capacity of their energy resource streams. The determination of a technical capacity for a given energy resource is based on environmental, and social and economic principles to assure that it is secure, renewable, accessible locally, affordable, and obtained with minimal impacts on the environment.

A sustainable-energy future is not far away if a region or nation can bring their consumption in line with the technical capacity of their energy resources along with providing the financial measures needed to achieve the goal. Governments and their business partners need to formulate their own sustainable-energy strategy through the development of their energy resources and consumption spreadsheet and allocate the necessary human resources and capital funds to develop the energy resources and energy distribution systems.

Appendix 1

California's energy consumption and energy resource projections 2005–2050 in quads/year[a]. For source of data, see text under respective resource heading

Type	2005	2010	2020	2030	2040	2050
State natural gas heat	0.2794	0.2794	0.2794	0.2794	0.2794	0.2794
Imports, US, NG heat	1.7347	1.4585	2.2393	2.4236	2.1447	1.9907
State gas/diesel[b]	1.3463	1.3463	1.3500	1.3500	1.3500	1.1000
Imports, US, gas/diesel[b]	0.7592	0.8200	0.8200	0.8200	0.6000	0.2000
Imports, foreign, gas/diesel[b]	1.5119	1.7500	1.7000	0.6000	0.0000	0.0000
State hydro. elect.	0.0991	0.0991	0.0991	0.0991	0.0991	0.0991
Imports, US, hydro. elect.	0.0454	0.0454	0.0454	0.0454	0.0454	0.0454
State coal elect.	0.0976	0.0976	0.0976	0.0976	0.0976	0.0976
Imports, US, coal elect.	0.1129	0.1129	0.1129	0.1129	0.1129	0.1129
State natural gas elect.	0.3584	0.3584	0.3584	0.3584	0.3584	0.3584
Imports, US, NG elect.	0.0450	0.0450	0.0450	0.0450	0.0450	0.0450
State nuclear elect.	0.1031	0.1031	0.1031	0.1031	0.1031	0.1031
Imports, US, nuclear elect.	0.0230	0.0230	0.0230	0.0230	0.0230	0.0230
Biomass elec.	0.0200	0.0250	0.0500	0.0500	0.0500	0.0500
Biomass – CH4/Ethan./BioD	0.0500	0.0750	0.3300	0.3400	0.3000	0.2400
Imports, US, ethanol bioD	0.0678	0.0800	0.1000	0.1818	0.3636	0.3636
State biomass to H2	0.0000	0.0000	0.0000	0.0350	0.0900	0.1600
Import biomass to H2	0.0000	0.0000	0.0000	0.0140	0.0280	0.0280
Solar elect.	0.0024	0.4292	0.1549	0.2948	0.9408	1.4048
Solar elect. to H2	0.0000	0.0000	0.3500	1.7500	2.7300	3.8500
Wind elect.	0.0145	0.0356	0.0220	0.1664	0.3017	0.4097
Wind elect. to H2	0.0000	0.0000	0.0350	0.1750	0.3220	0.4900
Small hydro elect.	0.0159	0.0172	0.0197	0.0223	0.0248	0.0272
Geothermal elect.	0.0478	0.0735	0.0247	0.0520	0.0520	0.0520
Geothermal elect. to H2	0.0000	0.0000	0.0077	0.0560	0.0560	0.0560
Prospective	0.0000	0.0000	0.0000	0.0000	0.0000	0.0000
Total	6.7344	7.2742	8.3672	9.4948	10.5175	11.5859
Baseline consumption (applies population increase numbers to 2010–2050)	6.7344	7.2732	8.3641	9.4515	10.4911	11.5402
Total electrical supply in quads/year	0.9851	1.4650	1.5485	3.4510	5.3618	7.2242
Total gas supply in quads/year (less amount going to electricity gen.)	2.0141	1.7379	2.5187	2.7030	2.4241	2.2701
Target supply (in quads/year) transportation fuels based on population growth	3.7352	4.0340	4.6391	5.2422	5.8188	6.4007
Total supply transport fuels in quads/year	3.7352	4.0713	4.6927	5.3218	5.8396	6.4876
Total renewable energy	0.2184	0.7355	1.0940	3.1373	5.2589	7.1313
Percent of renewable energy in total energy consumption	3.24%	10.11%	13.08%	33.04%	50.00%	61.55%

[a]One quad = 10^{15} BTUs = 10^{18} J
[b]"State," "federal," and "foreign" indicate source of crude oil. All refining is conducted in California

Appendix 2

Projections for concentrated solar power						
Total land mass in California[a]	155,959 sq. miles					
Parabolic trough, 6 h storage with slope less than 1%[b]	5,900 sq. miles		471,000 MW		1,640,000 GWh (5.599 quads BTUs)	
Concentrating PV, < 5% slope[b]	14,400 sq. miles		1,534,000 MW		3,558,000 GWh (12.147 quads Btus)	
	2005	2010	2020	2030	2040	2050
Solar Ca parabolic troughs, 6 h storage, <1 slope land[c] (quads)	0	0.746	5.599	5.599	5.599	5.599
Concentrating PV, <5% slope land[c] (quads)	0	0	12.147	12.147	12.147	12.147
CA electrical energy load[d] (quads)	0.9851	1.465	1.5485	3.451	5.3618	7.2242
CA all energy load[d] (quads)	6.7344	7.2742	8.33651	9.4925	10.4912	11.5408

[a] www.dof.ca.gov/html/fs_data/stat-abs/table/as.xls
[b] Stoddard 2006b
[c] Stoddard 2006c
[d] Appendix 1

Appendix 3

Projections for solar PV grid connected per navigant paper September 2004				
Solar panels	2010 MW[a]	Capacity factor[b]	Annual GWh	BTUs in quads
Residential[d]	20,132			
	2,237			
	22,369			
Total res.	44,738	0.179	70,151	0.240
Commercial[d]	16,915			
	1,879			
	18,794			
Total com.	37,588	0.164	54,000	0.184
			Total Quads	0.424
Solar panels	**2025 MW**[c]	**Capacity factor**[b]	**Annual GWh**	**BTUs in quads**
Residential[d]	28,794			
	3,199			
	31,993			
Total res.	63,986	0.179	100,333	0.343
Commercial[d]	27,899			
	3,100			
	30,999			
Total com.	61,998	0.164	89,069	0.304
			Total quads	0.647

Sources: Chaudhari et al. 2004
[a] Page 83 Technical market for PV (MWp) in 2010 – by state and segment
[b] Personal communication with author
[c] Page 82 Technical market for PV (MWp) in 2025 – by state and segment
[d] State of California data was in three segments, SCE, SDGE, PG&E (note: no Muni data)

Appendix 4

Biomass to electricity capital cost in $2006 dollars – model deployment

Year	Cost per trillion BTUs ($)	Technical capacity trillion BTUs per year	Cost ($)	Quads
2010	44,444,444	20	888,888,880	0.0200
2020	44,444,444	30	1,333,333,320	0.0300
2030	44,444,444	0	0	0.0000
2040	44,444,444	0	0	0.0000
2050	44,444,444	0	0	0.0000
Sum	44,444,444	50	2,222,222,200	0.0500

Source: Jenkins 2006b

Biomass to CH_4/ethanol/BioD/heating capital cost in $2006 dollars – model deployment

Year	Cost per trillion BTUs ($)	Technical capacity trillion BTUs per year	Cost ($)	Quads
2010	44,444,444	80	3,555,555,520	0.0800
2020	44,444,444	275	12,222,222,100	0.2750
2030	44,444,444	15	666,666,660	0.0150
2040	44,444,444	15	666,666,660	0.0150
2050	44,444,444	15	666,666,660	0.0150
Sum	44,444,444	400	17,777,777,600	0.4000

Source: Jenkins 2006b

Concentrated solar plant capital cost in $2005 dollars – model deployment scenario

Year	Cost/100 MWe ($)	Cost per watt ($)	Technical capacity MWe	Cost ($)	MGWh/year	Quads
2007	494,386,000	4.94	100	494,386,000	348	0.0012
2009	457,590,000	4.58	100	457,590,000	348	0.0012
2011	583,384,000	3.89	250	972,306,667	870	0.0030
2015	631,373,000	3.16	950	2,999,021,750	3,307	0.0113
2020	631,373,000	3.16	2,600	8,207,849,000	9,051	0.0309
2030	631,373,000	3.16	155,666	491,416,547,090	541,873	1.8500
2040	631,373,000	3.16	155,666	491,416,547,090	541,873	1.8500
2050	631,373,000	3.16	155,668	491,422,860,820	541,880	1.8500
Sum			471,000	1,487,387,108,417	1,639,551	5.5975

Source: Stoddard et al. 2006d

Solar PV capital cost in $2005 dollars – model deployment scenario

Year	Cost per watt ($)	Technical capacity MWe	Cost ($)	MGWh/year	Quads
2010	2.25	82,326	185,233,500,000	124,148	0.4238
2025	2.25	43,640	98,190,000,000	65,809	0.2247
2050	2.25	34,015	76,533,750,000	51,295	0.1751
Sum		159,981	359,957,250,000	241,251	0.8236

Source: Chaudhari et al. 2004

Wind capital cost in $2005 dollars – model deployment scenario					
Year	Cost per watt ($)	Technical capacity MWe	Cost ($)	MGWh/ year	Quads
2010	0.76	2,000	1,520,000,000	6,170	0.0211
2017	0.63	3,029	1,908,270,000	9,344	0.0319
2030	0.63	30,990	19,523,700,000	95,604	0.3264
2040	0.63	30,990	19,523,700,000	95,604	0.3264
2050	0.63	30,991	19,524,330,000	95,607	0.3264
Sum		98,000	62,000,000,000	302,330	1.0322

Source: Yen 2005

Geothermal capital cost in $2005 dollars – model deployment scenario					
Year	Cost per watt ($)	Technical capacity MWe	Cost ($)	MGWh/ year	Quads
2010	4	1000	3,620,000,000	7,538	0.0257
2017	4	1,995	7,002,450,000	15,038	0.0513
Sum		2,995	10,622,450,000	22,576	0.0771

Source: Sison-Librilla 2005

References

Achtelik G (2011) "California hydrogen highway network," CaH2 Net, California Air Resources Board, 8 Feb, p 2. http://www.hydrogenhighway.ca.gov/facts/progress.pdf. Accessed 4 Mar 2011

Al-Juaied M, Whitmore A (2009) Realistic costs of carbon capture. Discussion paper, Belfer Center for Science and International Affairs, Cambridge, MA

Baker DR (2010) Solar industry goes through shakeout. San Francisco Chronicle, 17 Apr 2010, p D-1. http://www.sfgate.com/cgi-bin/article.cgi?f=/c/a/2010/04/16/BUB91CUQPQ.DTL. Accessed 8 Feb 2012

Baldassare M, Han E (2005) CA 2025, "Its your choice: what kind of California do you want?" Section: the challenge, Public Policy Institute of California, p 7

Beberich S (2012) Reliable power for a renewable future – 2012–2016 strategic plan. California ISO, 250 Outcropping Way, Folsom 95630, p 6. www.caiso.com/documents/2012-2016strategicplan.pdf. Accessed 7 Feb 2012

Boer R (2009) Optimized core design and fuel management of a pebble –bed type nuclear reactor. Ios Press, Amsterdam. ISBN 13:9781586039660

Borenstein S (2011) The private and public economies of renewable electricity generation. Energy Institute at Hass, 2547 Channing Way, #5180, Berkeley, WP 221, pp 2, 24. http://ei.haas.berkeley.edu. Accessed 7 Feb 2012

Brundtland GH (1987a) Our common future, world commission on environment and development. Oxford University Press, Oxford/New York, p xi. ISBN 0-19-282080

Brundtland GH (1987b) Our common future, world commission on environment and development. Oxford University Press, Oxford/New York, p 8. ISBN 0-19-282080

BTAC (Biomass Technical Advisory Committee) (2002) Vision for bioenergy & bio-based products in the United States, p 9. http://www1.eere.energy.gov/biomass/pdfs/final_2006_vision. Accessed 8 Feb 2012

Byron JD (2009) 2009 Integrated energy policy report. California Energy Commission, CEC-100-2009-003-CTD, p 133

CA Assembly Bill 32 (2006) http://www.arb.ca.gov/cc/ab32/ab32.htm. Accessed 8 Feb 2012

CA Department of Finance (2011) http://www.dof.ca.gov/research/demographic/reports/estimates/e-6/view.php. Accessed 8 Feb 2012

CEC Oil Supply Sources to Ca Refineries (2005) http://www.energy.ca.gov/2006publications/CEC-600-2006-006/CEC-600-2006-006.PDF. Accessed 8 Feb 2012

Carpenter S (2010) A gas station at home? It's Honda's vision of the future. Los Angeles Times, 25 June 2010. http://latimesblogs.latimes.com/home_blog/2010/06/honda-clarity-hydrogen-fuel-cell-home-refueler.html. Accessed 8 Feb 2012

Chaudhari M et al (2004) PV grid connected market potential under a cost breakthrough scenario. Navigant consulting under contract to the Energy Foundation, EF, p 117373

Clark WW II (2009) Sustainable communities. Springer, New York. ISBN 978-1-4419-0218-4

Clark WW II (2010) Sustainable communities design handbook. Elsevier, New York

Clark WW II (2006a) A green hydrogen economy, Special issue on hydrogen. Energy Policy, Elsevier, vol 34, Fall, pp 2630–2639

Clark WW II (2006b) Partnerships in creating agile sustainable development communities. J Clean Prod, Elsevier Press

Clark WW II, Morris G (2002) Policy making and implementation process: the case of intermittent resources. J Int Energy Policy, Interscience. http://www.clarkstrategicpartners.net/documents/Journals/WindPolicyWWCGreg352-204.pdf. Accessed 8 Feb 2012

Clark C, Yin Y (2007) The sustainability of biomass energy in the Pacific Northwest: a framework for the PNW region of the Sun Gran Initiative. Biomass Feed Stock Partnership workshop, 27–29 Aug 2007, Oregon State University, Sun Grant, Portland

Department of Commerce (2004) Income, poverty, and health insurance coverage in the United States: 2004. Issued Aug 2005. http://www.census.gov/prod/2005pubs/p60-229.pdf. Accessed 28 Feb 2012

Desvaux M (2007) The sustainability of human populations: how many people can live on Earth? RSS Journal Significance 4(3):4–6

Dutzik T et al (2006) A new energy future: the benefits of energy and renewable energy for cutting America's use of fossil fuels. Environment California Research & Policy Center, Los Angeles, pp 15–18, Fall

Fisher J (2011) A green future for the Los Angeles department of water and power.Synapse Energy Economics. 22 Pearl Street, Cambridge, MA, 02139, p 2. http://www.labeyondcoal.org/uploads/7/5/9/0/7590173/synapse_consulting_-_green_plan_for_ladwp_-_may_2011_-_final.pdf. Accessed 8 Feb 2012

Forest Stewardship Council (2008) Full review of FSC principles and criteria. FSC newsletter, News and Views 6(9):1–2, 5 Sept 2008. http://fscus.org/news/. Accessed 8 Feb 2012

CEC Gross System Power (2004) http://www.energy.ca.gov/2005publications/CEC-300-2005-004/CEC-300-2005-004.PDF. Accessed 8 Feb 2012

Hass R et al (2008) Towards sustainability of energy systems: a primer on how to apply the concept of energy services to identify necessary trends and policies. Energy Policy 36(11):4020

Hatcher M (2012) PW 2012: fusion laser on track for 2012 burn. Optics.org. 29 Jan 2012. http://optics.org/news/3/1/37. Accessed 7 Feb 2012

Hicks B, Nelder C (2007) Profit from peak oil. Wiley, Hoboken, 111 River Street 07030

Highway Statistics (2004) State highway agency-owned public roads 2000 to 2004. Rural and Urban Miles; Estimated Lane-Miles and Daily Travel. http://www.fhwa.dot.gov/policy/ohim/hs04/htm/hm80.htm. Accessed 8 Feb 2012

Jacobson MZ, Delucchi MA (2010) Providing all global energy with wind, water, and solar power, Part 1: technologies energy resources, quantities and areas of infrastructure and materials. Energy Pol. doi:10.1016/j.enpol.2010.11.040

Jenkins BM (2006a) A preliminary roadmap for the development of biomass in California. PIER collaborative report, California Energy Commission, CEC-500-2006-095-D, Figure 1.5, p 27

Jenkins BM (2006b) A roadmap for the development of biomass in California. Informational meeting, California Energy Commission, 19 Sept 2006

Keese WJ (2003) Renewable portfolio standard: decision on phase 1 implementation issues. California Energy Commission 500-03-023F, p 4

Kessides IN (2010) Nuclear power: understanding the economic risks and uncertainties. Energy Pol 38:3849–3864

Knittel CR (2012) Reducing petroleum consumption from transportation. Energy Institute at HAAS, WP 227, pp 15–20. http://ei.haas.berkeley.edu. Accessed 7 Feb 2012

Lipman T (2004) What will power the hydrogen economy? Present and future sources of hydrogen energy. Report prepared for The National Resources Defense Council, Institute of Transportation Studies – University of California, Davis, 95616, Figure E-4, p x

Long JCS, John M (2011) California's energy future: the view to 2050. California Council on Science and Technology, 1130 K Street, Suite 280, Sacramento, 95814, ISBN-13 978-1-930117-44-0, pp 31, 42. http://ccst.us/publications/index.php. Accessed 7 Feb 2012

Marks M (2005) Natural gas assessment update. CEC report 600-2005-003, p 131, CEC California historical natural gas supply by source, CEC, 2005. http://www.energy.ca.gov/2005publications/CEC-600-2005-003/CEC-600-2005-003.PDF. Accessed 8 Feb 2012

McCarthy E (2005) Experts say nukes create greenhouse impacts. California Energy Circuit, Berkeley, CA, USA

McGhee L (2006) McMansions yanked from village menu. Sacramento Bee, 19 Oct 2006, B-1

Moniz EJ et al (2010a) "The future of natural gas," an interdisciplinary MIT study interim report. MIT Energy Initiative, Boston, p xi. ISBN 978-0-9828008-0-5

Moniz EJ et al (2010b) "The future of natural gas," an interdisciplinary MIT study interim report. MIT Energy Initiative, Boston, p 14. ISBN 978-0-9828008-0-5

Moniz EJ et al (2010c) "The future of natural gas" an interdisciplinary MIT study interim report. MIT Energy Initiative, Boston, p xiii. ISBN 978-0-9828008-0-5

MPR (2005) California Public Utility Commission, Mar 2006. http://docs.cpuc.ca.gov/published/Comment_resolution/54445. Accessed 8 Feb 2012

Murtishaw S et al (2005) Development of energy balances for the State of California, prepared by Lawrence Berkeley Laboratory, Environmental Energy Technologies Division, Berkeley, Tel: (510) 486–7553, CEC PIER report 500-2005-068, June, Table 2, p 11

Myers DB et al (2003) Hydrogen from renewable energy sources: pathway to 10 quads for transportation uses in 2030 to 2050. Draft report by Directed Technologies, for the Hydrogen Program Office, Office of Power Technologies, U.S. Department of Energy, Washington, DC, Figure 2, p 2

Nyberg MJ (2009) 2009 net system power report. California Energy Commission, 200–2009 010-CMF, Table 2, p 5

Optimum Population Trust (2002) www:optimumpopulation.org/opt.contact.html

Overbye TJ (2006) Building tomorrow's super grid. EnergyBiz Magazine, p 76, Sept/Oct

Perez P (2005) Platts Ethanol finance and investment. Chicago Conference, Chicago

Prodi R (2003) EU roadmap towards a European partnership for a sustainable hydrogen economy. Speech, Brussels, 10 Sept 2003

Pyle W (1998) Solar hydrogen chronicles. Wheelock Mountain Publications, ISBN 0-9663703-0-9. http://www.goodideacreative.com/shc.html. Accessed 8 Feb 2012

Rosenberg M (2007) The ultimate answer: fusion power's long take off. Energy Biz Magazine May/June, pp 36–40

Roundtable (2008) Roundtable on sustainable biofuels, Version, Zero, standard for sustainable biofuels, 13 Aug 2008, p 2. http://www.biofuels.nsw.gov.au/__data/assets/pdf_file/0003/105429/RSB_Principles_and_Criteria_v0.pdf. Accessed 8 Feb 2012

Schaffer MB (2011) Toward a viable nuclear waste disposal program. Energy Pol. doi:10.1016/j.enpol.2010.12.010

Schwarzenegger (2004) Executive order 2-7-04 by the Governor of the State of California. http://www.renewableenergyworld.com/rea/news/article/2009/09/executive-order-raises-california-rps-to-33-by-2020. Accessed 8 Feb 2012

Schwarzenegger (2005) Governor Arnold Schwarzenegger's California hydrogen highways network action plan. http://gov.ca.gov/executive-order/11072/. Accessed 8 Feb 2012

CA Senate Bill 1078 (2002) http://www.leginfo.ca.gov/pub/01-02/bill/sen/sb_1051-1100/sb_1078_bill_20020912_chaptered.html. Accessed 8 Feb 2012

Simmons G (2005a) Developing cost-effective solar resources with electricity system benefits. IEPR Committee Workshop, Pier Renewables, California Energy Commission, CSP: Economic Potential with WTLR

Simmons G (2005b) Developing cost-effective solar resources with electricity system benefits. IEPR Committee Workshop, Pier Renewables, California Energy Commission, Trough: Cost Model Inputs & Results, Tower: LCOE Values

Sison-Librilla E (2005) Geothermal strategic value analysis, CEC 500-2005-105-SD, June draft, p 1, Table 6, p 12

Sorrel S et al (2010) Global oil depletion: a review of evidence. Energy Pol 38:5290–5295

Stoddard L et al (2006a) Economic, energy, and environmental benefits of concentrating solar power in California. Black and Veatch, NREL Sub Contract Report SR-550-39291, Table 6–2, pp 6–4

Stoddard L et al (2006b) Economic, energy, and environmental benefits of concentrating solar power in California. Black and Veatch, NREL Sub contract report SR-550-39291, Table 3–1, pp 3–1

Stoddard L et al (2006c) Economic, energy, and environmental benefits of concentrating solar power in California. Black and Veatch, NREL Sub contract report SR-550-39291, pp 8–1

Stoddard L et al (2006d) Economic, energy, and environmental benefits of concentrating solar power in California. Black and Veatch, NREL Sub contract report SR-550-39291, pp 5–5, Table 5–1, pp 4–3, Table 4–1, pp 3–1, Table 3–1

USA Bureau of Economic Analysis (2007) Pre, capita real GDP by state 2007. http://www.bea.gov/newsreleases/regional/gdp_state/2008/gsp0608.htm. Accessed 8 Feb 2012

US Census Bureau (2005) Gross Domestic Product by state-sorted by 2000–2005 GDP per capita growth. http://www.economics-chart.com/gdp/GDP-Per-Capita-by-gpc-growth.html. Accessed June 2010

USDOE, IEO (2006a) International energy outlook, Report# DOE/EIA-0484(2006), Release Date: June 2006, web site: http://www.eia.gov/forecasts/archive/ieo06/index.html. Accessed 8 Feb 2012

USDOE, EIA (2006b) Annual energy outlook 2006 with projections to 2030. http://www.eia.doe.gov/neic/press/press267.html. Accessed 8 Feb 2012

van Vliet O et al (2011) Combining hybrid cars and synthetic fuels with electricity generation and carbon capture and storage. Energy Pol 39:258, Table 8

Von Weizsacker EU (1998) Factor four: doubling wealth, halving resource use. Earthscan, London. ISBN 9781853834066

Wing IS (2005) The synthesis of bottom-up and top-down approaches to climate policy modeling: electric power technologies and the cost of limiting US CO_2 emissions. Energy Policy 34:3847

Woody T (2011) Solar energy faces test on greenness. The New York Times, Business Day Energy and Environment, New York Edition, 24 Feb 2011, p B-1

Yen D (2005) Strategic value analysis: economics of wind energy in California. CEC 500-2005-107-SD, June Draft p 1, and Table 4 (with build out to Technical Feasibility levels by 2050)

Chapter 6
Achieving Economic Gains Through the Setting of Environmental Goals: The Case of California

Tracey Grose

Abstract Faced with volatile energy prices, rising resource costs, and income uncertainty, Californians (and the USA) are finding new ways to save money and conserve resources. Concerns about climate change are part of this. These changes not only serve as a recessionary buffer but also develop resilience toward growing external shocks for a company, household, or economic region as a whole. Contrary to conventional wisdom, well-crafted regulation can be a market driver and spur business and employment growth. California's experience, dating back to innovative policies following the 1970s energy crisis, demonstrates that economic growth and environmental improvement can be achieved together.

Introduction

California's economy today is undergoing a complete transformation. Faced with volatile energy prices, rising resource costs, and income uncertainty, businesses, households, and public entities are finding new ways to save money and conserve resources. Concerns about climate change are part of this. However, the rising demand for global resources and the resulting price hikes and increasing volatility are taking place independent of political influence. As people seek alternatives, these changes not only save consumers money, especially during an economic downturn, but also help to develop resilience toward growing external shocks for a company, household, or economic region as a whole. Improving resilience through resource efficiency is about improving resource productivity and competitive advantage.

T. Grose (✉)
Research Director, Institute for the Future, Palo Alto, CA, USA
e-mail: tgrose@iftf.org

The transformation taking place is developing at varying rates across the USA, and public policy plays a vital role. Contrary to popular conventional economics, in the arena of clean energy and resource efficiency, regulation can be a market driver and spur business and employment growth. Following the energy crisis in the 1970s, California put into place a series of innovative policies that have resulted in higher energy efficiency than in the rest of the nation. A commitment to conservation through the implementation of standards, incentives, and mandates at the state and local levels has served to create new business opportunities for the state's entrepreneurs and new job opportunities for the state's workforce.

While modern economic theory may treat environmental degradation as a non-measurable "externality", the reality in which individuals, businesses, and public policy makers operate is very much impacted by constrained resources and the multiple costs of pollution. Recognizing this fact of environmental, health, and security costs, public policy (e.g., efficiency standards, incentives for early adoption of clean energy technology, and public procurement standards) can serve as a market driver for goods and services that help a community meet its environmental goals which concomitantly serves to spur employment growth in related sectors, improve resource efficiency across the economy, and therefore spur employment growth across all sectors.

The sections below briefly lay out today's changing context in the world's economy, its drivers, and the new opportunities that are arising. There is already a strong business case for energy efficiency retrofitting, and smartly crafted public policy can play a powerful role in spurring technological innovation and the broad-based adoption of products that improve resource efficiencies. California's experience is presented as an example of how setting environmental goals can yield positive economic results. California's story suggests that investing in environmental improvement must not come at the cost of economic growth and that the implementation of innovative public policy can serve to spur technological innovation and economic growth.

Economic Transformation: New Drivers and New Opportunities

In response to multiple driving forces, the global economy is transforming in significant ways. In the area of energy and resource efficiency, public policy can play a large role in achieving environmental goals and spurring economic growth. The idea that environmental regulation can create new markets and drive innovation has been described by Porter and van der Linde:

> Properly designed environmental standards can trigger innovations that lower the total cost of a product or improve its value. Such innovations allow companies to use a range of inputs more productively—from raw materials to energy to labor—thus offsetting the costs of

improving environmental impact and ending the stalemate. Ultimately, this enhanced resource productivity makes companies more competitive, not less.[1]

Globally, the demand is rising for vital resources, like energy and fresh water, as the world's population grows and the standard of living in developing countries rises. Add into that equation volatile fuel costs and the real impacts of climate change such as diminishing sources of fresh water, threatened coastlines, and changing weather patterns, the impetus has arrived for a transformation of the economy away from one based on fossil fuels and waste to one based on clean fuels and the efficient use of all resources.

Growing Resilience and Opportunity in a Context of Increasing Volatility

The transformation away from a carbon-based economy is about growing economic resilience. In addition to the environmental gains achieved, this transformation results in higher levels of resource productivity and competiveness for a company as well as an economy. A useful analogy can be observed in the application of information technology (IT) across the economy over the last several decades which resulted in significant gains in labor productivity across the economy.[2] New opportunities for cost savings and new product development emerged across industries, and the IT industry continued to grow and diversify offering wider ranges of products, services, and employment opportunities. Similarly, improving resource productivity will help improve the capacity for the economy and the environment to respond to external shocks and negative impacts (i.e., raise resilience) and generate new economic opportunity as related industries grow.

Economic and Environmental Resilience

With growing global demand for all natural resources driven primarily by the unprecedented economic growth in Asia, prices for natural resources including energy are expected to continue rising. The volatility of fuel prices in recent years has driven consumers—business, households, and the public sector alike—to seek out cheaper alternative means of transportation and new methods for fuel conservation. By adopting

---

[1] Porter and Claas van der Linde (1995).
[2] Labor productivity averaged 1.46% annual growth between 1973:Q4 and 1995:Q4 and averaged 2.91% per year over the 1995:Q4-2005:Q3 period. Jorgenson et al. (2005). See also Atkinson and McKay (2007).

products and practices that help conserve natural resources, prevent pollution, and manage waste and by investing in the development of new forms of these products and practices, our communities stand to reap multiple benefits. These actions not only reduce the overall consumption of scarce resources and the generation of greenhouse gas emissions, they also help stimulate new markets and economic opportunity.

In addition, public incentives and new regulations help spur innovation and the growth of new markets by lowering the cost of cleaner alternatives and increasing the cost of harmful fuels, products, and practices. Similarly, the lack of sustained political commitment and accessible project funding will hamper growth. Driving innovation in raising resource productivity offers a viable strategy for contending with the new set of circumstances we face today as businesses, households, and public policy makers. For the purposes of this chapter, "green innovation" is the process of finding new ways of doing things, through technological advance, new practices, and new forms of public policy, which generate both environmental and economic benefits.[3]

As resource costs rise, markets for alternatives open and present new business and employment opportunities. Also the adoption of green products becomes more widespread, these products improve in quality and price. Then, as these products and practices spread across the economy, businesses and households enjoy the cost savings that come from improved energy and resource efficiency. This cost savings can be reinvested in capital upgrades or new hires. While renewable energy generation systems may take longer, the return on investment for energy efficiency technologies and practices is almost immediate and frees up resources for other uses or investment.[4] This also means that businesses become more cost-effective which boosts their competitive advantage and improves the energy and resource productivity of the region as a whole. The region then benefits not only from environmental improvement (e.g., reduced pollution and demand for natural resources) but also from greater economic resilience and energy independence. The use of the term "economic resilience" is meant as the ability of an economy to withstand or recover from the effects of adverse exogenous shocks (e.g., volatile fuel prices, global financial crisis, or natural disaster).[5]

New Economic Opportunity: The Core Green Economy

There are different aspects to the changes currently taking place in the economy and the new opportunities arising as a result. At the core of these developments are the businesses which provide the products and services that enable the transformation across the entire economy (e.g., other businesses as well as households and public entities).

---

[3] Henton et al. (2008a).
[4] Farrell et al. (2007).
[5] See Briguglio et al. (2009).

This is called the "*core green economy*," and it consists of businesses that provide products and services that do the following:

- Provide alternatives to carbon-based energy sources
- Conserve the use of energy and all natural resources
- Reduce pollution (including GHG emissions) and repurpose waste

A larger part of the story is the *adaptive green economy*, which consists of businesses as well as the public sector, households, and nonprofit organizations that are using the products and services of the core green economy in order to improve the resource efficiency of their own operations. The economic actors of the adaptive green economy are altering their processes to improve sustainability, reduce costs, or anticipate regulatory changes. These institutions are reexamining their processes and investing in fundamental changes in their operations, as well as encouraging their suppliers to do likewise. Examples include the efforts of large USA corporations such as Staples, Walmart, and FedEx to significantly improve their own energy and resource efficiency and to set standards for their suppliers to follow suit. An added component of these efforts, besides the public relations value for the company, is the public awareness that is raised about actionable cost-effective measures. The actions of these companies and others demonstrate that transitioning away from business-as-usual and to the adaptive green economy is good for the bottom line.[6] This is not only evident in the growing activities of business associations but also in the growing course offerings related to sustainability at business schools.[7]

Additionally, included in the adaptive green economy are new businesses founded on principles of sustainability. From the outset, these companies develop their products with consideration for the entire product lifecycle. Examples include Tom's of Maine toothpaste and method cleaning products. The success and resiliency of companies in the adaptive green economy will signal to other companies in the overall economy to consider their own transformation to sustainable business practices. The jobs in the adaptive green economy are an important aspect to the overall transformation, because as with IT, jobs using IT are far greater in number than the jobs creating IT.

Then there is the *rest of the economy*, which consists of companies, households, and organizations that are committed to business-as-usual practices, and unless they adapt, they will be priced out of existence as prices for energy and all natural resources continue to rise.

The Role of Public Policy in Stimulating Innovation and Adoption

Public policy plays a significant role in stimulating technological innovation by supporting research, setting standards, and lowering the cost barriers to early adoption of new technology. In addition, in order to support the development of new

---

[6] Business Roundtable (2010); Forbes (2011).
[7] K. Galbraith (2009).

markets and economic growth, successful policies require a long-term commitment so that businesses and investors can plan strategically. Meaningful impact can be realized at all levels of government from federal and state policy to local policy.

Innovation is key to improving efficiency in our economy and creating new sources of value.[8] The research priorities set particularly by the federal government can play a significant role in spurring essential research and private sector investment. Since the global energy crisis in the 1970s, technology innovation in fields related to renewable energy sources and energy efficiency has taken place in waves. These waves reflect changes in public policy such as in research priorities set for federal funding (e.g., solar in the 1970s) as well as technological advance which spurred innovation in battery technology for small, remote devices like laptops and cell phones in the 1990s.

For example, in the early 1970s, the federal and California governments implemented incentives for the development and deployment of renewable energy.[9] This spurred research activity related to solar energy as illustrated in Fig. 6.1, and it also spurred business activity as entrepreneurs positioned themselves in the emerging market supported in part by incentives for early adopters of solar systems. However, when the price of oil fell and political winds changed in the early 1980s, the support for solar research and deployment stopped and the nascent market for solar technology was snuffed.

Federal policy can be especially effective in the definition of national standards thereby normalizing markets for product manufacturers as well as in the investment in research and development. However, at the local level where policy makers are often insulated from national political pressures, state and local policy makers can also have a significant impact in supporting the deployment of clean energy products and the development of businesses that provide these products.

Local policy makers have the ability to place their communities on the map as centers of early green/clean technology adopters. With this, the communities (i.e., households and businesses) reap the multiple benefits of resource efficiency in the form of cost savings and economic stimulus. As a center of early adopters, these communities become attractive to businesses that provide and manufacture green products and clean technology. For example, in recent years, five solar manufacturers have moved to the San Diego area primarily because California is the fastest-growing US market for solar.[10]

In a smaller geographical area, local and state policy makers can contribute significantly to the deployment of clean energy products and the development of businesses that provide these products. In particular, local policy makers can speed the deployment of clean energy technology by streamlining the permitting processes

---

[8] Extensive research has examined the compound economic benefits of an innovative economy and the key characteristics of an economy that creates new ideas and products and can commercialize them. See Romer (1990); Ian Morrison (1996); Utterbach (1994); Baumol (2002); Moon-Lee et al. (2002).

[9] Bezdek and Wendling (2007).

[10] Allen (2011).

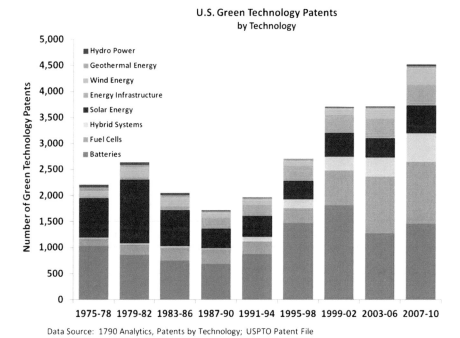

Fig. 6.1 Waves of innovation in green technology (Data Source: 1790 Analytics, Patents by Technology; USPTO Patent File)

related to the installation of renewable energy systems and energy efficiency retrofits. This approach is especially effective when municipalities cooperate to standardize requirements in the broader region.

Local policy makers can spur the demand for clean energy and green products by setting standards for energy efficiency (e.g., buildings and appliances) and water efficiency (e.g., low-flow fixtures) and by offering incentives such as rebates on products and retrofit services. Through the power of public procurement, localities can also support the development of new markets, and through bulk purchasing, they can also benefit from lower costs.

The California Example: Progress and Promise

The California example demonstrates that environmental policy can serve as a market driver and spur business and employment growth. Following the energy crisis in the 1970s, California put into place a series of innovative policies (e.g., standards and incentives instead of just regulations) that have resulted in higher energy efficiency than in the rest of the nation, based primarily on conservation and efficiency of energy usage.[11] A commitment to conservation through the creative realignment of incentives, implementation of standards, and setting of mandates at the state and local levels has served to create market certainty for new business and employment opportunities. Results provide evidence that environmental improvement need not come at the cost of economic growth and that the implementation of innovative public policy can serve to spur technological innovation and economic growth.

Policy Innovations Since the 1970s

For decades, California has been a national leader in innovative environmental policy. States have long been seen as the laboratories for new policies, and in the realm of environmental policy, California's innovative approaches are replicated in other states and used as a model for federal legislation as well as for other countries. According to the Congressional Research Service, "California has served as a laboratory for the demonstration of cutting-edge emission control technologies, which, after successfully demonstrated there, were adopted in similar form at the national level."[12]

Since the 1970s, the state has boldly set standards, designed incentives, enforced disincentives, and influenced major drivers of market dynamics toward improving energy efficiency and protecting natural resources and public health.

These policy innovations have been the product of combined efforts by public leaders, business leaders, grassroots organizations, and the State's cutting-edge technology innovation community. The OPEC oil embargo in 1973 served as a major force in spurring policy and technology innovation relating to energy efficiency. The next year, the State established the California Energy Commission to implement energy policy and planning. Meanwhile, a team of physicists at the Lawrence Berkeley National Laboratory, located at University of California, Berkeley, and operated by the University of California System for the US Department of Energy founded the Center for Building Science to research means for improving energy efficiency. In an early contribution to the cause, the center developed a computer program that calculated the energy performance of buildings. This program established the basis for the path-breaking legislation on energy efficiency standards for appliances and buildings (Title 20 and Title 24). Enactment in California was followed by the enactment of similar standards across the United States and other countries. By 1987, a uniform national standard for efficiency in appliances was in place.

A pioneering effort led by a group of efficiency advocates, utilities, and public leaders led to the realignment of investor-owned utilities' financial incentives from expanding consumption to investing in efficiency. This was made possible through the implementation of a decoupling mechanism of electricity and natural

---

[11] Clark and Bradshaw (2004).
[12] McCarthy (2007).

gas providers in 1982. This policy innovation removes the financial disincentive for utilities to encourage energy efficiency and conservation by making their profits independent of their sales.[13] Following California's lead, other states and countries are pursuing similar mechanisms to unlink economic incentives from environmental degradation.

The energy crisis in 2000 and 2001 provided another major force in spurring policy and technology innovation relating to energy efficiency. A result of the failed attempt at utility market deregulation, rolling blackouts characterized the 2-year period.[14] As in 1973, this crisis provided a fresh impetus for policy and technology innovation targeting improved energy efficiency in California. Ensuing policy innovations include broad-based energy efficiency campaigns, incentives for renewable energy sources, investment in technology research, and standards that reduce greenhouse gas emissions.

Innovative policies and approaches are surfacing in the State Capitol as well as in California's cities, counties, and regions. Examples of landmark legislation include the following state and local actions:

California Renewables Portfolio Standard was established in 2002 with the goal of increasing the percentage of power generation from renewable energy sources in the state's electricity mix of investor-owned utilities (IOUs) to 20% by 2017. This goal was accelerated to be achieved by 2010, and according to the California Public Utilities Commission, "Collectively, the large IOUs reported in their March 2011 Compliance Filings that they served 17.9% of their electricity with RPS - eligible generation in 2010, up from 15.4% in 2009."[15] In November 2008, Governor Schwarzenegger signed an executive order to accelerate the RPS target to 33% by 2020. In April 2011, Governor Brown signed this into law, requiring all public and privately own retail sellers of electricity to source at least 33% from renewable sources by 2020.

California's Clean Cars Law of 2002 (AB 1493) requires carmakers to reduce global warming emissions from new passenger cars and light trucks beginning in 2009. First in the world to reduce global warming pollution from cars, this law has now been adopted by 11 other states. Affecting nearly one-third of the US market, global warming emissions in 2020 will be reduced by more than 64 million tons of carbon dioxide a year. After implementation was held up at the federal level,

---

[13] Utility revenues have historically been tied to sales volumes, so companies were rewarded for selling more power and penalized for selling less. Therefore, there was a strong disincentive for utilities to encourage energy efficiency and conservation. The implementation of "decoupling" removes this barrier by assuring investor-owned utilities a fixed amount of revenues regardless of sales volumes. "The result of this simple, but profound, change has been that utilities have been free to aggressively help consumers reduce energy usage without doing financial harm to their business. As a direct result of decoupling and the programs it made possible, California's per capita energy usage has remained flat over the past 30 years, compared with an increase of 50% for the rest of the country." Pacific Gas and Electric Company (2006, p. 5).

[14] Clark and Bradshaw (2004); Clark (2003).

[15] California Public Utilities Commission (2011).

California was granted a waiver from the US Environmental Protection Agency in July 2009 to pursue these stricter vehicle emissions standards.

The California Global Warming Solutions Act of 2006 (AB 32) is the first law in the nation to comprehensively limit greenhouse gas (GHG) emissions at the state level. Five Western states (Washington, Oregon, Utah, Arizona, and New Mexico) have joined California to combine efforts toward reducing GHG emissions with the Western Regional Climate Action Initiative. As of November 2009, the California Air Resource Board reported that 591 of the 605 facilities that emit at least 25,000 metric tons of carbon dioxide per year had reported their greenhouse gas emissions for 2008.

AB 811 was passed into law in 2008 allowing property owners to receive public financing of renewable energy generation, energy efficiency, and water efficiency projects. Known as Property-Assessed Clean Energy (PACE) Programs, property owners enter into contractual assessments with public entities and repay the loan through increased property taxes. This loan structure allows property owners to immediately reap the gains of new technologies while not leaving them tied to the property while the loan is paid off. This creative financing model was first implemented by the City of Berkeley in January 2009 and has been replicated across the state in 23 counties. With 100 million dollars in financing, the Sonoma County Energy Independence Program (SCEIP) is the largest PACE program in the nation. The statewide program, CaliforniaFIRST, was awarded $16.5 million by the California Energy Commission for similar project financing.[16]

However, PACE does have an unresolved issue surrounding efforts to create municipal liens on commercial and residential properties that would trump existing mortgage holders. In the current system, municipal interests are secondary to mortgage holders, subjecting taxpayers to possible risk. As a result of the mortgage crisis, PACE programs for residential buildings across the nation have been brought to a halt. Fannie Mae and Freddie Mac, the nation's largest underwriters of home mortgages, have declared that they will not accept loans for homes using PACE programs due to these concerns. While Sonoma County has continued its program, efforts are currently underway to bring PACE back nationally.

Energy Efficiency Gains Are Economic Gains

Spurring the green economy raises resiliency and competitiveness. Improving efficiencies in the consumption of energy and all natural resources will boost the competitive edge of a company as well as an economy. In addition to new savings on resources not consumed, a company increases its resilience to external shocks (such as volatile fuel costs), thereby improving its competitive edge over other less resilient companies. The same is the case for a state or regional economy.

---

[16] See CaliforniaFirst program. http://www.californiafirst.org/.

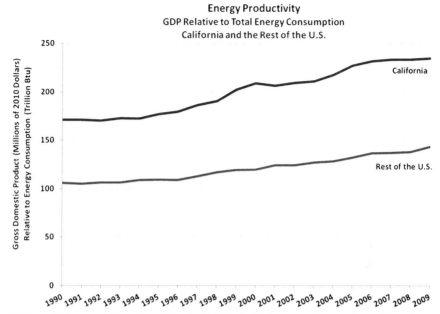

Fig. 6.2 Energy productivity (Data Source: U.S. Department of Energy, Energy Information Administration; Department of Commerce, Bureau of Economic Analysis)

California's energy productivity is 64% higher than that of the rest of the USA (Fig. 6.2). Measured as gross domestic product divided by total energy consumption, Californians generate 64% more economic value from a single unit of energy than the rest of the country. Since 1990, energy productivity increased 37% in California and 35% in the rest of the USA.

Much of this progress can be explained by California's successful realignment of financial incentives of investor-owned utilities (i.e., utility decoupling) and its sustained commitment to improving energy efficiency. California's per capita energy use has dropped at a faster rate than the nation. Total energy consumption is 45% higher for the state and 39% higher for the nation when compared to consumption in 1970 (Fig. 6.3). Per capita consumption in the USA has remained near 1970 levels, but in California, it is 22% lower and continues to fall. Total energy consumption includes petroleum, natural gas, electricity retail sales, nuclear, coal and coal coke, wood, waste, ethanol, hydroelectric, geothermal, solar, and wind energy.

When considering California's relatively high energy efficiency standards, it is important to note that while electricity rates are higher in the state, the average monthly residential bill is 14% lower than that of the rest of the nation (Fig. 6.4).

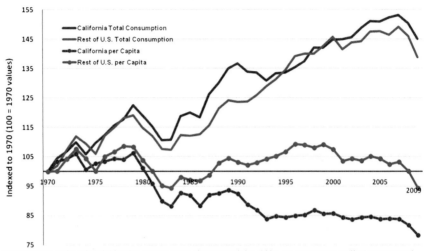

Fig. 6.3 Total energy consumption (Data Source: Energy Information Administration, U.S. Department of Energy; Population Division, U.S. Census Bureau; California Department of Finance)

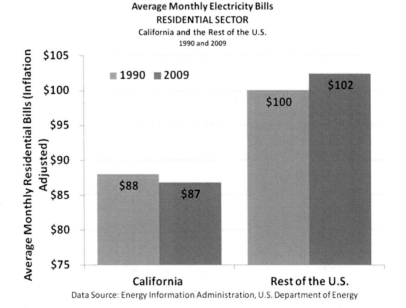

Fig. 6.4 Average monthly electric bill (Data Source: Energy Information Administration, U.S. Department of Energy)

| Venture Capital Investment in Clean Technology |||||||
| Values are inflation adjusted |||||||
	2005	2006	2007	2008	2009	2010
U.S.	$ 1,290,357,521	$ 3,120,002,801	$ 3,989,101,648	$ 7,741,304,014	$ 6,512,948,135	$ 6,595,810,720
California	$ 533,632,229	$ 1,373,965,703	$ 1,930,463,390	$ 3,602,821,454	$ 2,307,982,906	$ 3,891,865,000

Data Source: Cleantech Group™, LLC (www.cleantech.com)
Analysis: Collaborative Economics

Fig. 6.5 Venture capital investment in Cleantech

Business and Employment Growth in the Green Economy

California's leadership in technology, business, and public policy innovation has fueled the state's green economy and will determine its sustained success. As a result, California is home to many companies that are driving technological advance in products and services that will enable the entire economy to transition to clean energy sources, improve resource efficiencies, and reduce pollution. Further, the state attracts the largest percentage of total US cleantech venture capital investment (Fig. 6.5), helping drive technological advance in the state. Many of these California companies produce tried and tested products with mature markets outside the USA where efficiency standards are more stringent. Therefore, raising US standards would open up vast new domestic markets for California companies in the core green economy. California's core green economy is diverse, distributed across the state, and is growing at a faster rate than the economy as a whole.[17]

The Core Green Economy Defined

The core green economy provides the products and services that enable the transformation toward a cleaner, more efficient, and more competitive economy. As described earlier, the core green economy consists of businesses that provide products and services that do the following:

- Provide alternatives to carbon-based energy sources
- Conserve the use of energy and all natural resources
- Reduce pollution (including GHG emissions) and repurpose waste

The core green economy is composed of 15 segments. The broad scope of these segments reflects the many different factors associated with mitigating the sources and impacts of climate change. These segments were based originally on the cleantech segments defined by the Cleantech Network; however, while cleantech's focus is on new technology, the definition of the core green economy is broader in order to encompass all products and services that meet the criteria described above.

---

[17] Henton et al. (2012).

The lack of standardized industry data with information on "green" products, services, and occupations has resulted in the development of multiple approaches to defining "green jobs" and the green economy. The definitions vary largely depending upon the character of the data underlying the analysis. Some approaches focus on the activities of occupations and are based on job postings or employer surveys. Other approaches focus on businesses that operate in a "green" manner regardless of the end products and services they sell. While different approaches are valid and contribute different perspectives on changes under way, the approach presented here represents the most comprehensive analysis of businesses and employment in the emerging green economy.[18]

The fifteen segments of the core green economy		
Green segment	Description	
1. Energy generation	• Renewable energy generation (all forms of solar, wind, geothermal, biomass, hydro, marine and tidal, hydrogen, cogeneration)	• Renewable energy consulting services
	• Research and testing in renewable energy	• Associated equipment, controls, and other management software and services
2. Energy efficiency	• Energy conservation consulting and engineering	• Alternative energy appliances (solar heating, lighting, etc.)
	• Building efficiency products and services	• Energy efficiency meters and measuring devices
	• Energy efficiency research	
3. Transportation	• Alternative fuels (biodiesel, hydrogen, feedstock-neutral ethanol infrastructure)	
	• Motor vehicles and equipment (electric, hybrid, and natural gas vehicles, diesel technology)	
4. Energy storage	• Advanced batteries (e.g., Li-Ion and NiMH)	• Fuel cells
	• Battery components and accessories	

(continued)

---

[18] The Green Establishments Database is a composite database that draws information from multiple sources (including the Cleantech Group™, LLC, and New Energy Finance) for the identification and classification of green businesses and also leverages a sophisticated internet search process. CEI designed the parameters of the internet search platform which was engineered by a series of developers of business intelligence tools. The National Establishments Time-Series (NETS) database based on Dun & Bradstreet business-unit data was sourced to extract business information such as jobs.

This methodology was originally developed on behalf of Next 10, a California nonprofit, as part of the 2008 *California Green Innovation Index*. Since then, the methodology has been further developed and refined and resulted in the following published analyses: *California Green Innovation Index* (2009a, 2010, 2012), Next 10's *Many Shades of Green* (2009, 2011, 2012), Joint Venture: Silicon Valley Network's *Index of Silicon Valley* (2009, 2010), and *Cleantech and California's Growing Green Economy* (2008b) prepared for the California Economic Strategy Panel. The analysis completed for the Pew Charitable Trusts' 2009 *US Clean Energy Economy Report* provided the most comprehensive accounting of the reaches of the growing clean energy economy across the fifty states and the District of Columbia.

The fifteen segments of the core green economy		
Green segment	Description	
5. Air and environment	• Environmental consulting (environmental engineering, sustainable business consulting)	• Emissions monitoring and control
		• Environmental remediation
6. Recycling and waste	• Consulting services	• Recycling machinery manufacturing
	• Recycling (paper, metal, plastics, rubber, bottles, automotive, electronic waste, and scrap)	• Waste treatment
7. Water and wastewater	• Water conservation (control systems, meters and measuring devices)	• Research and testing
		• Consulting services
	• Development and manufacturing of pump technology	• Water treatment and purification products/services
8. Agricultural support	• Sustainable land management and business consulting services	• Sustainable supplies and materials
		• Sustainable aquaculture
9. Research and advocacy	• Organizations and research institutes focused on advancing science and public education in the areas of renewable energy and alternative fuels and transportation	
10. Business services	• Environmental law legal services	• Green staffing services
	• Green business portals	• Green marketing and public relations
11. Finance and investment	• Emission trading and offsets	• Project financing (e.g., solar installations, and biomass facilities)
	• Venture capital and private equity investment	
12. Advanced materials	• Bioplastics	
	• New materials for improving energy efficiency	
13. Green building	• Design and construction	• Site management
	• Building materials	• Green real estate and development
14. Manufacturing and industrial support	• Advanced packaging	• Industrial surface cleaning
	• Process management and consulting	
15. Energy infrastructure	• Consulting and management services	• Cable and equipment

Key Trends in California's Core Green Economy

California's core green economy is diverse and growing at a faster rate than the economy as a whole. Over the recent economic downturn, the state's core green economy reflected greater resilience than the economy as a whole. In 2010, California's total employment suffered losses of 7%, while the core green employment dropped by 3%. Over the longer term, 1995–2010, employment in California's core green economy increased 53%, while total statewide employment grew 12%. (Job numbers reported by year are current as of January of the reported year.)

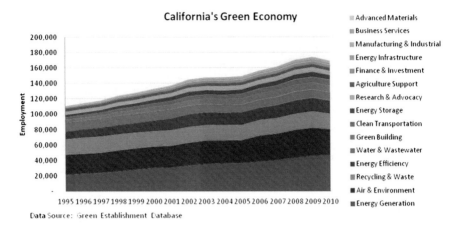

Fig. 6.6 California employment by green segment (Data Source: Green Establishment Database)

While aggregate core green employment contracted by three percent, specific segments witnessed continue growth in the most recent observable 12-month period. Employment expanded by 1% in the state's largest segment, energy generation, as well as in clean transportation (e.g., vehicles, components, and fuels). Smaller segments also grew: energy infrastructure (14%) and advanced materials (4%) (Fig. 6.6).

In addition to viewing the core green economy by green segment, that is, by the field of application of products and services, businesses establishments can be viewed by their primary functions such as research and development, manufacturing, and services. California's core green economy consists largely of high-value services and manufacturing. Employment in businesses that primarily offer services account for 40% of all jobs in California's core green economy. Manufacturing represents 27% of all green employment; by contrast, in the state economy as a whole, manufacturing accounts for 10% of total employment. Research and Development is strong particularly in transportation, energy generation, and air and environment.

Conclusions

Nationally, there is much that could be accomplished under ideal circumstances such as setting meaningful building and appliance efficiency standards, a renewables portfolio standard; encouraging utility revenue decoupling; and placing a price on carbon. However, in the absence of action at the national level, there is much that state and local policy makers can accomplish through technological and policy innovation to simultaneously meet environmental goals and grow economic opportunity.

The California example demonstrates that contrary to popular economics, in the arena of clean energy and resource efficiency, regulation can be a market driver and spur business and employment growth. Following the energy crisis in the 1970s and early 2000s, California put into place a series of innovative policies that have resulted in higher energy efficiency than in the rest of the nation. A commitment to conservation through the implementation of standards, incentives, and mandates at the state and local levels has served to create new business opportunities for the state's entrepreneurs and new job opportunities for the state's workforce.

The analogy of the IT industry is useful. The full economic impact of the growth of the IT industry over the last 30 years has manifested itself well beyond the immediate sectors of software and hardware. More significantly, the economy-wide penetration of IT has increased productivity across all sectors and transformed the entire economy. As a result, new markets, businesses, and occupational opportunities have emerged, and new value has been generated outside of the core IT sectors. Similarly, the widespread application of products and services that enable higher resource efficiency and reduced negative environmental impacts will have a transformational impact across the entire economy by increasing productivity and creating new markets and job opportunities.

Public policy can play a central role in this development, and California's experience demonstrates this. While modern economic theory may treat environmental degradation as an externality, the reality we as individuals, businesses, and public policy makers live in does not. The transformation of the economy toward a cleaner, more resilient economy can be spurred by implementing policy that encourages the adoption of the products and services that enable this transformation. This is not about the public sector choosing the "winning" technology. Instead, this is about the public sector aligning incentive structures with economic and environmental goals. By implementing similar efficiency standards, incentives for early adoption of clean technology, and targets for renewable energy generation at the federal level, the nation could benefit from improved resource productivity, competitiveness, and economic growth.

References

Allen M (2011) Manufacturing: employment declines, but clean energy creates jobs. Economic Trends 2011, San Diego Bus J. http://www.sdbj.com/news/2011/jan/03/brighter-outlook/

Atkinson R, McKay A (2007) Digital prosperity: understanding the economic benefits of the information technology revolution. The Information Technology & Innovation Foundation, Mar 2007

Baumol W (2002) The free market innovation machine. Princeton University Press, Princeton

Bezdek RH, Wendling RM (2007) A half century of US federal government energy incentives: value, distribution, and policy implications. Int J Glob Energy Issues 27(1):42–60

Briguglio L, Cordina G, Farrugia N, Vella S (2009) Economic vulnerability and resilience: concepts and measurements. Oxford Dev Stud 37(3):229–247

Business Roundtable (2010) Enhancing our commitment to a sustainable future. Business Roundtable's S.E.E Change and Climate Resolve initiatives

California Public Utilities Commission (2011) Renewables portfolio standard quarterly report, Mar 2011. http://www.cpuc.ca.gov/NR/rdonlyres/62B4B596-1CE1-47C9-AB53-2DEF1BF52770/0/Q12011RPSReporttotheLegislatureFINAL.pdf

Clark W II, Lead Author (2003) California's next economy. Governor's Office Planning & Research, Sacramento

Clark W II, Bradshaw T (2004) Agile energy systems: global lessons from the California energy crisis. Elsevier Press, Oxford

Farrell D, Nyquist S, Rogers M (2007) Making the most of the world's energy resources. The McKinsey Quarterly McKinsey & Company (Number 1, 2007)

Forbes (2011) Top ten green companies in fortune 500

Galbraith K (2009) Sustainability field booms on campus. New York Times, 19 Aug 2009

Henton DJ, Melville, Grose T et al (2008) California green innovation index. Next 10. http://www.coecon.com/Reports/GREEN/CAGreenIndex.pdf

Henton D, Melville J, Grose T et al (2008) Clean technology and the green economy: growing products, services, businesses and jobs in California's value network. California Economic Strategy Panel, California Regional Economies Project. http://www.labor.ca.gov/panel/pdf/FINAL_Green_Economy_March_2008.pdf

Henton DJ, Melville, Grose T et al (2009) California green innovation index. Next 10. http://www.coecon.com/cagreen.html

Henton DJ, Melville, Grose T et al (2009) Many shades of green diversity and distribution of California's green economy. Next 10. http://www.coecon.com/manyshades.html

Henton DJ, Melville, Grose T et al (2010) California green innovation index. Next 10.

Henton DJ, Melville, Grose T et al (2011) Many shades of green diversity and distribution of California's green economy. Next 10. http://www.coecon.com/manyshades.html

Henton D, Melville J, Grose T et al (2012). Many Shades of Green. Next 10

Joint Venture: Silicon Valley Network (2009) Climate prosperity: a green print for Silicon Valley. http://www.coecon.com/climpropserity.html

Jorgenson DW, Ho MS, Stiroh KJ (2005) Productivity volume 3: information technology and the American growth resurgence. MIT Press, Cambridge

McCarthy J (2007) California's waiver request to control greenhouse gases under the clean air act. CRS Report for Congress. Congressional Research Service, 20 Aug 2007

Moon-Lee C, Miller WF, Hancock MG, Rowen HS (eds) (2002) The Silicon Valley edge: a habitat for innovation and entrepreneurship. Stanford Business School Books, Stanford

Morrison I (1996) The second curve: managing the velocity of change. Ballantine Books, New York

Pacific Gas and Electric Company (2006) Global climate change: risks, challenges, opportunities and a call to action

Pew Charitable Trusts (2009) The clean energy economy: repowering jobs, businesses and investments across America. http://www.coecon.com/cleanecon.html

Porter M, Claas van der Linde (1995) Green & competitive. On competition. Harvard Business Review Book, p 352

Romer P (1990) Endogenous technological change. J Polit Econ 98(5):S71–S102

Utterbach J (1994) Mastering the dynamics of innovation. Harvard Business School Press, Boston

Chapter 7
Social Capitalism: China's Economic Raise

Woodrow W. Clark II and Li Xing

Abstract Social capitalism has emerged as a new form of economics, historically rooted in the Nordic countries, Germany, and more recently China. Due to the growth of China's population, the nation has needed more infrastructures for its growing cities and regions. The demand, however, has been with a strong concern for the people and the environment. Thus, like northern Europe, China has been aware that if it is to meet the needs of a nation over 1.3 billion people, then, it must also provide basic infrastructures ranging from energy, water, waste, transportation to WiFI for community.

Historically, China has a long cultural tradition for the support of societal needs. Under Mao and since the cultural revolution, China has had five-year plans which scope out what the central and regional governments are going to do. Critical to any plan is the need for finances to support it. China is now in its 12th five-year plan which focuses heavily on renewable energy with a strong commitment to billions of US dollars to implement their plans. With this plan and future ones in place (all of the past five-year plans have achieved their goals), China has "leapfrogged" and now leads the green industrial revolution which is growing around the world.

Woodrow W. Clark II, Ph.D. Managing Director, Clark Strategic Partners, Los Angeles, CA, USA at: wwclark13@gmail.com and Li Xing, Ph.D. Associate Professor at Aalborg University, Denmark at: xing@ihis.aau.dk with contributions from Jerry Jin, Ph.D. in Beijing; David Nieh in Shanghai; and ML Chan Ph.D. in Silicon Valley.

W.W. Clark II, Ph.D. (✉)
Qualitative Economist, Academic Specialist, Cross-Disciplinary Scholars in Science
and Technology, UCLA and Managing Director, Clark Strategic Partners, California, USA
e-mail: www.clarkstrategicpartners.net

L. Xing, Ph.D.
Aalborg University, Aalborg, Denmark
e-mail: xing@ihis.aau.dk

Background and Cultural History

The historical transformation from Maoist social-communist China to Dengist capitalist China since the end of the 1970s represents a sharp contrast as to China's national policy objectives, political agenda, economic growth, and more importantly sustainable development strategy. The Dengist China took the capitalist mode of economic development based on privatization of ownership and the means of production and distribution, to the marketization and allocation of resources including the total acceptance of economic inequities and political privileges.

The basis of this form of capitalism was initially to place emphasis on western market-oriented economics and advanced technologies as the essential productive forces, along with the promotion of the interests of the privileged, professional, and entrepreneur classes, to include the commercialization of welfare and social security (Chan 2009). The 1970s was the period in which western capitalism, promoted by the UK and USA, became the norm for over three decades (Economist 2009).

More significantly, what China has witnessed in the past three decades is an economic growth path based on increasing demand for energy consumption rooted in an annual GDP growth of 10%. By the end of the first decade of the twenty-first century, China had outpaced the USA in clean technology investments (Scientific American, April 5 2011). According to M. L. Chan, China at the end of 2010 ranked number #1 in clean technology investments, while the USA fell to #3 (Chan 2010). However, China also in 2011 become the number #1 nation in carbon and air pollution surpassing the USA even though the per capital air pollution in China is less than that of the USA.

By the end of 2008, China ranked fourth in the world in wind turbine manufacturing and installed 12 GW (gigawatts) of power (Zhen et al. 2009). Today, Chinese companies control over half of the annual USD $65 billion wind turbines market (Rosenberg 2010). Could China become the number one economy in the world, surpassing the USA? (Time 2011) With its GDP at $5.88 trillion in 2010, China passed the former number economy, Japan GDP of $5.47 trillion. The GDP per capital of $7,518 for the Chinese versus $33,828 for the Japanese is much lower, but some economists predict that by 2030, China will have a GDP of $73.5 trillion compared to the USA $38.2 trillion and Japan in third place with $8.4 trillion. The world economics has changed. Some Americans are skeptical and fearful of the Chinese economic challenge. A recent series of meetings in the US Congress have focused on "China's energy and climate initiatives" with a subtitle of "successes, challenges, and implications for US policies" (US Congress 2011). The presentations were varied and presented no conclusions.

During the socialist period (after WWII), due to historical reasons, China took a development course with a commitment and goal by emphasizing human capacity and economic equality, thereby mobilizing social and economic resources in pursuit of a self-reliance development strategy. No matter how the socialist strategy is interpreted and assessed from today's perspective, it was historically the only possible option if China wanted to sustain its economic development, national security, and independence. The strategy of self-reliance emphasized the primacy of internally generated *independent* development, not only at the national level but also at the provincial, regional, and local levels. The national Five-Year Plans started under Mao and continue today. These internally generated *independent* development plans were reflected in the institutional structure of each unit,[1] where the multifunctions of party leadership, production planning, medical service, and resource demand and supplier controls were closely integrated. Self-reliance under such a socioeconomic structure in which politics, economics, service, and supply existed together at the unit level and created a sustainable society based on their own resources and needs, under the overall guidance of the state's Five-Year Plans.

China's Five-Year Plans and Finance

The Five-Year Plans or *The Five-Year Plans for National Economic and Social Development* are national goal-setting economic and policy position papers derived over a year or two from high-level and local committees. At the national government level, the Five-Year Plans outline key construction projects and infrastructure plans and administer the distribution of productive, manufacturing and business sources, and growth of academic and educational institutions along with individual sector contributions to the national economy. The plans also provide large sums of national funding for implementation. Aside from giving the nation, business, government, and foreign interests a "road map" about Chinese policies, the plans map the general direction of future development including specific measureable policies and targets. The last Five-Year Plan was from 2006 to 2010 and officially called the *11th Five-Year Development Guidelines* (The Development and Reform Commission, PRC (2007). The current 12th Five-Year Plan or 12-5-yr Plan came out officially in March 2011 and now the 13th Five-Year Plan is being developed especially with the new national administration in office.

In the context of energy, particularly for homes and buildings, self-reliance also means that the focus for renewable energy was primarily on technologies that could help in the heating and cooling of buildings. Solar thermal systems were developed by state operated companies in the 1990s and then spun off into private-public-owned firms. Sundra (Sundra Solar Corporation, 2011) is a case in point with solar thermal systems that appear all over China on homes (Kwan 2009). Now their market has grown worldwide as examples from colleges in the USA illustrate (Eisenberg 2009).

The Chinese choice of a self-reliance and self-sufficiency development path was projected as a potential "ideological threat" by the capitalist western nations, because the central goal of the socialist politics, with specific plans from the central government, were seen as an attempt to challenge the capitalist ideology of competition that would reduce, costs but unfortunately also lead to an inequitable hierarchy in the world order solar panels (Downs 2000, 2004, 2006; Jiahua et al. 2006; Kaplinsky 2006; Liu 2006; Konan and Jian 2008; Li 2010; IEA, 2010, US Congress

---

[1] A unit, in Chinese "*Dan Wei*," refers to any functional organization, for example, a ministry, a university, a company, a factory, etc. A "Dan Wei" was self-managed "mini-welfare state" combing supply, demand, and welfare.

2011 among others). Seen from the interpretation of world system theory, the socialist self-reliance and self-efficiency policy aimed at transforming the basic logic of capitalism into "social capitalism" (Clark and Li 2003). However, in reality, it was actually designed toward a nationwide mobilization for industrialization for the purpose of catching up with the western capitalist countries that reward people with money, no matter where the funds or by means the funds were acquired. This is because socialist states were still operated within the capitalist world economy, and the dynamics of capitalism was capable of distorting and limiting national economic planning, leading to the constraints of their policy options (Chase-Dunn 1982, 1989). Nevertheless, such a socialist program for national plans was based on self-reliance existed outside the US-led capitalist economic system. In other words, it was merely an ideological challenge without being able to construct a sustainable alternative model (or paradigm) based on recent historical evidence to replace the capitalist system.

In addition, the Chinese self-reliance policy was also a response both to the internal constraints of socioeconomics and the external pressure by the US-led economic trade embargo after WWII and then the Korean War. Furthermore, the failure of China's dependent relations with the Soviet Union since the Republic was founded in 1949 and the Cold War, prevented China from actively using western economics. What should be pointed out here is that China's emphasis on self-reliance and independence in the Chinese energy industry was primarily derived from the lessons learned due to the Sino-Soviet split in which China was deprived not only of the Soviet technicians and specialists that were helping China develop its industries, but also of around 50% of its oil supplies that were imported from the Soviet Union (Downs 2000: 11–12). The discovery of the Daqing[2] oil field at the end of the 1950s was declared to mark the end of China's external oil dependence except for defense and civilian applications. Such was the case in the modern market-oriented economy after Mao in the 1980s.

The post-Mao transformation of policy orientation from socialism to opposite extreme of market capitalism with its objectives, have their roots in the change of the regime's perceptions of the external international political economy starting with the USA and UK since the 1980s and especially due to the end of the Cold War in the early 1990s. The perceptions were generated from the conceptualization of international relations that (1) the superpowers, including their respective alliances, were exhausted in their endless competition, leading to a situation where no major serious conflict was likely in the future even with the decade long US military involvement in the Middle East and the 2011 military presence in Northern Africa due to social unrest and change.

On the other hand, there was the emergence of nonconventional security challenges whereby the Chinese government-controlled fossil fuel companies bought international oil and gas producing and transport companies (Peyrouse, 2007 and Shoumatoff 2008); (2) economic development became the key objective for all

---

[2] Daqing was the largest oil field in China, and it is located in Heilongjiang province. It was discovered in 1959 and the production started in 1960.

nations, and economic power thus emerged to become more important and relevant than traditional military strength. Thus, soon, China had to face economic interdependence by increasing its global economic presence in acquiring fuel supplies; (3) the post-Cold War US-based and controlled world order, which is an American-centric new world order, would likely remain for a unknown period of time. Therefore, China should, in the words of Deng, "observe calmly, secure our position, cope with affairs calmly, hide our capacities and bide our time," and (4) there would be growing global competition for natural resources hence for energy security (Intriligator, 2013). It was this last area that began after the turn of the twenty-first century, when China's unknown global challenges were still being defined and hence significant tasks needed to accomplished moving from totally government-controlled industries to ones that were collaboration or joint ventures with foreign companies and often owned by a majority of Chinese workers (Clark and Isherwood 2010).

China Leapfrogs into the Green Industrial Revolution (GIR)

Driven by the above new strategic understanding, China has been pursuing a global foreign economic policy that was directed at creating a stable and peaceful environment for its economic growth through active engagement with the West and with the surrounding Asian nations (Clark and Cooke 2011). This strategy has become China's globalization focus for a new or next economy (Li and Clark 2009). China grasped opportunities for increasing international trade and foreign direct investment and, more importantly, for securing access to natural resources and energy supplies through its own international trade and investment in the resource-rich regions such as Africa, Latin America, and Southeast Asia and, in recent years, Central Asia. China's global policy strategies under an active role of the state have been seen as effectively making it one of the "globalization's great winners" (Thøgersen and Østergaard 2010).

In short, China no longer is an "emerging nation" and founding member of BRIC (Brazil, Russia, India, and China) but has leapfrogged technologically, economically, and politically into the Green Industrial Revolution (3IR) which moves an economy off the dependence on fossil fuels to renewable energy, smart green grids, advanced storage technologies to sustainable communities that reverse carbon emissions and reduce pollution. This GIR started in Asia and the EU over two decades ago (Clark and Cooke 2011).

China, by the end of the first decade of the twenty-first century, moved actively into the GIR. The Pew Charitable Trusts (SJBJ 2011: 8) was quoted in April 2011 with global data. For example, China reported $54.4 billion in clean tech investments for 2010, which was an increase of 39% from the year before and thus led all nations. Germany was second with $41.2 billion with the US third at $34 billion or 51% over 2009. Italy was third with $13.98 billion or up 124% from 2009 while the rest of the EU ranked fifth with $13.48 billion, which was off 1% from 2009.

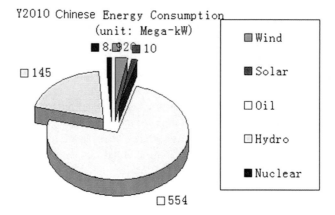

Chart 7.1 China Energy Demand and Supplies (2010)

China's remarkable achievement in economic growth (US DOE, 2011) was made possible by its growing involvement in the capitalist world system. Steven Chan verified this fact as he told the story of SunTech becoming the world's number #1 solar manufacturing company in 2010 (Chan 2011). But China remained in charge with caution and intense controls from the central government. It did not, for example, experience the deep 2008 global economic recession. In other words, China's economic growth is inseparable from its increasing dependence on global markets, with some estimates suggesting that more than 40% of its GNP is derived from international trade (Chun 2010). In other words, China's rapid economic growth has been driven by exports with the assistance of foreign investments and joint ventures that have dominated the most dynamic sectors of the economy. Its market-driven growth encourages more concessions to induce capital flows and growth in unlimited possibilities of expansion and more structural changes to meet the demand of the overwhelming pursuit of external markets and resources (Lo 2011). In addition, its integration with the world market is followed by overdependence of the productive forces on the fluctuations of the world market. The most affected area is China's energy demand and supply. Chart 7.1 illustrates the problem in China today with most of its energy demand coming from the supplies of coal, natural gas, and nuclear power. To summarize part of a recent series of presentations note in the US Congress (April 2011):

> China is now the global manufacturing leader of most renewable energy technologies, and the largest user of clean energy. In 2010 alone, clean energy investment in China topped $154 billion, while approximately 77 gigawatts of old fossil fuel power was shut down.

The Debate over Energy and Consumption-Based Economic Growth

During the past three decades, China's GDP has enjoyed an average growth rate of 9–10%. Since 2002, China's energy consumption has been growing at a faster rate than its GDP growth. From 2000 to 2005, China's energy consumption rose by

60%, accounting for almost half of the growth in world energy consumption (Downs 2006: 1). In 2004, China contributed 4.4% of total world GDP, whereas China also consumed 30% of the world's iron ore, 31% of its coal, 27% of its steel, and 25% of its aluminum. Between 2000 and 2003, China's share of the increase in global demand for aluminum, steel, nickel, and copper was, respectively, 76%, 95%, 99%, and 100% (USDOE, 2011). On a global scale, an increase in the rise of personal car ownership alone could mean an extra billion cars on the road worldwide within the next 10 years (NYT 2010). As a report by Chinese Academy of Social Sciences in 2006 predicted accurately on the mounting worldwide impact of China's resource consumption:

> China is currently the world's third largest energy producer and the second largest energy consumer. In 2002, China accounted for 10% of total world energy use and is projected to account for 15% of global energy use by 2025. China is the world's largest coal producer accounting for 28% of world coal production and 26% of total consumption. China is the third largest consumer of oil and is estimated to have the world's sixth largest proven reserves of oil. China has roughly 9.4% of the world's installed electricity generation capacity (second only to the United States) and over the next three decades is predicted to be responsible for up to 25% of the increase in global electricity generation. China emitted 10.6% of global carbon emission from fossil fuels in 1990 (second only to the United States) and 14.2% in 2003. This share is projected to rise to 22.2% by 2020 (CASS 2006: 28).

This situation has given rise to problems of net energy imports, environmental pollution and ecological destruction, cross-border pollution, and mounting carbon dioxide (CO_2) emissions:

> The domestic environment has deteriorated rapidly, with some 70% of urban population exposed to air pollution, 70% of seven major water systems heavily polluted, over 400 cities short of water, and 3,400 km2 (equivalent to Japan's Tottori Prefecture) turning into desert every year. Cross-border pollution, notably acid rain and sandstorms, have reached the Korean Peninsula and Japan. Global environmental problems: China is the world's second largest CO2 producer after the U.S. (Li 2003: 1)

By early 2010, China moved ahead of the USA in being the number emitter of carbon in the world, alto per capital, the USA is still the leading nation.

China's escalating energy consumption is placing heavy pressure on the world's energy prices. Chinese energy demand has more than doubled during the past decade. According to the study of Konan and Jian (2008), China will consume about 41% of global coal consumption and 17% of global energy supply by 2050. Metal prices have increased sharply due to strong demand, particularly from China which has contributed 50% because of the increase in world consumption of the main metals (aluminum, copper, and steel) in recent years. Due to its rapid growth and rising share in the world economy, China was expected just after the turn of the twenty-first century to retain its critical role in driving commodity market prices (*World Economic Outlook*, September 2006). It did so, according to almost every study by the end of the first decade of the twenty-first century (ACORE 2011; IEA 2011; US Congress 2011). China is willing to offer above world market prices for purchasing raw materials, which attributes great comparative advantages to the developing world, but also threatens the west who seek to purchase and control those fossil fuel supplies (Tseng 2008).

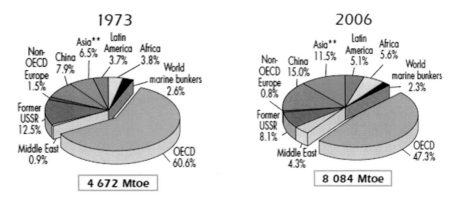

Fig. 7.1 Shares of total final consumption 1973 and 2006 (Source: International Energy Agency 2008, p. 30)

Historically, the world has been already burdened by the high energy consumption by the West, particularly by the United State. Today, China's growing appetite for international trade drives its mounting demand for resources to sustain its economic growth and to fuel its countless development projects. China has already become the world's largest importer of a range of commodities, from copper to steel and crude oil (Silverstein 2011). The phenomenal rise of commodities prices worldwide in recent years is claimed to be attributed to China's growing import demand. Some even worried that there would not be enough resources in the world, for example, gas and oil, to satisfy the ever increasing demand driven by China's economic growth. Furthermore, taking China's neighbor – India – into consideration with its population of one billion, it will add additional pressures on the demand for the same resources.

After a short period of self-sufficiency in energy supply especially in coal and petroleum, China became a net importer of petroleum in 1993, and it took only a over a decade later than that China became the second largest oil importer and consumer after the United States in 2009 (USDOE 2011). China's energy profile used to be heavily weighted toward fossil fuel technologies (petroleum and coal) at a time when reductions are urgently needed to stabilize global climate change. Based on the 2008 statistics from the International Energy Agency, the growth rate of China's energy consumption and its share of the global total final consumption are comparably much higher than the rest of the world (Fig. 7.1).

It is foreseeable that China's resource imports of oil and gas will continue to increase if its targeted economic growth is to be maintained and it does not move fast enough into the Green Industrial Revolution. The implication is therefore clear that not only global commodity price and international geopolitical power relations will be affected, but also China's international politics, such as its foreign policy rationale, international aid objective, arms sale consideration, and compulsory expansion of its long-range naval power projection capabilities, which are closely connected energy economic issues. The rise of China, as a key factor in energy consumption, is forcing the current international energy regime to adjust or modify its established rules of the capital market economics which provides China a challenge today. This is because the international energy regime "is influenced not only by economic, political, and social factors of resource-rich countries but also by international political factors, particularly change in the international balance of power, adjustment of relationships among countries and changes to international rules" (Xu 2007: 6). Events in North Africa and the Middle East in early 2011 may have changed all this however. Clearly, the price of fossil fuels will rise and remain high.

One of the key challenging questions raised here is whether the global ecological system is on the verge of reaching the limit and whether the expansion of global resource consumption is ecologically possible. The consequences facing China are very severe, and the Chinese growth model could face a fundamental challenge because the peak of resource exhaustion and the imperativeness of ecological sustainability would impose severe limits to its future economic growth, hence fundamental social changes will be inevitable (Li 2010).

Li and Mingi argue that "World oil production is projected to have peaked in 2008. World natural gas production is projected to peak in 2041. World coal production is projected to peak in 2029. Nuclear energy is projected to grow according to IEA's "Alternative Policy Scenario". Long-term potential of the renewable energies is assumed to be 500 EJ (12,000 million tonnes of oil equivalent). The world's total energy supply is projected to peak in 2029" (Li 2010:128).

Furthermore, China's coal production is projected to peak in 2030, oil production to peak in 2016, and natural gas production to peak in 2046. China's long-term potential of nuclear and renewable energies are assumed to be 1,000 million tonnes of oil equivalent. China's energy imports are assumed to keep growing from now to 2020 to sustain rapid economic growth. The Chinese economy is assumed also to keep growing at an annual rate of 7.5% from 2010 to 2020. By 2020, China's energy imports are projected to grow to near 700 million tonnes of oil equivalent, comparable to the current US energy imports. After 2020, China's energy imports will stay at 8% of the rest of the world's total fossil fuels production. China's total energy supply is projected to peak in 2033 (Li 2010: 130–131).

Clark and Isherwood (2010) found this same pattern of short-term fossil fuel energy supplies in China, while doing their study of Inner Mongolia Autonomous Region (IMAR) for the Asian Development Bank in 2007. IMAR is the second largest coal producing region in China and in the use of renewable energy (solar and wind in particular) to transition from the environmental problems caused by coal, China needed to provide public policies in its next Five-Year Plan for sustainable development with the financial resources. The 11th Five-Year Plan started that with over half of 1.5 GW (gigawatts) of power from global solar installation estimates being done in China (Chan 2011). Today, the IMAR is developing such renewable energy resources like wind and solar, while controlling its coal production through advanced coal technologies that are "cleaner." Almost monthly, multi-MW of wind power is being installed and operated in IMAR (Martinot and Li 2010) Wang et al. (2013) found that there was an over abundance of wind power in IMAR which has lead the local Government to move into local on-site or distributed energy in order to efficently use the renewable power surplus.

The Rise of China in the Context of Energy Dependency

In order to keep the economic growth rate, China has to make the access to adequate energy supplies as a national priority and to a great extent a national *security* priority (Constantin 2005; Huliq News 2008; Li 2010). China is perhaps one of the few countries that regard energy security as a vital component of their *national interests*. Currently, China is "the world's second largest consumer and third largest producer of primary energy" (Martinot and Li 2010). There is no sign that China's energy consumption will slow down; on the contrary, it will steadily increase. Thus, for energy consumption to keep pace with its targeted economic growth at a moderate rate of 8–9%, China will have to utilize every fuel source available including investment on renewable energy and the expansion of nuclear power. It is expected that China's import of energy resources will increase at a steady rate particularly from Russian natural gas and liquefied natural gas (LNG) shipped through Chinese seaports which are both difficult negative environmental options (Clark and Isherwood 2010).

China's growing interest in resource-rich regions such as Africa, Latin America, Middle East, Central Asia, and Southeast Asia is no doubt linked with its energy security consideration (Brautigam 2008). How will the rapidly increasing demand for energy, raw materials, and other natural resources shape Chinese policies toward its international relations especially with resource-rich countries? Can China afford depending on global energy markets, either via exclusive bilateral deals or direct investment in resource exploration in order to sustain its economic growth? What strategies will China use to secure its share of the global resource market? To find the answers to these questions, it is of importance to take an energy security approach to explore the geopolitical, economic, energy, and environmental implications behind China's rapidly growing energy challenges and to understand the Chinese anxiety and concern with issues of energy security in attempting to search for new sources of energy supply.

China's economic and foreign policy behavior is increasingly influenced by growing energy concerns. Andrew Chung, principal at Lightspeed Venture Partners in Silicon Valley, notes that the Chinese increase in clean (green) tech financing is reflective of the US drop: "First, a lot of the clean technologies are dependent on policy and government support to scale up. In some other parts of the world, you have more consistency in the way these types of funds are appropriated" (SJBJ 2011: 8). As the world's second largest economy and trading nation in 2010, China's search for energy and its global strategies for energy security have led to heavy debates and even in some cases have resulted in political conflict. China is predicted by some economists and members of the US Congress to be the number one economy and trading nation within the next decade (US Congress 2011).

The western nations have been expecting that ideally, China's energy vulnerability might drive it toward cooperation with rival oil consuming nations through participation in multilateral organizations and other forums. Since energy security is no doubt playing a more decisive role in Chinese foreign policy, Beijing's relations with both the existing major energy-consuming powers and energy-exporting countries will shape its motivation and justification on energy issues as well as on nonenergy issues.

In recent years, China's "going-out" economic and foreign policy encouraged its national oil companies (NOCs) to try to acquire some western oil companies but still secure the control over the access to some overseas energy supplies including purchasing equity stakes in foreign oil companies (US Congress 2011). This strategy has been regarded as "mercantilist" in the West and particular in the United States where the attempt of a Chinese NOC to buy out the American oil corporation UNOCAL in 2005 triggered political backlash in the US Congress causing the final withdrawal of the Chinese company. The incident indicates the lack of trust of the USA in China's energy diplomacy, because the USA politicians felt that the Chinese move was to undermine American energy security. Hence, the USA oil and gas company Chevron bought UNOCAL. Now almost a decade later, the same series of issues have arisen with the shale oil discovered in Canada as well as the sale of oil and gas refineries in the USA.

In the studies of China's energy security with its economic and foreign policy, a number of geopolitically vital areas cannot be disassociated with China's efforts to maintain both energy security and good international relations within these regions and with the major western powers. China's energy diplomacy with the Middle East, Russia, Central Asia, Asia-Pacific, Africa, and Latin America has become a global topic, where Beijing's efforts toward greater energy security through multilateral organizations are discussed. It is still too early to predict whether the world will witness the evidence supporting the liberal hypothesis that economic interdependence promotes international cooperation or confirming the realist conviction that competition and power accumulation will eventually lead states to conflict and war as history has shown in the past. Energy demand is seemingly accelerating China's "peaceful" rise to global prominence and moderating the conflict aspects of Chinese foreign policy, while China establishes hundreds of new solar, wind, and other renewable energy companies. Chan (2011) stated that of the global installation of solar in 2011, over half of the 2.5 GW will be in China. Nonetheless, the social movements in the Middle-east along with issues over Itan, could place China in the role as mediator for all nations seeking an end to further conflicts.

China has been struggling to develop and promote good relationships with underdeveloped regions that contain potential energy reserves, such as Africa and Latin America, through its unique international aid system linking development aid and trade with energy suppliers. Recently China has aimed to prepare for technological advances and changes in the climatic that will bring maritime transport in the Arctic waters to make possible the linking of North Atlantic and the North Pacific into closer commercial relations. Some policy makers expect that China will increasingly strengthen its political economic strategies for international relations in the Arctic region and speed up its research through its polar research bases in the Antarctica. In addition, China is adopting different policy strategies and objectives to different regions around the world.

Currently, China is one of the key investors in Africa, and its trade and investment relations in Latin America are going to accelerate in the coming years

(Hanergy 2011). China's increasingly dynamic economic relations with these regions through long-term financing of infrastructures, renewable energy technologies, and smart grid systems are seen by some western critiques as challenging the traditional ties between these regions and their historical colonial ties with the western powers. Intensification in China-African and China-Latin American trade relations also accelerated the "neocolonialist" argument claiming that China's is imposing the regions with a renewed "colonial" relationship. However, despite the criticism on China's energy-oriented policy in its economic and political relations with the two regions, the Chinese approach and engagement to its aide policy and practice have indeed a far-reaching long-term and permanent realignment of power relations in the conventional international aid system that has already changed the system in many ways (Opoku-Mensah 2010).

Currently, the Iran nuclear issue is testing China's foreign policy orientation in the context of its energy security consideration. The China-Iran relationships has grown out of mutual need for products, ranging from technology to consumer goods to China's soaring need for energy supplies (Dorraj and Currier 2008: 70). Thus, it has been a painful foreign policy decision for China to lend support to the USA-led UN sanctions against Iran's nuclear program, fearing the grave consequence that this might lead to loss of one of its major energy suppliers.[3] China is being torn between the imperative need for energy on the one hand and the US pressure on its role as "responsible stakeholder" and "strategic reassurance."[4]

From an internal Chinese perspective, energy security has become the essential premise for China to achieve its national goal of quadrupling its gross domestic product (GDP) in 2020. There is a genuine consensus among Chinese leaders and scholars that energy has become a key strategic issue for China's economic development, social stability, and national security and that the realization of China's key national interests[5] is highly dependent on the access to sufficient energy resources (Liu 2006; Zhang 2006). China's now outdated "market economy" had locked itself in a "tiger-riding dilemma," that is, any slowdown in economic growth would put the country in a risky situation, leading to social unrest and political illegitimacy (Li and Clark 2009). China's government fears that domestic energy shortage and rising energy costs could undermine the country's economic growth and thus seriously jeopardize business and job creation (Lo 2011). Beijing increasingly stakes its political legitimacy on economic performance and rising standards of living for its people. Consequently, the threat of economic stagnation due to energy shortage

---

[3] According to data released by the General Administration Agency (GAC), Iran supplied 11.3% of China's energy consumption in 2009 (adapted from People's Daily Online, 10 February 2010).

[4] "Strategic reassurance," coined by James Steinberg, deputy secretary of state in a conference sponsored by the Center for a New American Security, states that "China must reassure the rest of the world that its development and growing global role will not come at the expense of security and well-being of others."

[5] China's national interests are defined by the government as including sustained economic growth, the prevention of Taiwanese independence, China's return to as a global power status, and the continuous leadership of the Chinese Communist Party (CCP). Today, energy security is defined as a core part of China's national interests.

represents real risks of social instability, which could in turn threaten the continued political authority of the state and the Communist Party. Energy security, hence economic stability, and sustainable development are basic strategic political concerns for the leadership.

In fact, some scholars of energy politics point out that state-led pursuit of energy supplies is often seen as the source of international conflicts (US Congress 2011). However, behind it, other sources of conflict – nationalism, geopolitical competition, and competing territorial claims – are most likely to have been at the root cause of those conflicts (Constantin 2005). One Chinese scholar of strategic studies clearly explains the reason why energy security has become a core component of China's national interest:

> With external trade accounting for almost 50 percent of China's economy, China is now highly interdependent with a globalized market. This shift also includes hard social, political, and geopolitical choices that deeply impact matters of national security. The more developed China becomes the greater its dependence grows not only on foreign trade but also on the resources to fuel the economy. With these complex and expanding interests, risks to China's well-being has not lessened but has actually increased, making China's national security at once both stronger and more vulnerable (Zhang 2006).

China's sensitivity on the confluence of geopolitics and resource politics is also derived from the fact that historically China has been a weak sea power. One of China's key weaknesses through centuries of its development and into the modern age is its lack of a strong navy to safeguard its global interest and is perhaps one of the major factors leading to China's massive investment on raising and modernizing its naval capabilities. China therefore has good reasons for acquiring an aircraft carrier to enable it to protect its national interests (Cole 2006). China has territorial disputes in the South China Sea over the Spratly Islands with neighboring countries; Taiwan remains a continuing issue; and protests in Japan and China over the small uninhabited islands. Even more significant for China the security of the major maritime transportation routes through China which transport the majority of its foreign trade, as well as its oil imports upon which country has become dependent. Based on the historical lessons, China has a clear understanding on the linkage between its energy security and international geopolitics, which is noted clearly by one scholar:

> The history of capitalism and its spread globally have shown that it is often accompanied by cruel competition between nation states. Those countries that lose out are not necessarily economically or technologically underdeveloped or those with a low level of culture. Rather, they are most often those nations who forgo the need to apply their national strength to national defense and therefore do not possess sufficient strategic capability. (Zhang 2006: 17)

Today, the rise of China is due in large part to its rapid emergence as a major force in world energy markets and energy geopolitics (Chan 2011; Lo 2011; and Intriligator, 2013). Beijing's booming energy consumption and heavy investment for energy security have raised a new range of contentious issues between China and other world powers that are adding a new layer of issues to already complex and dynamic relationships. China's economic growth is supported by three primary pillars: (1) export-led growth, (2) real property growth, and (3) government spending whereby exports have been the key engine driving its economic growth. The current global financial crisis (2008–2009) has already indicated a concern for the first pillar because European and especially American consumers can no longer consume at the debt-supported levels as they had in the past (Economist 2009). One of the perplexing questions is whether the sustainability of China's export-oriented development strategy can be counted on to be sustainable and reliable into the future. Current data (Economist, 2012) suggests that China has weathered this storm and become the new financial center for economic markets. Hence, businesses are coming to China to invest but also seek investments. The new Chinese export has become investment "capital" and financing (SJBJ 2011; Lo 2011).

New Policy: Change to Economic Growth Strategy by Promoting Sustainable Energy

In November 2005, Chinese Premier Wen Jiabao claimed at the Plenary of the Chinese Communist Party that "energy use per unit of China's GDP must be reduced by 20% from 2006 to 2010," and this declaration was turned into a policy goal set up by China's current 11th Five-Year Plan (2006–2010).[6] In this national policy, planning China's energy policies is defined to be "from a growth at any cost model" to "a sustainable, energy-secure growth path." In order to deal with the rising energy intensity, Chinese government has introduced a number of energy and emission saving policies as well as administrative plans and legal frameworks to strengthen energy conservation work (op. cite. Economist, 2012). According to HSBC in late 2010, the next 12th Five-Year Plan (enacted in March 2011) would focus on three key issues:

1. "Achieving more balanced and sustainable growth is the key"
2. Requiring "real reforms of income distribution, industrial regulations and fiscal system"
3. Taking "steps towards financial reforms … (that will) unleash the power of consumers and inland regions"

As Lo (April 2011) reports from the central government and his real estate company in Shanghai, China has aggressively begun doing just these three target areas plus more. Lo adds three other key elements to the 12th Five-Year Plan, however: (4) "strengthen environmental protection," (5) "enhance innovation," and

---

[6] Five-Year Plan, shortened for *The Five-Year Plan for National Economic and Social Development*, and even shorter for the plans are 11-5 year Plan, is a national goal-setting policy paper. At the macrolevel, it disposes national key construction projects, administers the distribution of productive forces and individual sector's contributions to the national economy, and maps the general direction of future development including specific policies and targets. The current Five-Year Plan for 2006–2010 is also called the *11th Five-Year Development Guidelines*. The 12th Five-Year Plan or 12-5-yr Plan came out in March 2011.

(6) "improve living standards" by changing the focus of Chinese "export-oriented" economic model to a "domestic demand-oriented" economic model (Lo 2011: 7–9). Some concrete policy measures were implemented in line with these macropolicy goals which after two years appear to be achieving even more than predicted (Wang, et al. 2012). In December 2007, China's Information Office of the State Council issued the country's first ever white paper on its energy conditions and policies:

> China's Energy Conditions and Policies. China's National Energy Administration (NEA) was set up in 2008 to coordinate energy issues concerning various ministries, commissions, and state-owned energy companies. In order to promote the development of emerging energy industries and meet the carbon emissions reduction targets of 2020, the NEA has compiled a development plan for emerging energy industries from 2011 to 2020 that will require direct investments totaling 5 trillion yuan according to China Daily.[7]

Consider the growth of renewable power systems in China. The Chinese government is to launch a series of policies to support new energy development through China's 12th Five-Year Plan, or 12th Five-Year Plan for National Economic and Social Development, from 2011 to 2015 (Lo 2011) which is focusing on new energy, including wind, solar, and nuclear power, and the plan is being under final review. The new energy policy will increase China's proportion of nonfossil energy in overall energy consumption from 12% to 13% by 2015, according to China's Energy Research Institute (US DOE, US-China Research Center Report, 2011). This development trend is noticed by international consulting organizations with data such as:

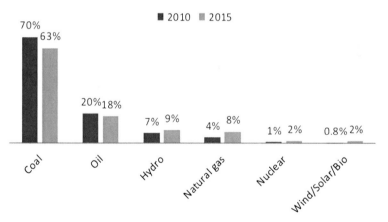

Chart: China's Estimated Energy Consumption
(US DOE, US-China Research Center Report, 2011)

Energy security and environmental problems in China should be resolved primarily through self-reliance efforts but also through international economic cooperation. In order to diversify access to energy supplies and reduce dependency

---

[7] *People's Daily* online, July 22, 2010, available from http://english.peopledaily.com.cn/90001/90778/90862/7076933.html

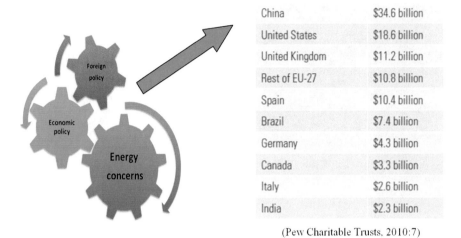

Fig. 7.2 China's ongoing and future policies connected with energy concerns (Pew Charitable Trusts 2010, p. 7)

on certain exporters, China is taking many political and economic measures and providing economic aid to strengthen its cooperative relations with resource-rich countries (Tseng 2008; Ziegler 2006 and US DOE, US-China Research Center Report, 2011; and Wang et al. 2012). However, the dynamic debates on energy security are still going on in China (Downs 2004). Certain indications can be drawn from the debates that many of China's analysts and policy maker are not fully convinced of the benefits of reliance on world energy markets. The political consensus today is to move toward "green" (renewable energy power generation) and integrated energy infrastructure systems that are sustainable.

In addition, China has been alert at the soaring demand in global energy in recent years and at the possibilities of long-term global energy shortage, called "peak oil" which now includes "gas" (US Congress 2011). Hence, China's energy security will be one of the most important parts of its broader foreign policy in the years to come. The world will soon focus on China's new economic and energy policies; its energy market reform; and its new strategies in meeting the political challenges of rising energy costs and environmental pollutions (Clark and Li 2003). Much global attention will be given on China's move toward technology development and innovation in generating clean coal, natural gas power along with new institutional developments. Chinese energy policies and each of China's steps along with practices bear significant implication on greenhouse gas emissions pollution and climate change.

Already being burdened with serious environmental problems and energy shortages, the continuing global economic downturn presents China with a historic opportunity to rethink its growth strategy in order to move ahead with a more stable and sustainable path. Today, a promising optimism is that China seems to be firmly committed to the creation of a largely self-sustaining innovation system as part of a knowledge-based economy of the future. China is sparing no effort to meet its 12th Five-Year Plan energy conservation goals, in which China will cut its per unit of GDP energy consumption by 20% from 2005 levels by the end of 2010 (Fig. 7.2) which it almost met except for the continued global economic downturn.

China's policy determination for clean and renewable energy can be clearly seen from its ambitious plan published in 2007 – "Middle and Long-term Development Plan of Renewable Energies" – which was approved by the People's Congress in 2008 as *The Renewable Energy Law*. The new policy is determined at moving the country toward renewable energy in order to reduce energy consumption and cut the surging carbon dioxide emissions. The target of this policy plan reflects another of China's policy concerns in coping with the environmental and economic challenges of climate change. The linkage of energy policy to climate change policy can be read from the policy document – *China's National Climate Change Programme 2007* prepared by one of China's key government institutions, the Development and Reform Commission. Some examples of China's successes can be seen in communities that are becoming sustainable (Wang and Li 2009; Kwan 2009).

Through the legal framework stipulated in the new laws, the Chinese government has set efficiency goals, imposed taxes and regulations designed to curb demand and reduce emissions of greenhouse gases. In addition, the government energy and environment institutions are imposed with defined guidelines and responsibilities. The new policy toward alternative energy is supported by financial incentives including direct subsidies and innovative policy measures, tax-related incentives, custom duties, and pricing incentives. Some concrete policy incentives are (1) connecting "intermittent"[8] sources of electricity like wind or solar to the national grid as well as providing battery and fuel cell storage; (2) connecting utilities mandated to open smart grid transmission lines to renewable generators, with ratepayers bearing part of the extra costs; (3) feed-in tariffs guaranteeing renewable energy producers a steady, high price for electricity so as to enable them to compete with coal producers; and (4) tax breaks, preferential loans, and other financial incentives encouraging investors to support renewable ventures (China FAQs 2010).

As is pointed out by Lo (2011) for the 12th Five-Year Plan and SGCC wrote a report on the 11th Five-Year Plan: "The strategy of building a world-leading strong and smart grid with ultrahigh voltage grid as its backbone and subordinated at various voltage levels featured as being IT-based, automated, interactive, based on independent innovation…Since 2009, SGCC has started 228 demonstration projects of 21 categories in 26 provinces and municipalities" (SGCC 2010: 1).

It is likely that China will meet and even exceed its renewable energy development targets for 2020 with applying other alternative energies including hydro, wind, biomass, and solar PV power. It is expected that more than one-third of China's households could be using solar hot water by 2020 if current targets and policies are continued (Martinot and Li 2010). China expects the policy objectives to be reached through the integration of a number of relationships: the responsibility of the state and the obligation of the public, institutional promotion and market mechanism, current demand and long-term development, and domestic practice and international experience.

---

[8] It refers to energy-generation installations which are not state owned.

In recent years, China has won the global recognition for its achievement in the development and application of alternative energy. China overtook the United States for the first time in 2009 in the race to invest in wind, solar, and other sources of clean energy. American clean energy investments were $18.6 billion last year which were a little more than half the Chinese total of $34.6 billion. Just a few years ago, China's investments in clean energy totaled just $2.5 billion (Tankersley and Lee 2010). In recent years, it is increasingly recognized that China's "green leap forward" policy has made it become the world's largest makers of wind turbines and solar panels surpassing western competitors in the race for alternative energy. As one USA newspaper points out:

> China vaulted past competitors in Denmark, Germany, Spain and the United States last year to become the world's largest maker of wind turbines, ... China has also leapfrogged the West in the last two years to emerge as the world's largest manufacturer of solar panels. And the country is pushing equally hard to build nuclear reactors and the most efficient types of coal power plants. These efforts to dominate renewable energy technologies raise the prospect that the West may someday trade its dependence on oil from the Mideast for a reliance on solar panels, wind turbines and other gear manufactured in China. (*New York Times*, January 30 2010)

Clean renewable energy strategy emphasizes a sustainable growth path based on equity is leading the transition to knowledge and information economy. When referring to China's alternative renewable energy policy, some studies have shown that China is facing both opportunities and challenges. The potential opportunities are plenty, such as solar energy, wind energy, biomass energy, small hydropower energy, geothermal energy, and ocean energy, where the challenges are apparent as well, such as the lack of coordination and policy consistency, weakness and incompleteness in incentive system, lack of innovation in regional policy, immature financial system for renewable energy projects, and the limited investment in research and development of renewable energy (Zhang et al. 2007). There is still a long way to go before China's renewable energy market becomes mature, socially and culturally embedded.

Conclusion Remarks: Challenges and Optimism Ahead

This chapter aims at providing a framework of critically understanding China's transformation from a self-reliance development path to a "market-driven" dependent growth strategy over the last decade to now a "social capitalist" economic system. This chapter's emphasis is on the economic and ecological consequence of China's insatiable demand for energy driven by the growth-based industrialization policy in the past decades. The chapter argues that since the beginning of the twenty-first century, energy has become a key concern on the agenda of China's economic and foreign policy-making calculations as it moves rapidly into The Green Industrial Revolution. Among China's core national interests – securing energy resources, generating national renewable companies and systems, gaining market access and political recognition – energy and economic security are at the top priority for developing a sustainable nation. It is expected with the rapid economic development and the improvement of people's living standard, energy demand in China will unavoidably continue to increase, which will be inseparable with its environmental problems, such as the emission of sulfur dioxide, carbon dioxide, and particulates among other issues of pollution and waste.

Above all, the western definition of "market-driven" economies in energy is questionable in China such that different definitions and meanings that are needed for "market" and therefore "capitalism." And that is what China has done: redefined capitalism so that it has a societal focus, direction, and set of policy along with financial strategies. For example, the rapidly emerging renewable energy industry in China has created a new market finance mechanism for long-term debt, which involves the Chinese business financing the entire sale, installation, and operations along with maintenance of the renewable energy technologies and products. In short, China may have discovered a new model of the World Bank.

Chinese policy makers understand the fact that due to a growing need for and even competition over energy resources and maritime transportation security, global resource-based competition, geo-territorial claims on sea areas and shelves will become harsh which could lead to armed confrontations. In order to understand the implication of the underlying dynamics of international political economics in resource-rich regions, it is of great importance to understand the source of international economic competition that can result in political conflict for access to energy and natural resources in order to understand the interactions between individual national interests. The international geopolitics and geoeconomics in the acquisition and distribution of states' wealth and power are manifested in their respective country's economic and foreign policies.

China's soaring demand for energy in connection with its export-oriented economy poses a variety of new challenges for its economic and foreign policy. Hence, the country will be more and more dependent on the purchase of natural resources abroad for sustaining its economic development. Any crisis to China's access to overseas resource and maritime shipping routes will have a negative impact on China's growth and its trade-dependent economy. China will endeavor to protect the strategic areas concerning its national interest. It has no choice, but may turn more internally in the future. In recent years, China's energy diplomacy in the context of the political economy of global energy developments has drawn the attention of the West, especially in connection with the sensitive regions, such as the Middle East, Central Asia, Latin America, and Africa. As one Chinese scholar bluntly states, "The determining factor shaping the rise and fall of a country ultimately is not just the size of its total economic volume but also the strategic ability of the country; that is, the ability to use national forces to achieve political goals" (Zhang 2006: 22).

However, despite the above global reality described by this realist perception, China's deep sense of its energy insecurity and vulnerability is changing its development policy toward clean and renewable energy. China is accelerating R&D on renewable energy supply and advanced energy conservation-based techniques and products; it is making necessary structural changes in industrial and agricultural sectors moving to nonenergy intensive industries. Furthermore, China is trying to rely primarily on domestic resources while strengthening mutually beneficial international energy cooperation. The optimism that China is presenting to the world is not groundless. China is not only one of the world's leading producers of renewable energy but also is overtaking more developed countries in exploiting valuable economic opportunities, creating green-collar jobs, and leading development of critical low-carbon technologies.

Such optimism in China's own "green revolution" is also confirmed by the front page of a report by Climate Group (2009), "As one of the world's major economic powers, China will have to be at the forefront of this journey. This report shows that it can be." Nevertheless, China still has a long way to meet its policy objectives on energy and environmental sustainability. Due to its size and population, the consequences of failure in China's case are much more serious than many other counties. China should not be left struggling alone on the road to optimism, and the whole world must pay more attention to China. World peace and a sustainable planet depend on global harmony and collaboration beyond convention competition over supply and demand.

References

ACORE (2011) American Council on Renewable Energy. US-China Quarterly Review, Mar 2011
APCO (2010) China's 12th Five-Year Plan: how it actually works and what's in store for the next five years. APCO Worldwide
Brautigam D (2008) China's Africa aid. The German Marshall Fund of the United States
Chan ML (2009) Building strong grid policy, state grid corporation. In: Proceedings of international conference on UHV
Chan ML (2010) Green White paper series, Technical Advisor, JUCCCE, Shanghai
Chan S, President of SunTech (2011) Global solar industry prospects in 2011. In: Proceedings of the SolarTech conference, Santa Clara. www.SunTechSolar.com
Chase-Dunn C (1982) Socialist states in the world-system. Sage, Beverly Hills
Chase-Dunn C (1989) Globalization: a world-systems perspective. J World-Syst Res 5:165–185
China FAQs (2010) Renewable energy in China: an overview. World Resources Institute
Clark WW II, Cooke G (2011) Global energy innovation. Praeger Press, Santa Barbara
Clark WW II, Li X (2003) Social capitalism: transfer of technology for developing nations. Int J Technol Trans 3:1–11
Clark WW II, Isherwood W (2007) Report on energy strategies for Inner Mongolia Autonomous Region, Asian Development Bank. doi:10.1016/j.jup.2007.07.003 in Clark WW II, Isherwood W, (Authors and Co-Editors) (2010) Utility Pol J. Special issue on line as China: environmental and energy sustainable development, Winter
Cole BD (2006) Chinese naval modernization and energy security. A paper prepared for the Institute for National Strategic Studies, Washington, DC
US Congress (2011) China's energy and climate initiatives: successes, challenges, and implications for US policies. From Energy Strategy Institute (EESI) and World Resources Institute (WRI), 5 April 2011
Constantin C (2005) China's conception of energy security: sources and international impacts. Working paper no. 43, University of British Columbia
Dorraj M, Currier CL (2008) Lubricated with oil: Iran-China relations in a changing world. Middle East Pol 15:66–80

Downs ES (2000) China's quest for energy security. RAND, Santa Monica

Downs ES (2004) The Chinese energy security debate. China Q 177:21–41

Downs ES (2006) China: executive summary. The brookings foreign policy studies energy security series, The Brookings Institution

Economist (2009) Special issue and articles on "collapse of modern economic theory", 16 July 2009

Economist, "Briefing, China's new leadership", The Man who must change China, October-November, 2013, pp. 21–24.

Eisenberg L (2009) "Los Angeles Community College District (LACCD)" and "Appendix A". Sustainable communities. Springer Press, pp 29–44

Hanergy Corporation. www.hanergy.prc

Hongbin Q, Sun JW (2010), Economics: from quantity to quality of growth, China's next Five Year Plan: HSBC Global Research, Oct 2010 (revised March 2011). www.research.hsbc.com

Huliq News (2008) China's energy demand increases global pressure to seek out new sources. Available at http://www.huliq.com/597/73705/chinas-energy-demand-increases-global-pressure-seek-out-new-sources

IEA International Energy Agency (2010) Key world energy statistics. IEA, France

Intriligator, Michael, "Energy Security in the Asia-Pacific Region", Contemporary Economics Policy, Special Issue with co-editors Woodrow W. Clark II and Michael Intriligator, "Global Cases in Energy, Environment, and Climate Change: Some Challenges for the Field of Economics" Western Economic Association International, Blackwell Publishing, due in 2013.

Jiahua P et al (2006) Understanding China's energy policy. A background paper prepared for Stern review on the economics of climate change, the Chinese Academy of Social Sciences (CASS), Beijing

Kaplinsky R (2006) Revisiting the revisited terms of trade: will China make a difference? World Dev 34:981–995

Konan DE, Jian Z (2008) China's quest for energy resources on global markets. Pac Focus 23:382–399

Kwan CL (2009) Rizhao: China's Green Beacon for sustainable Chinese cities. In: Clark WW (ed) Sustainable communities. Springer, New York, pp 215–222

Li, Z (2003) Energy and environmental problems behind China's high economic growth: a comprehensive study of medium- and long-term problems, measures and international cooperation. IEEJ: Mar 2003. http://eneken.ieej.or.jp/en/data/pdf/188.pdf

Li M (2010) Peak energy and the limits to economic growth: China and the World. In: Xing L (ed) The rise of China and the capitalist world order. Ashgate, Farnham, pp 117–134

Li X, Clark WW (2009) Globalization and the next economy: a theoretical and critical review. In: Li X (ed) Globalization and transnational capitalism: crises, challenges and alternatives. Aalborg University Press, Aalborg, pp 83–107

Liu, X (2006) China's energy security and its grand strategy. Policy analysis brief, The Stanley Foundation

Lo V (2011) China's 12th Five-Year Plan. Speech by Chairman. Shui On Land and President, Yangtze Council, Asian Society meeting, Los Angeles, 25 April 2011

Martinot E, Li J (2010) Powering China's development: the role of renewable energy. Worldwatch annual reports, Worldwatch Institute

NYT New York Times (2010) China leading global race to make clean energy, 30 Jan 2010

Opoku-Mensah P (2010) China and the international aid system: transformation or cooptation? In: Li X (ed) The rise of China and the capitalist world order. Ashgate, Farnham, pp 71–85

Pew Charitable Trusts (2010) Who's winning the clean energy race? G-20 clean energy fact book. The Pew Charitable Trusts, Washington

Peyrouse S (2007) The economic aspects of the Chinese-Central Asia Rapprochement, Silk Road paper. Central Asia-Caucasus Institute and Silk Road Studies Program: A Joint Transatlantic Research and Policy Center, Johns Hopkins University

Rosenberg M (2010) Global renewables war is on. U.S. missing in action. EnergyBiz, 17 Dec 2010

Scientific American (2011) Clean tech rising, 5 April 2011. www.scientificamerican.com/technology

SGCC (2010) State grid road map of China. State Grid Corporation of China. Framework and roadmap for strong and smart grid standards, 18 Aug 2010

Shoumatoff A (2008) The Arctic oil rush. Vanity Fair. Available at: http://www.vanityfair.com/politics/features/2008/05/arctic_oil200805

Silverstein K (2011) Energy avenues open with China. EnergyBiz http://www.energybiz.com/article/11/01/energy-avenues-china-open

SJBJ (San Jose Business Journal) (2011) Clean energy financing jumps to record $243B. The News, Data from Pew Charitable Trusts, 1 April 2011, p 8

Spears J (2009) China and the Arctic: the awakening snow dragon. China Brief IX:10–13

Sundra Solar Corporation, Beijing (2011) www.sundrasolar.com

Tankersley J, Lee D. (March 25, 2010) China takes lead in clean-power investments: US falls to No. 2 in funding for such alternative sources as wind and sun. Los Angeles Times

The Climate Group (2009) *China's* clean revolution ii. The Climate Group, London

The Development and Reform Commission, PRC (2007) China's National Climate Change Programme 2007, Beijing

Thøgersen S, Østergaard CS (2010) Chinese globalization: state strategies and their social anchoring. In: Li X (ed) The rise of China and the capitalist world order. Ashgate, Farnham, pp 161–186

Time (2011) World: now no. 2, could China become no. 1? 28 Feb 2011, p17

Tseng Y-H (2008) Chinese foreign policy and oil security. Int Asian Forum Int Q Asian Stud 39:343–362

UNCTAD (2005) Review of maritime transport 2005, UNCTAD

US DOE. US Department of Energy. Annual global reports. www.usdoe.usa.org

US DOE. US Department of Energy. US-China clean energy research center. http://www.whitehouse.gov/blog/2010/09/03/us-and-china-advancing-clean-energy-research-throughcooperation

Valsson T (2007) How the world will change with global warming? University of Iceland Press, Reykjavik

Wang W, Li X (2009) Ecological construction and sustainable development in China: the case of Jiaxing municipality. In: Sustainable communities. Springer, New York, pp 223–241

Wang, Xuan, Nan Wu and Tor Zipkin "Wind and solar energy systems in IMAR: the status quo and suggested paths" Paper from Cross-disciplinary Scholars in Science and Technology Program, UCLA, to be published in 2013.

www.WorldWatchReports.org

Xu Q (2007) China's energy diplomacy and its implications for global energy security, FES briefing paper 13, Friedrich-Ebert-Stiftung, Berlin

Zhang W (2006) Sea power and China's strategic choices. China Security, pp 17–31

Zhang P et al (2007) Opportunities and challenges for renewable energy policy in China. Renew Sust Energy Rev 13(2007):439–449

Zhen YZ, Hu J, Zuo J (2009) Performance of wind power industry development in China: a diamond model study. Renew Energy J, 34 Elsevier Press, pp 2883–2891

Ziegler CE (2006) The energy factor in China's foreign policy. J Chin Pol Sci 11:1–23

Chapter 8
The "Cheap Energy Contract": A Critical Roadblock to Effective Energy Policy in the USA

Michael F. Hoexter

Abstract The "Cheap Energy Contract" is offered as a characterization of how policymakers and economic policy advisors approach the pricing of energy in the USA. The demand that the unit costs of energy can be particularly inexpensive has inhibited policymakers from implementing policies that would accelerate the adoption of new cleaner energy sources and increase energy efficiency. Economic theories of the role of energy are reviewed for their contribution to an understanding of energy pricing. Conventional energy economics and pessimistic varieties of biophysical economics associated with the "peak oil" school of analysis, both insist that energy be very inexpensive. Conventional environmental economics and solutions-oriented biophysical economics, both allow for the unit costs of energy to become more expensive to meet environmental and technological challenges. Energy policy and future energy economics should both allow for consideration of the production costs of promising newer technologies, including renewable energy, in establishing price expectations for energy. The Cheap Energy Contract should thereby be replaced by an "Affordable Energy Contract" that eschews a reliance exclusively on "least-cost" solutions for solutions that lead to better and best practices in energy sourcing and use.

Introduction

Conventional wisdom in economics and in energy policy has played a role in reinforcing one of the more troubling social agreements of our time, what might be called the "Cheap Energy Contract." There is historically a pervasive belief that is

---

M.F. Hoexter, Ph.D. (✉)
Terraverde Consulting, 200 Davey Glen Rd., Belmont, CA 94002, USA
e-mail: Michael.terraverde@gmail.com

particularly strong in the United States and Canada, but extends through many parts of the world, that the unit costs of energy must be particularly cheap. And in some cases below what classical economists would call the "natural price" of produced energy, i.e., the price that would enable producers, particularly those undertaking new risks via implementing or selling new, cleaner technologies, to stay in the energy business without substantial government help.

These low prices, in addition, do not account for the much-discussed negative externalities of energy use, such as climate change and local pollution. Outside of temporary episodes of attention paid to energy during energy crises or energy-related environmental disasters, energy is called upon to be the almost silent support for economic activity and only rarely its focus. The conventional tools and assumptions of both energy economics and economics in general tend to reinforce a set of unexamined beliefs about the necessarily low profile and low valuation of this key economic input.

In this chapter, I explore both the reasonable and the unreasonable basis for the Cheap Energy Contract and why it must change in order for industrial and postindustrial economies to remain viable in an era of limited atmospheric and energy resources. The Cheap Energy Contract can be stated as follows:

1. Government, consumers, and energy producers are parties to the contract, with government acting as investor, mediator, and partial guarantor.
2. Total energy expenditures for consumers and industry must be negligible for all but the most energy-intensive industries (airlines, aluminum, or logistics).
3. Per unit energy costs must be low enough to allow temporarily doubled rates of energy use not to "bust the bank."
4. Real or artificial energy shortages are unacceptable despite energy's low cost.
5. Government is ultimately responsible for guaranteeing that energy is cheap and available; elected officials risk being voted out of office if energy prices rise substantially or energy availability is reduced through either government action or independent of government action.
6. Depending on which political ideology vis-à-vis regulation and government expenditure is currently dominant, government subsidy of energy may need to be hidden from public view.
7. Dominant players in energy markets sacrifice some freedom to set prices in exchange for political influence and subsidies: oil sellers have more pricing power though experience more competition on the retail level than electricity retailers that are regulated by government agencies.

The effects of the Cheap Energy Contract are observed in the behavior of US politicians with regard to the price and taxation of energy, as well as interpretations of electoral behavior by politicians. The reflexive (Soros 1987) nature of social and market processes intensifies the effect of the Cheap Energy Contract as fears and expectations about energy's cheapness reverberate and may depress nominal offering prices while not necessarily reducing overall costs, both paid and unpaid, requiring therewith subsidies. By this recursive social process, the insistence on "cheap" has an important role in codetermining the current nominal, but not necessarily the real price of energy, especially in the longer term.

In the United States, there are some fairly recent lynchpin political events that reinforce the effect of the "contract" on contemporary energy policy. The electoral defeat of Democratic President Jimmy Carter in 1980 by Republican Ronald Reagan occurred after a massive spike in oil prices in 1979–1980 in the wake of the Iranian Revolution. The defeat of Carter, the founder of the US Department of Energy, effectively ended the efforts of the US Federal Government to respond in a concerted manner to the challenge of the oil crises of the 1970s and to mounting alarm about the external costs of fossil fuel use.

One of the two major early policy failures of the USA Democratic Clinton Administration in the 1990s was the attempted levy of a BTU tax which may have contributed to the defeat of Democrats in the midterm elections of 1994, Afterwards, though not necessarily as a consequence, Republicans held a majority in both houses of congress for almost 12 years. In 1999–2000, oil prices rose again, though not as dramatically as in 1979–1980, a price spike which coincided with the defeat of incumbent Democratic Vice President Al Gore by Republican George Bush in the 2000 Presidential election. The very dramatic spike in oil prices in 2008, which subsided somewhat before the November 2008 election coincided with the defeat of the candidate of the incumbent President Bush's Republican Party, John McCain, by the current president, Barack Obama.

While it is impossible to know whether energy pricing was critical to the outcomes of these elections, it is clear that the price of energy is a concern for USA politicians going forward. Disowning gas tax hikes during the 2008 presidential campaign, Barack Obama showed a preference for fuel efficiency standards, a less effective policy strategy that transfers responsibility to auto manufacturers rather than raising gasoline taxes (Mankiw 2007; NBC News 2008). The preference of both Clinton and Obama administrations for cumbersome emissions trading (cap and trade) schemes, rather than for direct carbon taxation is, particularly in the United States, a sign of the strength of a Cheap Energy Contract by, again, shifting responsibility from the government to the seemingly impersonal "invisible hand" of carbon markets to raise the price of energy.

In March 2010, Obama's move to expand offshore oil drilling and, in May 2010, the reiteration of that stance in the face of criticism after the largest oil spill in US history, the Deepwater Horizon blowout, at first seemed puzzling given the president's nominal commitment to clean energy. Upon further reflection, looking toward the 2010 midterm elections, Obama may have been reckoning that a show of commitment to increasing energy supply in all forms, and therefore pledging further allegiance to the Cheap Energy Contract, might guard against opposition attacks and voter discontent about energy prices.

The 2011–2012 conflict surrounding the Keystone XL pipeline, slated to transport tar sands from Alberta to American refiners in the United States, is a critical test for Obama's commitment to clean energy as opposed to lowest common denominator support for the Cheap Energy Contract. A fairly strong movement that combines climate campaigners from 350.org and local environmentalists along the proposed pipeline route has emerged that is attempting to block the pipeline. Protest and civil disobedience have been part of the arsenal of the movement. Without mention of the issue of climate change, the Obama administration has, as of February 2012, seemed to err on the side of blocking the pipeline, on environmental grounds, as it would cut across the crucial Oglala aquifer. An extended environmental review might doom the project. Still, observers believe that the Obama administration has not turned decisively against intensified fossil fuel exploration and extraction; continued grassroots pressure may be essential to keep the tar sands in Canada (Battistoni 2012). This lack of commitment by the Obama administration has lost it the strong environmental and clean tech support that it once had in 2008.

While this chapter questions uncritical acceptance of the Cheap Energy Contract and its assumptions, rises in energy prices, as well as perceived and real political influence on those prices, do have real negative effects for politicians around the world. A recent example is illustrative of the dangers: In early 2010, one of the precipitating events of the riots that ousted the corrupt Bakiyev government in Kyrgyzstan was 400% rises in the cost of heating and 170% rises in the cost of electricity within a short period of time (Asman 2010). Kyrgyz energy consumers had come to rely on low subsidized prices for energy, a pricing regime which came to an abrupt end in 2009 and 2010. The Kyrgyz government was already known for cronyism and corruption which are considered by many observers the ultimate causes of its downfall (Cohen 2010); however, the abrupt multifold increase in energy prices greatly magnified long-standing political discontent.

Political influence upon energy prices via taxes or price controls does not necessarily lead to abrupt price spikes or to political unrest. Relative to energy price expectations in the United States, the sense of entitlement to cheap energy, or at least energy without a politically determined price component, is not nearly as powerful in Western Europe and Japan where over the last four decades, governments have implemented tax regimes that attempt to limit the use of imported or domestically produced petroleum as well as limit the size of cars via taxation schemes (Johannson and Schipper 1997). Though the high price of petroleum is viewed as onerous by drivers in these countries, there is also apparently willingness to accept these taxation regimes by European electorates to date (Sterner 2007). More recently, European governments have set wholesale prices for renewable energy that incentivize the building of wind turbines, solar installations, and other renewable electric generators, which, in many cases, will raise retail electricity prices incrementally. The so-called Porter hypothesis (Porter 1990) that environmental technology innovation and market leadership can be the result of a system of environmental fees seems to have been borne out by the experience of Europe relative to the USA in the area of both fuel efficient vehicles and renewable energy deployment in the last decades.

A prospective view of what must be achieved in the next decades puts the inadequacy of the Cheap Energy Contract as a guide to pricing energy in high relief. To achieve virtual carbon neutrality within a few decades and additionally to escape the effects of oil shocks in the near term will mean the additional investment of tens of trillions of dollars in the energy and transportation sectors over this period worldwide, estimated by the IEA as $10.5 trillion in additional investment by 2030 just for climate mitigation in a scenario that still assumes predominant use of fossil fuels (IEA 2009). The IEA's figure is also premised on continued growth of use of fossil resources that may be based on unrealistic assessments of the cost and availability of these resources in the next two decades; therefore, the $10.5 trillion figure for worldwide expenditures may be conservative. Expenditure on electrified train systems, electric vehicles batteries, battery charging and changing networks, renewable energy generators, new electric transmission, and new generations of nuclear power plants will all involve costs and expectations of profit on investment that require revenue, some of which will originate in the sales of energy or taxation that finances energy-related investment. Sitting astride this view of the future of energy, the Cheap Energy Contract mandates as a foundation of our economies that this revenue stream must come largely from miniscule payments for energy or payments that are no more than that to which we become accustomed in an era of cheap fossil energy.

The Cheap Energy Contract and the Marketing of Energy

The influence of beliefs about the necessarily low price of energy can be found in the marketing of new energy technologies as well. Even in businesses that would and do profit from higher energy prices, the mantra of low energy prices continues to be repeated over and over again. Marketers of renewable energy generators often vastly underestimate their costs in public discussions or leave open the distinction between production costs and selling prices. In the last few years, a number of solar module sellers claim to be at or near the $1/W mark in terms of costs (Kanter 2009); however, this number refers not to a price that buyers pay, let alone the higher installed cost of the system, but to production costs. Current low selling prices for modules are around $1.50/W, and installed costs would be then somewhere closer to $2.50–4.00 W depending on system size. Almost never are these price distinctions discussed in public communications, as the lowest possible cost number is chosen to represent the price of solar panels. By contrast, televisions are not usually sold with reference to their low production costs, but to either their full sale price or a discounted but still retail price.

While there are some rational marketing reasons why sellers of energy conversion devices would discuss costs rather than price to give off the impression of being cheaper than they are, the rush to publicly cite cheap energy costs has had the effect of maintaining price expectations for buyers and consumers below the actual total cost of producing any energy, let alone, higher quality energy that is most useful. In another example, the venture-funded start-up in the area of concentrating solar thermal electric generation, Ausra, now AREVA Solar, chose to develop a technology (linear Fresnel concentrating solar thermal power) that from projections would lead to lower costs, but ultimately less useful energy for electricity generation (though sufficient for industrial process heat) because of the technology's lower temperature steam output than competing technologies (Clarke 2010). The company advertised itself heavily as though it were the low cost leader in solar thermal electricity generation until it had to refocus its business to produce instead solar industrial process heat with lower steam temperature requirements. In conforming to the dictates of the Cheap Energy Contract, Ausra had allowed cost to determine its selection of technology. Meanwhile price expectations for the reliable electricity generation via concentrating solar were set below what could be produced in the near future.

The 2010–2012 shake-out in the photovoltaic industry can be traced directly to the demand that energy prices be as cheap as possible and often below the cost of production. From the perspective of consumers, the current price war, a direct result of capacity build-out by heavily subsidized Chinese panel producers and low polysilicon prices, is a boon, many producers of new technologies that have initial higher costs are being pushed out of business, as conventional silicon solar cell manufacturers are pushing out more innovative but higher cost competitors. While those who believe in cost as the primary determinant of economic success will cheer this result, the merits of emerging technologies in photovoltaics are overlooked by the race to the bottom in price. Technologies with potential greater future merit, particularly thin film technologies, may no longer reach market and inhibit the development of production facilities of scale, which might enable them too to compete with the now 60-year-old crystalline silicon technology.

Is Energy a Unique Good?

In the dominant neoclassical economic tradition as well as in many schools critical of mainstream economics, the physical characteristics of goods and services are treated as secondary or nonexistent for the purposes of economic analysis (e.g., Borenstein et al. 2002). Energy or individual energy products are viewed strictly as commodities, goods with interchangeable qualities that are sold on markets primarily via the price and quantity available. In this view, the predominant energy-economic institutions are then a (commodity) market for oil, a market for natural gas, and a market for electricity, and the predominant reality is the record of prices paid in the past on these markets.

Critics of neoclassical economics have suggested that this dominant school of economics has an unrealistic view of the place of the production of goods and services themselves in their price evolution as well as containing a model of the physical world that is unrealistic (Mirowski 1989; Clark and Fast 2008). A secondary, less privileged discourse of business and engineering analyses of the cost structure of energy production (e.g., Blair et al. 2008) is more in tune with the business and technological realities of energy but is not absorbed into the dominant economic model of energy markets.

Biophysical Economics

The attribution of a pivotal causative role for energy in economics comes largely from outside of economics' mainstream. Early twentieth century biophysicist Alfred Lotka (1925) drew a distinction between exosomatic and endosomatic energy. Endosomatic energy is food energy for human beings, while exosomatic energy is energy used by work animals and machines to do useful work for human beings. What we call "energy" is mostly exosomatic energy, though with the introduction of a biofuel industry, an area of overlap between exo- and endosomatic energy has been reintroduced and with it potential competition between end-use markets. Basing his view of the economy upon energy and thermodynamics (Soddy 1926), Nobel-winning chemist Frederick Soddy campaigned during the 1920s for innovations in the economic system which have now become commonplace features of macroeconomic policy (Zencey 2009). More recently, the influential heterodox economist Nicolai Georgescu-Roegen (1971) reinterpreted economics as an application of the laws of thermodynamics in biological systems.

Ecological economists in the tradition of Georgescu-Roegen have studied how exosomatic energy use correlates with higher levels of worker productivity, as people can through the use of energy conversion devices, turn mostly fossil energy into work that replaces manual labor. David Pimentel and Mario Giampetro (1990) observe that in societies which do not use powered machines the ratio of exosomatic to endosomatic energy is approximately 4–1, with work-animal feed consumption plus biomass burning representing all exosomatic energy. By contrast, as of the early 1990s, developed countries consumed exosomatic energy at an average of ten times the rate per capita of nonindustrial societies or 40–1. The ratio in the United States is approximately 90–1, twice that of other industrialized nations. 85% of exosomatic energy now comes from fossil sources (IEA 2009).

The "peak oil" school of energy economics that operates largely outside academic economics (Heinberg 2003) points out how depletion of the most valuable fossil source of exosomatic energy, petroleum, will undermine economies and civilization as we know it through skyrocketing prices and severe resource conflicts. While the very specific and dire predictions of peak oil theorists seem at times tinged with Schadenfreude, their attention to exosomatic energy contrasts with the homogenization of energy as just one commodity among many: while rice can substitute for wheat as a food commodity, much of the fixed capital of developed societies becomes less useful and economical with the growing expense and potential shortages of oil. Since the First World War (Sassi 2003), government policies and political events throughout the world have reflected the critical nature of oil for economic development and the maintenance of governments' military effectiveness in an era of mechanized warfare.

Closely associated with extra-academic peak oil theorists, biophysical economics locates itself within the academy between the biological discipline of ecology and economics (Hall et al. 2006). Biophysical economists propose an "energy theory of value" which suggests that economic value for humans ultimately stems from biologically or socially available energy for use by humans, which has a phylogenetic relationship with the motivational systems and motivated behavior of all biological organisms and communities. Biophysical economists are scathingly critical of the dominant neoclassical paradigm (Hall et al. 2007).

Neoclassical economists would argue that energy is just one of many commodities, which, because of economic development and increased demand within advanced market economies, leads to higher consumption of energy. Neoclassical economics starts from the assumption that markets determine the availability of energy more than any physical constraints (Clark and Fast 2008). Biophysical economists believe that the availability of abundant exosomatic energy is a physical prerequisite, one of the most important causes of economic development.

Biophysical economists point out that dominant neoclassical economics works with the assumption of an unlimited physical world, where scarcity is simply a matter of insufficiency of suppliers rather than physical supply. Growth then becomes the default assumption of most economic projections, a "pro-growth" bias (Cleveland 2003). Biophysical economics points to the equal potential for decline or shrinkage of economic activity especially when constrained by real physical limits. The collapse of civilization that many biophysical economists and peak oil theorists predict is based both on examining the historical record of growth and decline of past human civilizations (Roman, Mayan) as well as the observation of animal populations that have exhausted their food supply after periods of exponential growth in population (Diamond 2005).

Environmental Economics and Energy

While biophysical economics places energy at the center of economics, more conventional environmental economics views energy as a commodity traded on markets but puts it in a class of goods and services that create negative externalities for the environment (Owen 2004). Most climate policy instruments that hinge on carbon pricing are shaped using a model of an energy market with negative externalities. The negative effects of fossil fuel combustion upon the atmosphere are now recognized as one of a set of negative externalities that now and in the future require some form of economic policy management. This however puts greenhouse gas emissions into the same category as acid-rain pollution, ozone depletion, and local pollution of waterways or groundwater. Recently, in the United States, there is an acutely aware of the considerable external costs of fossil fuel extraction both with the largest oil spill in US history in the Gulf of Mexico and public awareness of pollution of groundwater via unconventional natural gas extraction via hydraulic fracturing in the northeast and midwest.

The concept of a "market failure" is key to environmental economics: the implication is that the normal functioning of markets yields good-enough economic results with the exception of certain areas. Conventional environmental economics has as its goal the internalization into markets of the negative environmental externalities or the positive externalities of different market activities (e.g., Longo et al. 2008). Energy is then treated as another, very important, market within which the production and sale of the energy commodity should be shaped via market signals to produce less damage to the environment and human health. There is a voluminous literature on and within environmental economics about the externalities of energy as well as other environmental problems

Analysts and Critics of Energy Subsidies

Other energy-economic schools of thought that operate both within and outside of academic economics are analysts and critics of energy subsidies as distortions of energy markets (Koplow 2010). One group of energy subsidy analysts point out that energy production with damaging effects on the environment, like gas, oil, and nuclear energy, is the recipient of government assistance in the form of subsidies and rulemaking that favors them over other sources of energy. These critics of subsidies suggest that the removal of these subsidies would favor cleaner, cheaper, and more environmentally friendly energy sources. The UN Environmental Program is one such critic of energy subsidies for fossil fuels (UNEP 2008). Another school of critics of energy subsidies from a libertarian perspective sees energy subsidies as departures from the ideal of free, unregulated markets, which should determine the prices of goods and services in almost all cases (Bradley 1997). This group of critics also argues that subsidies ultimately inflate the prices of energy, which is a primary concern in all product sectors for neoclassical economics.

Despite their range of preferences and values, critics of energy subsidies usually operate within the assumptions of neoclassical economics that unsubsidized and often unregulated markets provide the optimal economic benefits. Energy markets however have often operated as oligopolies or monopolies in reality, so the neoclassical ideal of competitive energy markets may hold out a false path. Keen (2004) finds that in electricity as many other sectors that overall economic welfare is not well served by trying to use the tools of neoclassical competitive market theory to price electricity and deregulate electricity markets.

Energy Price Controls

Another policy instrument that has received mostly negative attention within traditional energy economics, especially in an era or liberalized energy markets, is the use of energy price controls, which are viewed by some economists and advocacy groups as either subsidies or other forms of market distortion. Feed-in-tariffs are one example of a wholesale energy price control that is either intended as a price stabilization mechanism and/or an incentive system for, most often, different renewable energy generation technologies. Usually Feed-in-Tariffs offer above-market rates or more-stable-than-market rates over a period of 10–20 years to enable the financing of capital intensive renewable generation (Mendonca 2009).

Feed-in Tariffs are usually financed via a surcharge on the pool of all ratepayers within a jurisdiction but could, in some policy designs, include as well a government tax subsidy. Critics of Feed-in-Tariffs claim that these technology-specific price controls are market distortions and waste ratepayer money, calling them "subsidies" as a pejorative (Monbiot 2010; Frondel et al. 2009). Feed-in-Tariff proponents point out that feed-in-tariffs are used to support the development of critical infant industries that would otherwise never reach sufficient market penetration to enter the technology cost curve; well-designed Feed-in Tariffs are structured so as to decrease in successive years of the program to pressure the industry to become more cost-competitive (Laurent et al. 2010). The use of Feed-in-Tariffs for wind and solar photovoltaics over the last 20 years, driven by the policy decisions of German and other European governments, has had the effect of reducing the price of energy from these technologies by stimulating demand and thereby enabling the creation of economies of scale for component manufacturers.

Due to the stabilizing and stimulating effects on demand, feed-in-tariffs function in part as a type of pro-renewable energy industrial policy, and as industrial policies are considered to be a less efficient means of stimulating economic growth within neoclassical economics, they are passed over as the object of serious study; via acts of politicized academic misinterpretation, each and every "industrial policy" is labeled "protectionism" or as another departure from the free-trade, market ideal (Rodrik 2008). The categorical dismissal by academic and media analysts of any and all industrial policies keeps mainstream neoclassical economics at a considerable distance from the reality of optimizing industrial development in any number of industries, as well as the choice of which industries to support.

Other price controls are those that reduce the retail cost of energy, which must be paired with energy subsidies to enable purchase of the energy supply from producers at market prices (Jha et al. 2009). With the expansion of energy market deregulation in many parts of the world in the 1990s and 2000s, subsides of the retail price of energy are more likely to now occur in developing countries, where high fuel or electricity costs can stifle development or injure key economic sectors. The cost of these subsidies is very high for governments with already lower levels of revenue and dependent upon the vagaries of oil markets.

Energy, Public Goods, and Infrastructure

On the border of discussions of energy subsidies is the role of infrastructure in economic life. Infrastructure is for the most part treated as an externality in neoclassical economics, albeit a positive one, which in turn reinforces a tendency to neglect infrastructure building and maintenance, especially in political regimes where neoclassical ideas are ascendant. Exosomatic energy use, first via draught animals and wind, has, since ancient times, involved transport either by land or by sea, which has benefited from ancillary services or structures that enable smooth passage and off- and on-loading of cargo. Governments have most often supplied this infrastructure, often collecting use fees that finance the building of these structures and their maintenance. An argument can be made that road building is an example of a very large government subsidy for the oil industry (ICTA 1998) and internal combustion engine vehicles, both in the demand generated for asphalt, a petroleum product, and the provision of a key positive externality for these two related markets, transportation and fossil fuels.

While institutionalist economics with its focus on the specifics of individual economic contexts and historical development can easily accommodate discussions of the role of infrastructure, neoclassical economics has tended to treat infrastructure as either an externality or just another good or service to be bought and sold on a par with other goods and services. Adam Smith in *Wealth of Nations* (1776) acknowledged the role of government in providing infrastructure for commerce, though this aspect of his work did not make it into the dominant neoclassical synthesis. The introduction of the notion of public goods into economic discourse by Samuelson (1954) offered a means of discussing infrastructure within the neoclassical framework. Nevertheless the concept of "infrastructure" implies that these goods are not only public but compulsory for economic functioning, which has not become part of the theory of public goods.

Must Energy Be Cheap?: Five Views

Neoclassical Economics in Theory

Because neoclassical economics tends to subordinate production costs and factors to market exchange, the differential costs of emerging or new energy technologies would tend to be subsumed to existing energy pricing and price expectations. For neoclassical economics, the tendency in theory is to accept the market price determined by supply and demand as the "right" price for energy. Any efforts to raise or lower the price of energy by price controls or subsidies would be considered to be "inefficient" and distortions of the "true" price of energy as determined by market exchange. That being said, energy is treated as just one factor of production or cost and there is no a priori demand that it be less expensive than other factors of production. In theory, neoclassical economics would describe how other cheaper inputs, for instance energy efficient devices, would substitute for energy if it became too expensive.

Nevertheless, neoclassical theorists would say that any good or service must be as cheap as possible, otherwise the sellers would be realizing "rents" or larger than "efficient" profits on their sales of a good or service. So, in neoclassical theory, any and every good must be cheap(er), though energy must not be exceptionally cheap. Surprisingly, despite its nominal commitment to capitalism and the profit motive, neoclassical economics tends to suggest that profits are often inefficient "rents" that must be reduced via competition on markets. Monopoly power or other distortions of the supposed norm of perfect competition, also known as "market power," are the only insurance of, theoretically disparaged, profits within neoclassical theory. Thus the gross structure of neoclassical theory supports a bias toward buyers/consumers rather than sellers/producers which has tended to correlate with the tendency toward deindustrialization in political regimes where neoclassical ideals are most assiduously pursued (the United States and Great Britain from the 1980s to the present).

Applied (Neoclassical) Energy Economics

In practice, most applications of energy economics draw from the tools and tradition of neoclassical economics but add in the "Cheap Energy Contract" that energy must be particularly inexpensive. Analyses of most climate proposals are evaluated not so much for how rapidly they will reduce emissions or incentivize production of new technologies but their effects on energy prices (US EPA 2009). New energy sources, like renewable energy, are discussed largely in terms of their higher costs rather than their benefits, their contribution to industrial capabilities, or the internal cost trajectory of the technology. Concern for impacts on consumers is not balanced by attention to the contribution of policies to nascent energy industries. In this framework, the price of energy can rise due to market factors like diminished supply or increased demand but not to accommodate the future-looking demand for cleaner energy that may have an origin in environmental concerns that preoccupy politicians and the public.

Contemporary energy economics is hampered by its neoclassical roots because neoclassical theoretical assumptions subordinate factors related to the production of goods and services to the history of exchanges of those goods on markets (Mirowski 1989). As existing market prices in energy are determined by the currently dominant fossil sources (85% of supply), the production costs of newer sources represent a discontinuity, a break, from expectable costs. Therefore, energy pricing is defined by still relatively cheap, though price-volatile, fossil fuels, which gives incumbent industries and technologies an appreciable advantage in energy-economic analyses. Because of the structure of neoclassical economic thought, these arguments about the price determination of energy end up being circular, as the real trajectories of production costs and methods of either fossil fuels or renewable energy remain largely "external" to the main focus of analysis.

By focusing on keeping nominal energy prices low, conventional energy economics may necessitate the subsidy of energy production and energy use, not just out of the intention to realize profit but also via recognition of the inter-networked systems (infrastructure) that energy demand requires. If energy prices must be set at a level that is not an "all-in" price, inclusive of costs and profit, then alternative sources of revenue need to be found to sustain energy services and the energy industry as a viable line of business.

(Neoclassical) Environmental Economics

While strict neoclassical economics, sometimes applied within an economic policy regime now called "neoliberalism," see the imposition of environmental regulations by government as inefficient or a step on the "road to serfdom" (von Hayek 1944), mainstream environmental economics share with neoliberal economics a focus on exchange and markets as the paradigmatic institution within economics. Environmental economics however then seeks to apply to the market price of energy the quantifiable externalized costs of energy production and use (mostly from fossil fuels), disadvantaging them as compared to energy sources with much lower external costs such as renewable energy. Carbon pricing is the most commonly discussed example of how a negative externality can be brought into the market.

Alternatively, though this has not been the trend, environmental economists could also reward cleaner energy for its positive externalities by offering a wholesale price premium to producers of, for instance, renewable electricity. This could only work in a multi-fuel electricity system where this price premium would not disadvantage cleaner sources on the market in competition with polluting sources. If this were in the form of a generic feed-in-tariff premium, clean energy would be sold into the wholesale market at one premium wholesale price rather than differentiated by technology.

In both cases, environmental economics does not subscribe to the Cheap Energy Contract but suggests that energy can have a price based on its market price plus or minus some penalty or premium.

(Pessimistic) Biophysical Economics

Many biophysical economists warn that the age of cheap fossil energy was a one-time bonanza which will rapidly and, for the current world economy, catastrophically come to an end, as oil rapidly depletes under increased worldwide demand (Kunz 2009). Centrally concerned with the process of energy production, biophysical economists link the cost of energy in broad social terms as well as market price to its EROI, or energy return on investment. If a lot of work, energy expenditure, and money is required to achieve a certain output of energy, then the price of that energy goes up, lowering its EROI, and with non-renewable resources, stocks (reserves) of that resource have already begun to or will soon diminish. The EROI ratio is, in addition, for biophysical economists, an indication of how close one may be coming to the end of abundant fossil resources.

A recent study (Hall et al. 2009) suggests that at a minimum, energy production should have an EROI of 3 or more (return two times more energy in surplus of what has been invested) in order for just a reduced, lower-energy version of our own civilization to continue. To maintain the current industrial civilization, some EROI analysts project that economies require an energy return of 8 or 9 on energy invested (Mearns 2008). Extracting oil from tar sands currently has an EROI somewhere between 2 and 10. Most often cited by the more pessimistic, "peak oil" economists are past EROIs of 30 or 100 that were achieved when oil could be either easily pumped or would itself gush from the ground. These higher EROIs helped facilitate cheaper market prices for oil.

Biophysical economists theorize that economic growth itself depends on high EROI energy sources, which means devoting minimal energy and investment to the extraction of an adequate energy supply to fuel the development of a sophisticated, differentiated economy. Kunz (2009) suggests that doubled unit energy costs endanger the business case for increasing the energy efficiency of mechanized production. Peak oil "doomers" are skeptical of the claims that energy efficiency increases substantially via technological development. Citing pessimistic numbers for the EROI of renewable energy (2–8), at least for liquid biofuels, the conclusion is that we are facing a "net energy cliff" (Mearns op cit). The current financial and debt crisis is also for peak oil "pessimists" of one piece with the depletion of oil and a further sign of the imminent collapse of industrial civilization. As in their analyses monetary resources are so closely linked to cheap energy resources, there are not sufficient funds to finance the development of alternatives to oil and other depleting fossil fuels.

This school of biophysical "pessimists" point out that no ready substitutes exist for oil and natural gas with EROIs of 15 and above and we will see shrinkage of the economy rather than intensified investment in clean energy. The most rational policy responses for them are adaptation to a lower-energy, lower technology, less complex, and more localized society.

Pessimistic biophysical economists are paradoxically supportive of the Cheap Energy Contract, as they feel that industrial economies are inflexibly wed to the cheap oil and gas of the past, oil and gas that was both much easier to extract and also for which, as is still the case now, we do not pay for its external costs. They see human beings as ultimately unable to modulate their energy use and with the capacity or desire to plan for a future without the concentrated energy of fossil fuels.

(Solutions-Oriented) Biophysical Economics

Others within biophysical economics feel that there is a window of opportunity to move to a non-fossil-fuel-dependent society, largely via a transition to electrified transport and work processes fueled eventually by renewable energy. While accepting the account of the historically reduced EROI and depletion of fossil fuels, these less pessimistic analysts have more positive assessments of the following components of a future energy system:

1. The current EROI of renewable electric generators (assessed as 5–35 rather than less than 5–6, as do "pessimists")
2. The capacity of current societies for large scale social cooperation to meet the challenges of peak oil and climate change
3. The capacity of electric transportation to take over from petroleum-powered transport in a timely manner
4. The flexibility of energy consumers to accept more expensive energy either temporarily or permanently
5. The flexibility of drivers to transfer to group or public transit and/or temporary compromises in the speed, load capacity, or range of electric vehicles
6. The effect of energy efficiency on the total energy expenditures for economic actors and make higher energy unit costs affordable in some applications

Table 8.1 Attitudes towards the Cheap Energy Contract by Economic School

Economic School	Stance Towards Cheap Energy Contract (CEC)	Production costs are determinative
Neoclassical theory	All goods and services must be cheaper	No
Conventional energy economics	Supports CEC; norms energy pricing on fossil prices	No
Environmental economics	Opposes CEC; energy must pay its imposed external costs	No
Pessimistic biophysical economics	Supports CEC; energy costs are normed in age of cheap fossil fuels	Yes
Solutions-oriented biophysical economics	Opposes CEC; energy costs could go higher to develop sustainable society	Yes

Those in the pessimistic school of biophysical economics call this position a "techno-optimist" one. An assessment of historical technology change (Greenwood 1999) reveals that in fact, technological change and energy change have often taken decades to occur but not usually accompanied with devastating collapse. Whether peak oil or another resource shortage will cut this transition short is to solutions-oriented biophysical economists a matter, in part, of the desire of both key actors and masses of people to work out energy and transportation solutions.

As biophysical economics places production and physical reality at the center of economic analysis, biophysical economists of the non-pessimist variety would be supportive of industrial policies, like feed-in-tariffs, technology research, public investment in deployment, and tax incentives, that develop non-fossil energy and transportation in a targeted manner.

Solutions-oriented biophysical economists then are not wedded to the Cheap Energy Contract though would maintain that a minimum EROI is necessary for any energy generation technology, and therefore some energy cost ceiling is necessary for the continued existence of complex societies (Table 8.1).

Conclusion: Energy Pricing Based on Physical, Economic, and Social Reality

The Cheap Energy Contract arose in an era of plentiful and relatively easily accessible fossil fuels, where additionally, the externalized costs of fossil energy were not paid by consumers or by taxpayers. In the USA, with plentiful supply of fossil fuels throughout the first and second industrial revolutions, a particularly durable sense of entitlement to cheap fossil energy has developed and remains. It would seem to be a low-probability event that all positive attributes of fossil energy (its plentitude, portability, energy density, and low cost) could be immediately and painlessly be transferred to cleaner sources of energy without a period of transition (Greenwood 1999). For economists and policymakers to continue to suggest that the needed transition to cleaner sources must immediately and smoothly conform to cost parameters and payment schemes, inclusive of infrastructure costs, established during the mid-twentieth century, would appear to be unreasonable and highly unrealistic.

The Cheap Energy Contract should be replaced by an "Affordable Energy Contract" that encourages better and best practices in the area of energy procurement and energy use. Controlled but higher levels of expenditure on energy, on a per unit basis, will encourage the deployment of energy efficient technologies leading to temporarily elevated energy-related expenditures but greater long-term energy security. Payments for energy itself should remain within the same order of magnitude if governments, investors, and energy consumers are enabled to make strategic investments in cleaner energy. Rather than leave energy pricing to the "luck of the draw" in fossil energy discovery or clean energy innovation breakthroughs, a policy of strategic investment on all levels will insure the future affordability of energy.

In reviewing five perspectives on the price of energy, some schools of thought suggest that uninterrupted access to uniformly cheap energy is a prerequisite for civilization and economic strength, while others suggest that there is flexibility in our tolerance for higher energy costs, especially during periods of transition between energy sources. Both conventional energy economics and the pessimistic version of biophysical economics are focused on the necessity for energy to be one of the lowest cost factors. Environmental economics and a solutions-focused version of biophysical economics suggest that we have some flexibility with regard to our demand that energy be very inexpensive, though neither recommends that energy costs be "expensive." EROI provides a meaningful physical and measurable parameter for evaluating the soundness of energy investments and the cost trajectories of various energy technologies.

While economic theory is still a "work in progress," the durable teaching of most schools of economics is that "incentives matter." The demand from the side of politicians, consumers, and, as a marketing strategy, purveyors of cheaper, more polluting energy that energy must be particularly cheap is one of the main cultural and policy barriers to spurring a revolution in the production and use of energy. Without incentives to build new cleaner energy production facilities, the clean energy future that we require to meet the dual challenges of climate change and fossil resource depletion will remain stalled. To undertake this risk, producers of clean energy and clean energy technologies must be rewarded to attract further investment in this area. Suggestions by advocates and economists that public expenditures on energy research and development will solve most of our energy problems are often also efforts to circumvent this dilemma; they are looking for an energy "Hail Mary pass" in the way of a cheap and clean energy innovation breakthrough that will allow continuity in energy pricing from the fossil era to the post-fossil era. This would seem to be a "low-probability" solution if viewed as the primary strategy to solve our energy challenges in a timely manner. Our current miserly way with energy pricing stands in the way of a rapid energy transition as does a recent (1980–present) and surprising lack of political imagination with regard to the power of concerted public efforts to face massive social and economic challenges.

In addition, countries and regions will need to accept that for some energy-related projects, government subsidies are required to rapidly move to cleaner generation and a cleaner transportation system, which are part of the lifeblood of our economy and society. Large- and medium-scale infrastructure projects require the government to backstop risks and provide funding. Because subsidies have been viewed as a violation of an economic taboo, we have not developed transparent and publicly discussed monitoring criteria for the expenditure of public funds. An open acknowledgement that some projects are in our common interest will enable us to develop means to assess the progress of these projects and keep them on time and on budget.

Creating an economic theory that supports and informs the tasks ahead is one crucial step. Reflexively and continually proffering the chimerical ideal of competitive markets must be replaced with a readiness by economists to confront real world problems with tools and goals that are appropriate to the physics, psychology, and sociology of specific economic problem domains as they actually exist. The derogation of industrial policy must be curbed in favor of strategic and often time-limited support for critical industries and public works projects with instruments that are performance based and transparent. Attention to the production costs of crucial technologies and creating pricing instruments that enable the growth of these technologies while pushing their production toward greater efficiency should be part of the main edifice of a new energy economics, not an afterthought.

References

Asman L (2010) Kyrgyzstan: utility price hike squeezes citizens. Eurasiannet.org 7 Feb 2010. http://www.eurasianet.org/departments/insightb/articles/eav020810.shtml. Accessed on 30 May 2010

Battistoni A (2012) Obama punts the keystone pipeline. Salon.com 18 Jan 2012. http://www.salon.com/2012/01/18/obama_postpones_the_keystone_pipeline/. Accessed on 8 Feb 2012

Blair N, Mehos M, Christensen C (2008) Sensitivity of concentrating solar power trough performance, cost and financing with the Solar Advisor Model. Presentation at 2008 14th biennial CSP SolarPACES symposium 2008, Las Vegas

Borenstein S, Bushnell JB, Wolak FA (2002) Measuring market inefficiencies in California's restructured wholesale electricity market. Am Econ Rev 92:1376–1405

Bradley R (1997) Renewable energy: not cheap, not "green". Cato Pol Anal 280, 27 Aug 1997. Accessed from http://www.cato.org/pubs/pas/pa-280.html

Clark WW, Michael Fast (2008) Qualitative economics: toward a science of economics. Coxmoor, Longborough

Clarke E (2010) Hovering in the wings: linear Fresnel technology. CSPToday, 14 Jan 2010. FC Business Intelligence: London. http://social.csptoday.com/industry-insight/hovering-wings-linear-fresnel-technology. Accessed on 29 May 2010

Cleveland CJ (2003) Biophysical constraints to economic growth. In: Al Gobaisi D Editor-in-Chief. Encyclopedia of life support systems. EOLSS, Oxford

Cohen A (2010) Kyrgyzstan's corruption instigated revolution. Forbes.com 9 Apr 2010

Diamond J (2005) Collapse: how societies choose to fail or succeed. Viking, New York

Frondel M, Ritter N, Schmidt CM, Vance C (2009) Economic impacts from the promotion of renewable energy technologies: the German experience. Ruhr Economic Papers: 156

Georgescu-Roegen N (1971) The entropy law and the economic process. Harvard University Press, Cambridge, MA

Greenwood J (1999) The third industrial revolution: technology, productivity and income inequality. Econ Rev, Federal Reserve Bank of Cleveland, issue Q II: 2–12

Hall CA, Klitgaard K (2006) The need for a biophysical-based paradigm in economics for the second half of the age of oil. Int J Transdiscipl Res 1:4–22

Hall CA, LeClerc G (2007) Making world development work: scientific alternatives to neoclassical economics. University of New Mexico Press, Albuquerque

Hall CA, Balogh S, Murphy DJ (2009) What is the minimum EROI that a sustainable society must have? Energies 2(1):25–47

Heinberg R (2003) The party's over: oil, war and the fate of industrial societies. New Society Publishers, Gabriola Island

ICTA (1998) The real price of gasoline. International Center for Technology Assessment: Report #3

International Energy Agency (2009) World energy outlook 2009. OECD/IEA, Paris

Jha S, Quising P, Camingue S (2009) Macroeconomic uncertainties, oil subsidies, and fiscal sustainability in Asia. Asian Development Bank Economics Working paper series: no. 150

Johannson O, Schipper L (1997) Measuring the long-run fuel demand of cars. J Transp Econ Policy 31(3):277–292

Kanter J (2009) First solar claims $1-a-Watt 'Industry Milestone'. New York Times Green (Blog) 24 Feb 2009, 4:30 pm

Keen S (2004) Deregulator: judgment day for microeconomics. Util Policy 12:109–125

Koplow D (2010) EIA energy subsidy estimates: a review of assumptions and omissions. Earthtrack, Cambridge, MA, March 2010

Kunz H (2009) Energy and globalization: a fairy tale – no happy ending. 2nd biophysical economics conference, 16 Oct 2009, Syracuse, NY

Laurent C, Rickerson W, Flynn H (2010) FITness testing: exploring the myths and misconceptions about feed-in tariff policies. World Future Council, Washington, DC

Longo A, Markandya A, Petrucci M (2008) The internalization of externalities in the production of electricity: willingness to pay for the attributes of a policy for renewable energy. Ecol Econ 67:140–152

Lotka A (1925) Elements of physical biology. Williams and Wilkins, Baltimore

Mankiw G (2007) The Obama Bush Plan. Greg Mankiw's Blog, 7 May 2007. http://gregmankiw.blogspot.com/2007/05/obama-bush-plan.html. Accessed 30 May 2010

Mearns E (2008) The global energy crisis and its role in the pending collapse of the global economy. The oil drum: Europe, 3 Nov 2008. http://www.theoildrum.com/node/4712. Accessed 30 May 2010

Mendonca M (2009) Powering the green economy: the feed in tariff handbook. Earthscan, London

Mirowski P (1989) More heat than light: economics as social physics, physics as nature's economics. Cambridge University Press, Cambridge, UK

Monbiot G (2010) Are we really going to let ourselves be duped into this solar panel rip-off? UK Guardian, 1 Mar 2010

NBC News (2008) Barack Obama interview. Meet the Press, 7 Dec 2008

Owen AD (2004) Environmental externalities, market distortions and the economics of renewable energy technologies. Energy J 25:127–156

Pimentel D, Giampietro M (1990) Assessment of the energetics of human labor. Agric Ecosyst Environ 32:257–272

Porter M (1990) The competitive advantage of nations. Harv Bus Rev 68(2):73–93

Rodrik D (2008) Normalizing industrial policy. Commission on growth and development working paper no. 3, Washington, DC

Samuelson PA (1954) The pure theory of public expenditure. Rev Econ Stat 36:387–389

Sassi M (2003) The emergence of the French oil industry between the two wars. Business and economic history on-line business history conference. http://www.h-net.org/~business/bhcweb/publications/BEHonline/2003/Sassi.pdf. Accessed 30 May 2010

Smith A (1776) An inquiry into the nature and causes of the wealth of nations. W Strahan and T. Cadell, London
Soddy F (1926) Wealth virtual wealth and debt. George Allen & Unwin, London
Soros G (1987) The alchemy of finance: reading the mind of the market. Wiley, Chichester
Sterner T (2007) Environmental tax reform: the Swedish experience. Eur Environ 4:20–25
United Nations Environment Programme (2008) Reforming energy subsidies. UNEP, Paris
United States Environmental Protection Agency (2009) EPA analysis of the American clean energy and security act of 2009 H.R. 2454 in the 111th congress 23 June 2009. Accessible from http://www.epa.gov/climatechange/economics/pdfs/HR2454_Analysis.pdf
von Hayek F (1944) The road to serfdom. University of Chicago Press, Chicago
Zencey E (2009) Mr. Soddy's ecological economy. New York Times, 11 Apr 2009

Chapter 9
Economic-Environmental Performance of Micro-wind Turbine in Mediterranean Area

Nicola Cardinale, Gianluca Rospi, Giuliano Cotrufo, and Tiziana Cardinale

Abstract Through the study of wind resources for the site of the "Murgia Materana" Park, located in the Murgia plateau between Basilicata and Puglia, we deduced the potential of wind energy in this area of southern Italy, a typical Mediterranean area. The measurements, carried out over a period of 365 days, allowed to characterize and assess the windiness of the site from an energetic, economic, and environmental point of view for four different types of micro-wind turbines: two horizontal axis turbines with power of 6 and 20 kW and two vertical axis turbines of equal power. This technology, characterized by a low environmental impact, can be used to supply loads, even in areas subject to environmental constraints, with very low costs of installation and maintenance. The analysis on energy productivity, the equivalent hours of operation, the return on investment, and environmental benefits in terms of emissions of CO_2, NO_x, SO_x, and TEP, compared to conventional sources (power plant), allowed to determine which of the turbines is the one that best suited to site studied. The last issue was to assess the cost per kWh produced and compare it with other energy sources. The value obtained was competitive equaling the cost of the kWh produced by a power plant or a third-generation nuclear power plant with significant environmental benefits such as zero emissions of CO_2 and storage of nuclear waste.

DICEM – Department of European and Mediterranean Cultures: Architecture, Environment, Cultural Heritage, University of Basilicata, Via Lazazzera n.c, 75100 MATERA (ITALY). Contact e-mail: nicola.cardinale@unibas.it

N. Cardinale (✉) • G. Rospi • G. Cotrufo • T. Cardinale
University of Basilicata, Basilicata, Italy
e-mail: nicola.cardinale@unibas.it

A Short Historical Introduction on Use of Wind Energy

The principle of wind energy working is among the oldest in the world, and it is one of the easiest forms to produce energy because it is achieved by using a very old system. The wind always existed, but it has been employed in order to simplify the life of man since a few thousand years. Wind power has been widely used over the centuries for a variety of destinations, from sailing (2500 BC) to the ventilation and to the food products drying or as a driving force of mechanical utilized for many applications.

Although known since the seventh century in Iran and Afghanistan, the windmills appeared in Europe 500 years later. The first windmill was historically reported at the Caliph Umar I time (634–644) when a Persian claimed to be able to build one. The first Persian mills were realized with a vertical axis. A trunk carried by four to eight horizontal arms supported the vertical blades. The "wind energy" technology for centuries did not undergo great innovations because it never prevails over other sources of natural energy which are more easily identifiable and in any case "cheaper."

In Europe, they appeared much later, in medieval times around the twelfth century AD, as a result of the Crusades, and they were built with horizontal axis and with more complex and efficient mechanisms. In the later centuries, the windmills found in Europe had a variety of applications: grain milling, mills for crushing olives, water pumping, etc.

After the Middle Ages, when the technology moved from the Mediterranean to the northern countries, the great development of wind energy began. The blades became stronger and more resistant to withstand stronger winds present in those areas, with a sharper impact surface to the outside to balance the different tangential speeds of the blade. The tower, which supported the blades, was raised on a tripod, so as to better capture the wind action. It was well anchored to the foundation and was movable only in the upper part, where there are the supporting beam and the blades axis.

Over the years, the windmill spread throughout Europe, for example, in St. Mary at Swineshead in Lincolnshire (1,170 approx), in Normandy (1,180 approx), in Weedley in Yorkshire (1,185) and in Buckingham at the Oseney Abbey (1,189). In 1192, thanks to the crusaders, the windmill was introduced into Palestine, where it was stated as a strategic machine able to grind corn even in the besieged cities. At the beginning of the thirteenth century, there were more than 120 windmills near Ypres in Flanders.

In regions of Iran and Afghanistan, around the tenth century, a windmill with a vertical axis was present. The blades, simple wooden plates connected to the rotation axis, are enclosed within a tower. The wind, entering through windows asymmetrically arranged, has an effective action only on one of the faces and thus is able to impress the motion to the axis. From these mill towers, in the thirteenth century, Chinese windmills were developed.

Going back to Europe, the windmills assume a very important role indeed in the Francesco di Giorgio Martini studies. He described five windmills: two with blades with horizontal axis and three with impeller with vertical axis. From the drawings, the blades of these wind turbines seem to be made by wood, and the overall structure of the machine appears unrealistic. The mill assumes a helical shape with horizontal axis, with canvas paddles adjustable when it has to withstand strong winds. The shape differs somewhat from one of northern Europe mills, and probably, it did not find practical realization.

It was around 1600, however, that the engineers introduced the most sophisticated technologies. First of all, blade profiles that best exploit the lift were applied. It is not the perpendicular impact of the wind with the blades rotation plane providing more power (strength), but the side force that uses aerodynamic profiles (lift) of the blades themselves.

The invention of the dynamo by the Belgian Gramme, in the middle of the nineteenth century, opened new horizons for the use of wind and hydropower energy, and in 1887, the French Duc de La Peltrie built the first wind turbine made in Europe applied in the electricity production. It had blades of 12 m in diameter, a dynamo, and a storage battery

After the first applications, the exploitation of wind energy in industry began. In the same period, the United States realized the possibility of producing electricity from wind, and in Ohio in 1890, Charles F. Brush created a windmill to produce electricity. We can estimate that in the early years of the last century in Denmark, about 30,000 windmills of various sizes were running, with an installed power of 200 MW.

The first wind turbines to generate electricity were built in the first decade of the 1900 with machines of power between 3 and 30 kW, often poor in the design and implementation. A big technological development was made to the turbines in the period between the two World Wars, when in the United States machines for more than one MW were built. In Denmark turbines spread to cover most of the domestic electricity needs by second decade of the 21st Century. Only after the 70s years, with restrictions on petroleum products that shook the economies of industrialized countries and thanks to further research about materials and in the aeronautical field, there was a decisive revaluation of wind technology for producing electricity. Recently, there was another push towards the appreciation of the wind potential thanks to the growing attention to the environmental problems, which led many countries to make intelligent use of energy resources, especially those with reduced or zero emission (Marchis 2010).

Introduction to Technical Analysis

Wind energy is characterized by high variability in space and time. So to install wind energy conversion systems, we need to consider two fundamental aspects: the evaluation and characterization of the wind resource which are different for each

Fig. 9.1 The "Murgia Materana" site

site (Garcia et al. 1998). However, the wind energy represents, among renewable sources, the one with the highest potential of use, as it is an absolutely free resource exploitable by using a simple turbine without the high cost of installation. In Italy, the installed wind power capacity in 2009 had an increase by approximately 40% compared with 2008, reaching an overall power of 4,898 MW. The Italian wind farm currently consists of over 4,250 turbines primarily concentrated in southern Italy, which represent 88% of national installed capacity (Statistical report wind 2010).

The wind turbine is an open flow fluid dynamics machine that converts the kinetic energy of a flow of air into rotational mechanical energy; it can be with a horizontal axis (HAWT = Horizontal Axis Wind Turbine) or with a vertical axis (VAWT = Vertical Axis Wind Turbine). The efficiency of an ideal machine is always less than 60% and, to this nonviscous loss, you have to add dissipative losses due to friction phenomena and wake vorticity. As regards the aerodynamic power output by the blades, this is obtained through the following formula $P_r = \frac{1}{8}\rho_0 C_p \pi D^2 V_0^3$ (Pallabazzer 2004). In practice, to optimize the energy production from a wind machine and reduce the power generation cost is essential to describe the wind variation at the project site; this is usually described by the Weibull mathematical model (Shabbaneh and Hasan 1997; Mayhoub and Azzam 1997).

The research described in this chapter concerns the study of the wind resource for the site of "Murgia Materana" Park, located within the municipal territory of the Matera city in southern Italy (Fig. 9.1). The wind monitoring, done at a height of 20 m above the ground, lasted 365 days. Through the data detected in the measurement campaign, we described the speed trend obtaining the Weibull distribution (Pallabazzer 2004). Later it was possible to derive the wind energy potential of the site studied.

Methodology of Analysis

The area of investigation

The analysis, conducted through the installation of a monitoring station at 20 m height from the ground (average height of the micro-wind turbines) composed of a multi logger and an anemometric probe, had a duration of about 365 days with a data acquisition interval of 10 min. The data processed were approximately 52,560, and the elaborations enabled us to evaluate the wind resource of the site, the energy producibility, the environmental impact, and the economic cost-benefit analysis of four different turbines. The turbines analyzed were four: two with horizontal axis with a power of 6 and 20 kW and two with vertical axis of the same power.

Analysis of the Wind

Considering that the wind resource is characterized by a random distribution in time, to make a detailed anemological analysis, it is necessary to use a statistic analysis. The measurements were carried out through the installation of an anemometer at a height of 20 m and considering an interval of acquisition of 10 min. The speed data obtained in situ, in number of a 52,560, were then processed using a spreadsheet. In this way, it was possible to obtain the hourly average speed v_h, the daily averages v_d, the monthly averages v_m, and the annual average v_y; finally, for periods of many years, we calculated the historical speed v, which represents the global magnitude characteristic of the site. Then we derived the average values the of the wind direction and of the wind rose. The average speed, for a continuous function in time, defined by the integral of the instantaneous velocities v_i measured in a given time interval, can be summarized by following formula:

$$v_m = \int_0^T v_i dt \qquad (9.1)$$

The average v_m can be hourly, daily, monthly, or yearly depending on the time T. In this case the v_m shows the hourly average. In reality, the average speed is calculated by a certain finite number N of average data v_j calculated in a finite range of time:

$$v_m = \frac{1}{N}\sum_{j=1}^{N} v_j \qquad (9.2)$$

The average speed v_m is not an exact value, but an approximate value, whose accuracy is inversely proportional to the Δt range. But it is not a sufficient parameter to define the state of a wind site, and for more accurate analysis we must derive the probability distribution function f(v).

The probability that the wind blows at a speed between v_{min} and v_{max} of the potential range is given by the following formula:

$$f_{v\min<v<v\max} = \frac{N_j}{N} \quad (9.3)$$

where Nj represents the number of times in which the speed is included in the range of v_{min} and v_{max} and N represents the total number of reliefs. The diagram is built by splitting the wind speed domain in ranges of a certain amplitude (usually 0.5 or 1 m/s). In the case described here, a range of 0.5 m/s was considered, and for each interval, we calculated the number of times which a wind of that intensity occurs in the time period T.

From the probability distribution f(v), it was possible to obtain the frequency of all the speeds that are smaller or greater than a given value v_o (v_o is the cut-in speed); this is obtained in the first case summing all speed frequencies which respond to the condition $v \leq v_o$ and in the second case summing all the speed frequencies which correspond to the condition $v \geq v_o$.

Repeating the calculation for each value of speed, you can get two histograms: the cumulative distribution of the frequencies C(v) and the distribution of duration D(v) which provide for each value of v the probability that the speed is smaller or greater than v.

Through the study of these two functions it was possible to verify the behavior of the wind turbines as a function of wind because they provide the intensity wind percentage smaller than the cut-in speed and greater than cut-out speed of the turbine.

$$C(v) = \frac{K}{V}\left(\frac{V}{V_m}\right)^k \Gamma^k\left(1+\frac{1}{k}\right)\exp\left[\left(\frac{-V}{V_m}\right)^k \Gamma^k\left(1+\frac{1}{k}\right)\right] \quad (9.4)$$

$$D(v) = 1 - \exp\left[\left(\frac{-V}{V_m}\right)^k \Gamma^k\left(1+\frac{1}{k}\right)\right] \quad (9.5)$$

Another important parameter, to obtain a good anemological classification of the site, is the power density, defined by the formula:

$$p = \sum \rho f_i \frac{v_i^3}{2} \quad (9.6)$$

The power density represents the flow average power per unit of rotor swept area. Part of this power is converted from the rotor into mechanical power available to the axis of the generator for conversion to electrical energy (Caffarelli et al. 2009). To obtain a correct simulation of the statistical probability distribution of the wind, the Weibull mathematical model was used. Fixed a range of a certain amplitude, the probability that the wind speed appears between the minimum and maximum values in the considered range is given by the following equation:

$$f(v) = \frac{k}{c} \left(\frac{V_i}{c}\right)^{k-1} e^{\left(\frac{-V_i}{c}\right)^k} \qquad (9.7)$$

In the expression, v is the central value in the range between v_{min} and v_{max}, while the parameters k and c are called respectively shape parameter (dimensionless) and scale parameter (m/s) (Deaves and Lines 1997). Both k and c are indicators that vary as a function of windiness. In fact k reaches values close to 1.5 in mountainous areas, close to 2 in coastal areas and temperate climates, and close to 4 in monsoon areas (areas with regular winds). The more the value of k is close to 2, the more the shape of the curve will be that of a Gaussian.

The more the values of scale parameter c are smaller, the more the curve will be concentrated around its peak value. Once parameters k and c are known, you can express analytically the functions C (v) and D (v) and the average speed function through the expression:

$$v_m = \frac{1}{T}\int_T v_i \, dt = \int_0^\infty v f(v) \, dv = \frac{k}{c}\int_0^\infty v \left(\frac{v}{c}\right)^{k-1} \exp\left[-\left(\frac{v}{c}\right)^k\right] dv \qquad (9.8)$$

This integral can be solved through the gamma function defined as follows:

$$\Gamma(x) = \int_0^\infty y^{x-1} e^{-y} \, dy \qquad (9.9)$$

Knowing the values of average speed and indicating the variable x as follows:

$$x = \left(1 + \frac{1}{k}\right) \qquad (9.10)$$

it is possible to know the value of the scale parameter c, evaluated through the expression:

$$c = \frac{v_m}{\Gamma\left(1 + \dfrac{1}{k}\right)} \qquad (9.11)$$

The equations containing the gamma functions can be approximated by some empirical expressions (Pallabazzer 2004). This method provides to calculate the value of relative standard deviation σ_r by the relation:

$$\sigma_r = \left[\frac{1}{N}\sum\left(\frac{v_i}{v_m} - 1\right)^2\right]^{\cdot} \qquad (9.12)$$

Calculated the value of the relative standard deviation σ_r it is possible to calculate the value of shape parameter k from the following relationship:

$$k = 0,9847\sigma_r^{-1,0966} \qquad (9.13)$$

The value of the gamma function is calculated by the formula:

$$\Gamma\left(1+\frac{1}{k}\right) = \left[0,568 + \frac{0,434}{k}\right]^{\frac{1}{k}} \qquad (9.14)$$

while the value of irregularity or fluctuation is calculated in the following way:

$$k_f = 0,9794 + 0,11339\sigma_r + 0,7068\sigma_r^2 \qquad (9.15)$$

The previous expressions evaluate with good approximation the actual values, in the field 1<k<4, which covers virtually all the real cases.

Calculated the scale parameter c is possible to calculate the cumulative distribution C (v) and the duration distribution D (v).

Analysis of Energy Wind Turbines

Our research studied the producibility of four small wind turbines, two with horizontal axis (power 6 and 20 kW) and two with vertical axis of the same power. All four turbines are classified as micro-wind turbines and can be connected directly to the national electricity grid. The analysis calculated the gross and net energy and the equivalent hours for each wind turbine in order to determine which machine is the best for the selected site. The total amount of energy E produced by a wind generator is obtained by the following formula:

$$E = \int_{v_i}^{v_o} W(v) f(v) dv \qquad (9.16)$$

where the integration limits are the cut-in speed (v_i) and the cutoff speed (v_o) that define a range of wind speed inside which the wind turbine operates, W(v) is the turbine power that varies in relation to the speed, and f(v) is the Weibull function, which expresses the time in which the wind has been blowing at a certain speed and that depends on the actual operating conditions of the turbine.

Another method to determine the amount of energy produced, if you have access to a discrete number of measurements in the field, is to schematize the area subtended by the curve of the Weibull function as a summation of many small rectangles of height equal to the number of speed values and base equal to the sampling interval of themselves. In our case study, a base equal to 0.5 was considered. By adopting this system, the equation can be transformed in the following formula:

$$E = \Delta v \sum_{j=v_i}^{v_o} W_j n_{hj} \qquad (9.17)$$

where Δv is the amplitude of the speed range and W_i is the power developed by the wind turbine for the given wind speed. The calculation is much more accurate the smaller the sampling range of the wind speed.

The previous energy production was gross. Then to consider the net production, the previous value was reduced of 10% due to different losses (Bartolazzi 2005).

The last analysis was to calculate the value of the equivalent hours, defined as the period of time in which you assume that the turbine operates in nominal conditions; this value is obtained by dividing the kWh generated by the nominal power of the turbine. The value thus calculated is useful to evaluate the return period of a possible investment.

Economic and Environmental Analysis

This type of analysis was done considering an average time of 20 years for the life-cycle turbine. The value takes into account the physical duration of the machine and the technological and commercial life of the product. The calculation was made considering a cost of installation and ordinary maintenance of 200.00 €/year and the revenues from the energy sale in the case of a fixed sales price for 15 years of 0.3 €/kWh (value of government incentive called all-inclusive tariff) and a price of 0.18 €/kWh from 16° to 20° year (revenues obtained by the mechanism of locally energy exchange)(http://www.gse.it/attivita/Incentivazioni%20Fonti%20Rinnovabili/Servizi/Pagine/Tariffaonnicomprensiva.aspx).

The interest rate r used for the purposes of the financial calculations derives from the combination of the money cost i (fraction of capital invested in relation to a given period of time, determined by the laws of supply and demand) and the inflation rate j. In the economic calculation, we considered a value of interest rate equal to 1%, percentage detected in September 2010 from the official website of the Central European Bank. The equation is the following:

$$r = (i + j) = 1\% \qquad (9.18)$$

The economic analysis was performed by calculating the Net Present Value (NPV), and Pay Back Period (PBP). The NPV returns the cash flows discounting them on the basis of return rate after a period of 20 years, while calculating the PBP it is possible to know how long after the initial investment falls (Awerbuch 2003; Martinot 2003).

After performing the economic analysis it was possible to calculate the environmental benefit that different wind turbines produce, compared to the same energy produced from conventional sources (e.g., thermoelectric power plants). The environment indicators calculated are avoided emissions of carbon dioxide CO_2, nitrogen oxides NO_x, and sulfur oxides So_x and saved tons of equivalent oil TEP.

Results Discussion

Wind Analysis

Table 9.1 shows the mean, median, mode, maximum, and minimum values for the different months of the year. As we can be seen from mode value, the wind resource is more distributed in the winter months, reaching values of around 7 m/s in January and December.

From the analysis of the frequencies distribution as a function of different classes of wind speed, we can deduce that higher frequencies correspond to classes that are from 2 m/s to 4.5 m/s.

The frequencies values can be read in Fig. 9.2 as a function of speed and wind hours. From observation of the figure, you can note which are the most present speeds during the entire observation period (range between 2 and 4.5 m/s).

Figure 9.3 indicates, however, the interpolation of the speed cumulative distribution curve $C(v)$ and the duration cumulative distribution curve of $D(v)$. These functions are useful to study the behavior of wind turbines because they provide the percentage of wind intensity smaller than the cut-in speed or greater than the cut-out speed and the characteristics of the machine.

Table 9.1 Analysis of statistical data

Statistical analysis					
Month	Mean	Median	Mode	Minimum	Maximum
January	5.6	5.6	6.7	0	15.1
February	5.2	5.2	4.8	0	16.2
March	4.5	4.1	4.3	0	14.6
April	4.1	3.6	2.5	0	13.6
May	4.2	3.9	2.2	0	16.3
June	4.1	3.7	2.2	0	12
July	3.5	3.3	1.3	0.1	11.7
August	3	2.7	1.7	0.1	12.8
September	3.8	3.3	2.5	0	13.2
October	4.2	3.8	3.7	0	17.3
November	3.4	3	1.9	0	12.9
December	5.2	5.2	7	0	14.8

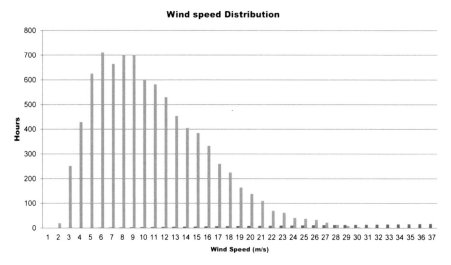

Fig. 9.2 Wind speed distribution

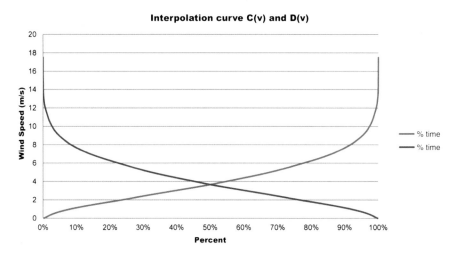

Fig. 9.3 Interpolation curve C(v) and D(v)

The curve cumulative gives us the time percentage in which the turbine does not work because the wind speed is lower than the starting speed of wind turbine, while the duration curve gives us the percentage of nonoperating time because of too high speed. Table 9.2 shows the percentages of time in which the different wind machines do not work. These percentages refer to the available time and are not energy percentages.

Table 9.2 The percentages of time in which the wind machine does not work

Turbine (kW)	Percentage of time with V < V cut-in (%)	Percentage of time with V > V cut-out
Vertical axis (energy power 6)	40	Regular
Vertical axis (energy power 20)	40	Regular
Horizontal axis (energy power 6)	32	Regular
Horizontal axis (energy power 6)	48	Regular

Table 9.3 Weibull coefficient

	Shape parameter (k)	Irregular coefficient (k_f)	Scale parameter (c)
January	2.18	1.2	6.37
February	2.09	1.21	5.96
March	1.7	1.3	5
April	1.8	1.3	4.6
May	1.8	1.3	4.7
June	1.8	1.3	4.6
July	1.93	1.25	3.9
August	1.83	1.27	3.42
September	1.6	1.35	4.2
October	1.53	1.37	4.71
November	1.48	1.39	3.79
December	2.07	1.22	5.88
Year	1.66	1.32	4.65

Even if we talk about time percentages, through Table 9.2, it is possible to obtain a first estimate on the producibility of single wind turbines. It is easy to see that the turbine with a horizontal axis of power 6 kW works more hours in a year.

The analysis of the Weibull model allowed to schematize the power of the wind resource in the site. Through Eqs. 9.13, 9.15, and 9.17, it was possible to calculate the Weibull parameters: the scale parameter c, the shape parameter k, and the irregularities coefficient k_f, summarized in the Table 9.3 for each month of the year.

The calculated average annual value of k was equal to 1.66, and this sums up the orography of the site studied, characterized by temperate climate and internal hilly area.

The annual average value of c is resulted equal to 4.65 m/s; this is representative of a curve with a flat average trend, preventing the achievement of high peak speed.

As for the mode values the winter months are characterized by high values of k, which are around 2, and c, which is around 6 m/s, allowing in this case the achieving of higher peak speed.

The Fig. 9.4 indicates the trend of the Weibull function, obtained from the calculation, in the different months of the year. From the analysis of single curves, it was possible verified that the two seasons that have a Weibull probability distribution

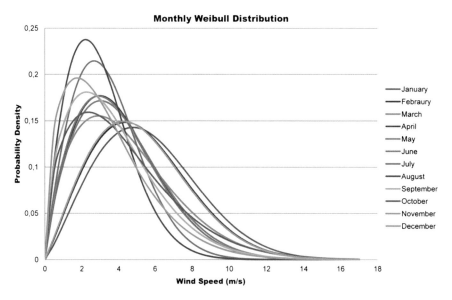

Fig. 9.4 Monthly Weibull distribution

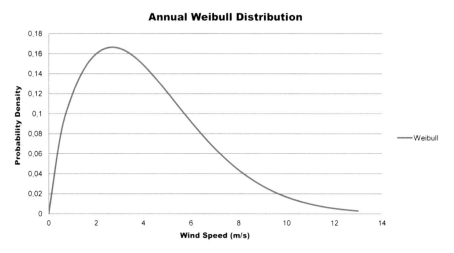

Fig. 9.5 Annual Weibull distribution

that covers high enough speed values are the spring and the winter, while autumn and summer seasons have a peak speed focused on lower wind values. It is important to specify that the annual average parameters are placed in an average situation relative to the four seasons (Fig. 9.5)

The final analysis concerns the calculation of the wind. This indicates the number of velocity data contained in different speed classes and the sectors in which the azimuthal quadrant was divided. Through this test, it was found that calm values are equal to 4.2% of total values.

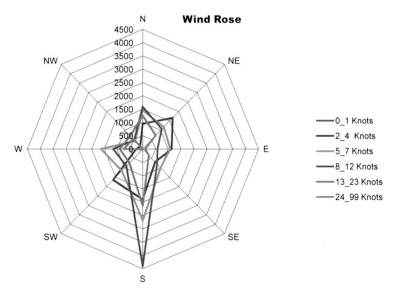

Fig. 9.6 Wind increases in strength

The Fig. 9.6 is a diagram of the wind rose in which you can see the absence of dominant winds, since there was no intensity greater than 20 m/s. Finally, by the same diagram, you can see that the prevailing winds are those from the southeast quadrant with speed of 8–12 knots.

Energy Analysis

Thanks to the elaborations of experimental data taken in situ, it was possible to calculate the energy production for each turbine. The turbines studied were 4, two with a vertical axis with power of 6 kW and 20 kW and two with a horizontal axis with a power of 6 kW and 20 kW, characterized by the power curves represented in Figs. 9.7 and 9.8.

The calculation of the producibility was done by multiplying the relative frequencies of the speed classes for the corresponding power value generated by the turbine. An example of calculation is shown in Table 9.4: it is representative of the turbine with a horizontal axis with a power of 6 kW, where the frequencies values, expressed in minutes, were first transformed into hour and then multiplied by the respective power values obtained from the machine power curve. In this way, it was possible to obtain directly the energy value in Wh for each speed class. The production in Wh is gross, so to consider the net production, the final value was reduced by 10%.

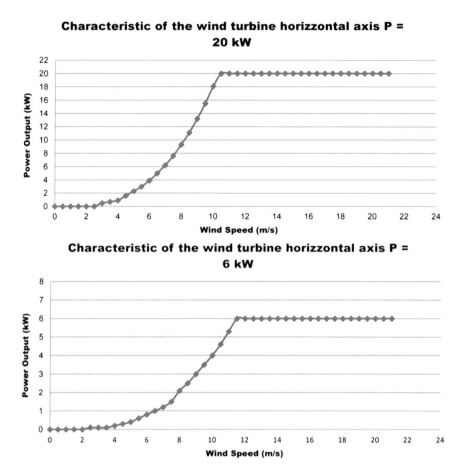

Fig. 9.7 Characteristic of the wind turbine with horizontal axis (20 and 6 kW)

Table 9.5, instead, summarizes the gross producibility for each wind machine; from the table, you can see immediately that at the same power the producibility of horizontal axis turbines is considerably greater.

To evaluate the return on investment, the most important data is the number of equivalent hours, defined as the time period during which it is assumed that the turbine operates in nominal conditions.

Table 9.6 shows the value of equivalent monthly hours and the annual total for each turbine. Through this calculation, it was possible to obtain that the turbine with a horizontal axis of power 6 kW is one that has a greater number of equivalent hours, even almost double compared to others studied.

So we can say that the turbine which best suits the site studied is that with the horizontal axis and a power of 6 kW. This result is certainly due to the power curve and the low speed cut-in.

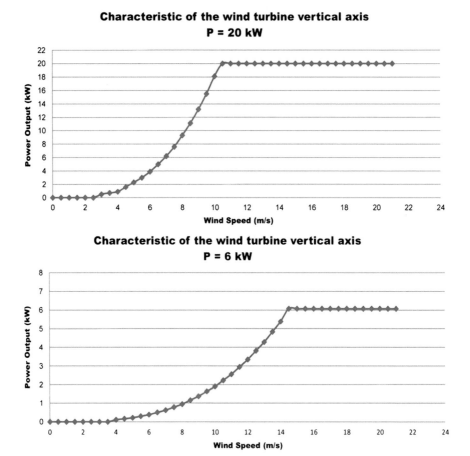

Fig. 9.8 Characteristic of the wind turbine with vertical axis (20 and 6 kW)

Economic and Environmental Analysis

This type of analysis was done considering a turbine average life of 20 years. The calculation was made considering the cost of original equipment and ordinary maintenance and revenues from energy sales. The economic analysis was performed by calculating the following indices: NPV and PBP, where the first returns the discounted cash flow after a period of 20 years and the second indicates the payback time.

Regarding the first installation cost, the average price supplied directly by several European manufacturers was considered. The values considered in the calculations are summarized in Table 9.7.

9 Economic-Environmental Performance...

Table 9.4 Wind energy generation from turbine with horizontal axis – 6 kW power

Wind speed (m/s)	Power (W)	Wind frequency in the site			Energy generated (Wh)
		%	min-wind	hour-wind	
0	0	0.23	119	19.83	0
0.5	0	2.93	513	252.17	0
1	0	4.99	2,576	429.33	0
1.5	0	7.28	3,756	626	0
2	0	8.27	4,269	711.5	0
2.5	100	7.73	3,989	664.83	66,483
3	200	8.16	4,210	701.67	140,334
3.5	300	8.15	4,205	700.83	210,249
4	450	7	3,614	602.33	271,048.5
4.5	700	6.77	3,495	582.5	407,750
5	1,000	6.17	3,184	530.67	530,670
5.5	1,300	5.29	2,730	455	591,500
6	1,550	4.72	2,435	405.83	629,036.5
6.5	1,800	4.48	2,311	385.17	693,306
7	2,100	3.87	2,000	333.33	699,993
7.5	2,500	3.03	1,562	260.33	650,825
8	3,000	2.62	1,355	225.83	677,490
8.5	3,450	1.92	989	164.83	568,663.5
9	4,000	1.61	831	138.5	554,000
9.5	4,500	1.28	663	110.5	497,250
10	5,050	0.83	426	71	358,550
10.5	5,500	0.73	376	62.67	344,685
11	5,950	0.48	250	41.67	247,936.5
11.5	6,200	0.44	228	38	235,600
12	6,400	0.39	203	33.83	216,512
12.5	6,450	0.27	140	23.33	150,478.5
13	6,400	0.16	85	14.17	90,688
13.5	6,300	0.11	56	9.33	58,779
14	6,200	0.04	23	3.83	23,746
14.5	6,150	0.02	12	2	12,300
15	6,100	0.02	8	1.33	8,113
15.5	6,050	0.01	6	1	6,050
16	6,000	0	1	0.17	1,020
16.5	6,000	0	2	0.33	1,980
17	6,000	0	0	0	0
17.5	6,000	0	1	0.17	1,020
					8,946,056.5

The cash flow for each case study returned the values listed in Table 9.8.

The results of economic analysis, shown in Table 9.9, calculated for each turbine, enabled us to evaluate the economic performance of different turbines and which turbine produces the best economic benefit.

Table 9.5 Comparison of different turbines producibility

	Net energy generated (Wh)			
	Vertical axis 6 kW	Vertical axis 20 kW	Horizontal axis 6 kW	Horizontal axis 20 kW
January	433.1	2,262.6	1,218.4	2,569.7
February	331.7	1,748.5	965.5	1,974
March	287.4	1,511.6	808.7	1,670.7
April	191.1	1,062.6	607	1,154.6
May	211.5	1,173.6	681.5	1,291.6
June	178	1,005.2	591.4	1,096
July	112.2	676.5	410.3	697.2
August	80.1	509	298.6	498.2
September	173.6	960.2	525.5	1,037.4
October	281	1,448.3	776.8	1,621.7
November	319.4	1,666.85	906.6	1,870.15
December	357.8	1,885.4	1,036.4	2,118.6
Year	2,956.9	15,910.35	8,826.7	17,599.85

Table 9.6 Equivalent hours

	Equivalent hours			
	Vertical axis 6 kW	Vertical axis 20 kW	Horizontal axis 6 kW	Horizontal axis 20 kW
January	72.2	113.1	203.1	128.5
February	55.3	87.4	160.9	98.7
March	47.9	75.6	134.8	83.5
April	31.9	53.1	101.2	57.7
May	35.3	58.7	113.6	64.6
June	29.7	50.3	98.6	54.8
July	18.7	33.8	68.4	34.9
August	25.5	25.5	49.8	24.9
September	48	48	87.6	51.9
October	46.8	72.4	129.5	81.1
November	53.2	83.35	151.1	93.5
December	59.6	94.3	172.7	105.9
Year	524.1	795.55	1,471.3	880

After performing the economic analysis, it was possible to calculate the environmental benefit in terms of avoided emissions of carbon dioxide CO_2, nitrogen oxides NOx, and sulfur oxides SOx and equivalent oil tons TEP saved in comparison to the same energy from conventional sources (steam turbine). The results were summarized in Table 9.10.

Table 9.7 Start-up cost

Turbine (kW)	Total cost system (€)
Vertical axis 6	20,060
Vertical axis 20	54,315
Horizontal axis 6	17,833
Horizontal axis 20	42,925

Table 9.8 Cash flow

Cash flow		
Vertical axis 6 kW	743.4	€/year
Vertical axis 20 kW	4,573.3	€/year
Horizontal axis 6 kW	2,448.3	€/year
Horizontal axis 20 kW	5,080	€/year

Table 9.9 Cost analysis

Turbine (kW)	PBP (year)	VAN (€)
Vertical axis 6	////	8,223
Vertical axis 20	14	20,231
Horizontal axis 6	8	21,941
Horizontal axis 20	10	39,917

Table 9.10 Avoided emissions

Turbine (kW)	CO_2 (Kg)	Nox (Kg)	Sox (Kg)	TEP/year
Vertical axis 6	1,812	4.2	3.8	419.1
Vertical axis 20	10,369.4	23.1	21.9	2,447.6
Horizontal axis 6	5,664.3	14.5	12	1,315.5
Horizontal axis 20	11,219.2	24.8	23.7	2,631.1

Evaluation of the Cost per Kwh

The final analysis was to evaluate the cost per kWh produced and compare it with other energy sources. Knowing the total cost of the system, it was possible to determine the average cost per kWh produced; this was calculated from the total energy produced by the plant in its lifetime (30 years), obtained by multiplying the annual production by the number of useful life years. Indeed, applying the formula described below, you can get the cost per kWh produced by the micro-wind source.

Cost per kWh generated = (system total cost)/(energy life cycle)

Table 9.11 shows total costs of each turbine, including maintenance costs and connection costs to the grid (equivalent to 30 Euros per year according to Italian AEEG Resolution 28/06).

Table 9.11 Cost per kWh

Turbine (kW)	Total cost system	Energy life cycle (kWh)	Cost per kWh generated (€/kWh)
Vertical axis 6	20,060	88,710	0.226
Vertical axis 20	54,315	477,310.5	0.114
Horizontal axis 6	17,833	264,834	0.067
Horizontal axis 20	42,925	528,000	0.081

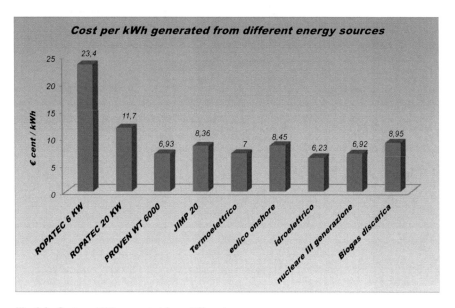

Fig. 9.9 Cost per kWh generated from different energy sources

From the analysis of Table 9.11, we note that the four turbines have a very different cost per kWh; it depends on the capability to exploit the wind resource of the site under study. The only turbine which cost per kWh produced is competitive compared to that produced by traditional methods (thermoelectric) is the horizontal axis turbine power 6 kW with a value of 6 € cents/kWh. Regarding the cost per kWh of renewable sources, the values obtained through the methodology SETIS of the European Commission were considered (https://odin.jrc.ec.europa.eu/SETIS/SETIS1.html).

The analysis of Fig. 9.9 allows us to do some important considerations. First of all, an appropriate choice of wind turbine allows us to obtain produced costs per kWh lowest of onshore wind turbines (turbine with a horizontal axis of 6 kW power) and also almost identical to those of thermal power plants and third generation nuclear plants. Another result is to have no emission of CO_2 into the environment compared to thermal power plants (thus no carbon tax) and no storage of dangerous waste due to the electricity production from nuclear power plants.

Conclusions

The wind analysis performed on the site of the "Murgia Materana" Park allowed to conclude that the site studied has a fairly good wind potential, and this is greater in winter, because of wind speeds greater than 5 m/s. The comparison made with the Italian wind atlas has allowed us to find that the site annual average speed is lower (value around 4 m/s), demonstrating that the analysis carried out on an annual basis are essential in order to obtain a correct characterization of the wind on the site.

The average-low value of the frequency distribution of wind speed makes the site more suitable for installation of micro turbines with low power. This is confirmed by the calculation of the equivalent hours, from which it was shown that the turbine with a horizontal axis with a 6 kW power is the best for the site studied.

The use of this technology, especially in natural reserves, can be useful for supplying loads with very low costs of installation and maintenance. In fact, choosing appropriate technologies, you can fall in investment in just 8 years.

Another result was to demonstrate that the turbines with horizontal axis have better performance than turbine with vertical axis; this occurs at expense of a greater visual impact, even if it is limited compared to large wind turbines.

The last analysis carried out was about the study of the energy cost produced by a micro-wind source. This allowed to verify that the turbines analyzed have a cost per kWh produced that varies considerably depending on the technology and the power. In addition, the turbine with horizontal axis, in particular with a 6-kW power, has a cost of kWh produced that is close to that produced by thermal power plants or nuclear third-generation power plants. It has the environmental benefit to remove both CO_2 emissions (compared to the thermoelectric power plants) and the storage of nuclear waste (compared to nuclear power plants).

References

Awerbuch S (2003) Determining the real cost – why the renewable power is more cost-competitive than previously believed. Renew Energy World 6(2):53–61

Bartolazzi A (2005) Renewable energy. Ed. Hoepli. (in Italian)

Caffarelli A, De Simone G, Stizza M, D'Amato A, Vergelli V (2009) Wind systems: design and economic evaluation. Ed. Maggioli. (in Italian)

Deaves DM, Lines IG (1997) On the fitting of low mean windspeed data to the Weibull distribution. J Wind Eng Ind Aerodyn 66:169–178

Garcia A, Torres JL, Prieto E, De Francisco A (1998) Fitting probability density distributions: a case study. Solar Energy 62:139–144

http://www.gse.it/attivita/Incentivazioni%20Fonti%20Rinnovabili/Servizi/Pagine/Tariffaonnicomprensiva.aspx (in Italian)

https://odin.jrc.ec.europa.eu/SETIS/SETIS1.html

Marchis V (2010) History of machines-Three millennia of technological culture. Ed.Laterza (in Italian)

Martinot E (2003) Renewable energy in developing countries – lessons for the market. Renew Energy World 6(4):102–113

Mayhoub AB, Azzam A (1997) A survey on the assessment of wind energy potential in Egypt. Renew Energy 11:235–247
Pallabazzer R (2004) Wind systems. Ed. Rubbettino. (in Italian)
Shabbaneh R, Hasan A (1997) Wind energy potential in Palestine. Renew Energy 11:479–483
Statistical report wind (2010). Gestore Servizi Energetici GSE. www.gse.it (in Italian)

Chapter 10
Energy Conservation for Optimum Economic Analysis

Stephen C. Prey[*]

Abstract This chapter looks at a systematic approach to determine energy conservation project cost effectiveness. It discusses various types of analysis methodologies. And in the analysis process of multiple conservation measures (ECMs), look for any relationships between the various ECMs.

This chapter then discusses how to preform a reduction analysis process on multiple ECMs to develop an optimized investment plan.

The last part of this chapter focuses on using life-cycle cost analysis (LCCA) to develop energy and cost savings data as they relate to converting high- pressure sodium (HPS) roadway and facility exterior lighting systems to an more energy efficient light-emitting diode (LED) technologies. The LCCA example looks at the costs and energy usage over twenty years from both the base case (normal business or the "do nothing" scenario versus conversion of all California State maintained roadway and maintenance facilities).

All statements and data contained in this chapter are based upon 40+ years of developing conservation training and successful implementation of various conservation projects and programs, in the private sector (Honeywell Inc., private consulting) and with the State of California (California Energy Commission, CalParks, and the Caltrans).

---

[*]Steve C. Prey, retired in 2012 as Statewide Energy Conservation Program Coordinator for the State of California Department of Transportation (Caltrans) since 1983, Chair "Emeritus" member of the California State Government Energy Policy Advisory Committee, technical advisor to the Sandia National Labs Special Projects Portable Hydrogen Generation Product team, and member of the National Academies of Science, Transportation Research Board Special Task Force on Energy and Climate Change. Prey can be reached at: steveprey@earthlink.net

S.C. Prey (✉)
Sacramento, CA, USA
e-mail: steveprey@earthlink.net

Introduction

This chapter looks at basic energy conservation analysis and how to optimize multiple conservation and efficiency opportunities that have direct relationships and interconnections. Where applicable, simple examples will be included at the end of each discussion section. All statements contained in this chapter are based upon 40+ years of developing conservation training and successful implementation of various conservation projects and programs, in the private sector (Honeywell Inc., private consulting) and with the State of California (California Energy Commission, CalParks, and the Caltrans).

Light emitting diode (LED) technology energy conservation, efficiency applications will be used as the focus for most of the sectional examples discussed in the chapter, and the principles discussed here can be applied to any conservation opportunity.

LED technologies have the potential to drastically change world lighting energy demands while, at the same time, offer better levels of service when artificial lighting application is required. LED lighting sources can be cycled on and off with no apparent degradation in life cycle. LEDs are instant on and have proven reliability by their deployment in traffic signals since the mid 1990s. Along with trends of higher light output per energy consumed, LED manufacturers can now design spectrum-specific products to meet greater number of applications.

An example of how Caltrans developed LED technology and applied it to traditional incandescent traffic signals is as follows: The research in how to apply LEDs to traffic signals started in Fresno California in 1991 when lack of reliable power at a remote intersection kept overloading utility circuits. Caltrans staff contacted a few vendors asking for technical assistance and potential solutions, and LED seemed the best option, rather those neon technologies which had fragility issues. By 1992, prototypes were being tested at Caltrans along with a number of other state/county/city transportation departments and a few electric utilities.

Background

Caltrans' Energy Conservation Program was brought into the project and funding was provided for human factor studies at the University of California at Berkeley along with large-scale red LED signal field tests. In 1995, the State Energy Commission and the US Department of Energy awarded the Caltrans' LED traffic signal project in Fresno with their respective annual energy awards. By 2000, Caltrans adopted performance standards for all traffic signal types and colors and proceeded to upgrade all state-owned and state-maintained intersection. It should be noted that at that time, the Caltrans performance specs were adopted by the LED manufacturer industry as the de facto "national product specifications."

Once statewide deployment occurred in the early 2000s, the energy/carbon footprint for the statewide owned/maintained traffic signal upgrade exceeded 93% reduction from the incandescent baseline. For Caltrans that meant about 13 MW of grid load was eliminated daily. Estimating that Caltrans has about 1% of the signalized intersections in the US, the national grid impact was at a level of one electric power plant available for other use. Return on investment (ROI) when utility rebate or incentives were applied often averaged less than 3 years or much faster in high electric rate service areas.

Side benefits meant longer intervals between lamp replacements (once a year out to 12 years), so limited resources could be re-tasked to other work as well as reduced service vehicle fuel usage. And, given much lower wattage use at the intersection, battery backup systems were installed statewide which allow for normal intersection operation during loss of utility power and the possibility to deploy signalized intersections in areas of no grid power through the use of renewable power systems.

So great was the impact of the LED application; many states outlawed the sale of incandescent lamps as early as 2003 such as in California until the passage of the US National Energy Act of 2005 which banned incandescent traffic lamp sales throughout the United States.

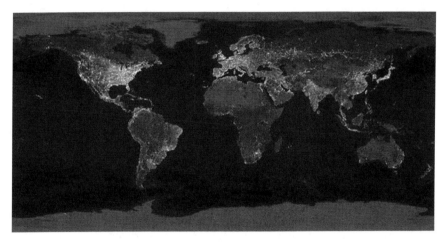

A typical example of one of many web sites would be http://apod.nasa.gov/apod/ap001127.html hosted by NASA

This was only the official start of applied LED technologies, finding their way to almost all levels of the world economies, new jobs, resource savings, and new opportunities for all those lights used in the world that light up the planet at night.

Since 2001, LED roadway lighting applications have been at various stages of development. Applications like roadway surface lighting, roadway lane delineation, flashing beacons, and sign lighting were all being investigated. Developing reliable product performance standards, human factor/vision research, higher efficiency LEDs (increasing light output and reducing energy consumed) and reducing the cost to deploy products are all elements needed in order to reach an optimum economic trigger point.

From 2011–2012 is the point in time where the economic trigger point for deployment of roadway surface lighting systems moved from the development and discussion phase to that of a deployment phase. During these 24 months, a number of state and local government Department of Transportation (or DOT) and utilities are funding deployment of LED roadway lighting projects. Caltrans was one of the early project developers. A number of years working with California electrical utilities, local government partners, national lighting standard committees, and lighting manufactures resulted in Caltrans adopting LED roadway performance specifications in December 2009. Funding analysis for statewide deployment started in the late 2000s through the summer of 2011 when costs to deploy large-scale projects became more realistic. Favorable funding options presented themselves in the winter of 2011/2012, resulting in departmental redirection of major funds in February 2012, procurement process starting in May, development of a qualifying products list (QPL) and request for bid (RFB) published in June, and bid opening in July with delivery expected to start in August 2012.

The examples discussed in the remainder of this chapter are extracted from those processes leading up to the "buy and deploy" decision by Caltrans management.[1]

Basic Energy Conservation Calculation Process

Conservation calculations have two primary components: the rate of consumption and the duration of the consumption, in short, "load" and "time." The process of determining the level of setting up the basic energy conservation calculation requires the user to fully understand the energy system and what is being analyzed. This needs to be done in order to determine how far the study needs to go when looking at conservation options.

Example
Changing out a light fixture to a more efficient high-pressure sodium lighting unit: do you just look for a higher efficiency lamp or ballast; look for alternative lighting technologies; or ask the most probing question "Why do you need the light in the first place, and why not just remove it all together?" This last question should always be asked *first* and more discussion a bit later.

Simple Loads

Simple loads are typically turned "on" and "off," so the load uses maximum energy when on and no energy when off.

Example
Examples might be light fixtures, pumps, fans, motors, computers, and equipment.

---

[1] For data updates on this project, please contact Gonzalo Gomez who is the LED roadway lighting project manager, as of July 1, 2012: gonzalo_gomez@dot.ca.gov.

10   Energy Conservation for Optimum Economic Analysis

Variable Loads

Another form of load gets a bit more complicated only from the fact that when a device is on or off, the amount of energy used can vary. For example, when a TV or computer is turned off, it still draws a little power to keep onboard electronics powered and operating. When turned on, the electricity use will vary depending on brightness of the screen, audio output levels, or activities of internal subsystems like hard drives or laser players. It is true for some heating, ventilation, and air conditioning systems (HVAC) where elements vary in output depending upon building demand. These types of energy loads typically require data monitoring equipment in order to quantify average rate of energy use.

Complex/Compound Loads

Complex/compound loads in energy systems are nothing more than a combination of loads, be they simple or variable. These types of loads could be viewed as the energy subsystem loads in a TV or computer, HVAC systems, or something much larger like a building, a complex, or a full power grid. It all depends on the level of analysis being requested at the time of a study. The size or scope of a complex/compound energy system can always be broken down into its basic components/elements if the need for total understanding of the load's operation is needed.

Time

Time is time, no matter the type of load, simple or variable, the amount of time the load uses energy determines the quantifiable resources.

Once the amount of energy consumed is calculated, economic value for the amount of used energy can then be determined.[2]

Understanding the Level of Analysis

As mentioned a bit earlier in this section, there is a need to determine the depth of understanding the level of study when looking for conservation options. "Why" is the system needed should be the first question asked. If not, then the option is to remove the system or at least turn it off until a real need is found.

---

[2] (Note: Economic value, for the energy used, is always done at the end of the analysis process. Since the amount of energy used is true given the same set of data, the value will vary by energy supply source (utility, gen set, or renewable source) at the time the economic analysis is performed. Also, over time (say 2001 vs. 2012) while the amount of energy remains the same, the energy value will change.)

If there is a need for the energy system, then the operation or use of the load should be studied to determine what kind of research is required in order to find a number of conservation options that can then be fed into the analysis process that would determine an optimum solution (more discussion on this later in the chapter).

Example (Maintenance Yard Lighting)

Currently, maintenance yard lighting for most government transportation department or electric utility service yards is on all night. The yard lighting systems use HPS types of fixtures or other lamp technologies that reflect the types of lighting systems used by a roadway/street owner's system. Doing so helps to reduce parts of inventory costs and simplifies stock carried on service trucks.

The reason for the lighting being kept on all night is due to not knowing when a crew might access the yard, and most lighting systems take from 10 to 20 min to come up to full brightness (so there is enough light for crews to safely access equipment and supplies). Many times, the crews are responding to an emergency, like a traffic accident or road spill, so the situation could be a matter of life protection. Crew safety takes priority over energy savings. So most times, energy conservation solutions focus on installing higher efficiency lighting components like solid-state ballasts.

The real question to be asked is what kind of lighting system could meet the lighting needs of the site and allow for the lights to be turned off when not needed. Before LED technology solutions were developed, an automatic lighting control system would need to work with only those lighting technologies that were "instant on" in nature like various kinds of incandescent technologies like halogen. While those fixtures were less efficient, if on time were reduced to 400 hours of operation per year as compared to existing 4,100 hours of nighttime operation, savings will result in an energy reduction. LED lighting fixtures are considered "instant on" and can be deployed in areas where you need "photons on demand," and other times the lights can be turned off and annual costs will be reduced. (4,100 h are the number of hours used by Caltrans' Energy Program to reflect fee structures supplied by California electric utilities when "flat rate"/no meter rates are calculated for roadway lighting accounts.)[3]

In the case of the maintenance yard lighting, there are two conservation projects that can be deployed. Install LED fixtures and control the lights so they only operate when needed. Upgrading to LED fixtures will yield about a 50% reduction in energy and green house gas/carbon footprint, and install a control system upgrade that would yield almost a 90% energy/carbon footprint. Both projects require LED fixtures to be installed to get optimum savings opportunities, yet not doing the LED upgrade would still allow a control system deployment along with fixture changes back to some form of incandescent technologies. The full conversion numbers will be included in a latter section of this chapter.

---

[3] (Note: Flat rate accounts with lighting controls being added would have to be converted to a metered account due to the change in operations. Typically, the conversion from flat rate to metered account costs is covered by the servicing utility. In California, most utilities are installing "smart meters" on accounts in order to optimize power delivery within the state's various service territories. In the case of this example, all maintenance service yards have metered accounts.)

10 Energy Conservation for Optimum Economic Analysis

Analysis Approaches

Analysis options range from simple payback to full detailed analysis over the product lifetime of operational costs, and when to choose the best approach will, most of the time, depend on who is your audience. This section will highlight several analysis approaches that could be used when determining cost effectiveness of energy projects along with typical results.

Example
Senior management may only want to see an estimate of costs and potential savings, while the budget office will want a full cash flow analysis that discusses all direct and indirect costs and benefits for the project. In addition to the complex analysis, the budget office may also look at the impacts of funding source options.

A Climate Change program manager may only be interested in greenhouse gas reductions and not want to know how the carbon footprint reduction was accomplished.

Simple Payback Analysis (SPA)

SPA looks at the cost to deploy the energy project and the economic benefits over a fixed period. Typical results are limited and do not yield a full overview of the project.

Example
If your project costs $1,000 and the annual energy savings is valued at $300, then the project will take 3.3… years to replace the costs. SPA does not contain info on side benefits, cost of money, and opportunity costs (could the money have generated a greater benefit if invested somewhere else or other factors that would be an outcome had other analysis paths selected).[4]

Function Analysis (Also Known as Value Engineering) and Systems Analysis (FA/VE/SA)

FA/VE/SA analysis processes look beyond the SPA process. These types of analyses look at the function or larger system that involves the conservation project. Typical results of these analysis options do ask the why/when/how questions that, when

---

[4] (Note: An SPA value for a project could be generated as an end product from a more complicated analysis process. This is an option for simplifying info and at the same time the more in-depth data is available upon request. A demonstration of this last point is found at the end of the chapter.)

addressed, allow for greater defining of direct and indirect costs and benefits for a project of any scale and complexity. Both functional and systems analysis approaches share some processes; the only real differences are mindset going into the project identification and defining phase of the process.[5]

Example
FA users would ask questions and identify basic functions of the project, for roadway lighting, ask why light the roadway, is it a safety issue, traffic control, driver assistance, and accident reduction and then look at alternative solutions for the identified functions, and non-energy functional solutions are a possible outcome. SA users would look at the whole lighting system and look at potential system tweaks like better lamps or products to light the roadway and some of the same questions might be asked as in the FA process.

Life Cycle Cost Analysis (LCCA)

LCCA has become the typical common number crunching tool used by many, including those who use the FA/VE/SA project analysis processes. LAAC has at least two analysis elements; firstly, a base case element that defines the current operational data conditions and, secondly, at least one alternative data set as defined by the project base case description. The LCCA's key difference with the SPA process is that LCCA adds operational time of use and those related costs and benefits data (including cost of money and inflation of costs and benefits). Direct and indirect benefits can be included in the calculation process as well as in any other data sets that have a relationship to the project like labor and maintenance cost/savings. LCCA then allows the user to compare and contrast any number of alternative project packages with a common base case scenario. Typical results from LCCA will include as much data as the user wishes to include.[6]

Example
Please refer to the LED roadway and maintenance facility example discussed later in this chapter.

---

[5] (Note: All state and local government transportation departments must use the FA/VE analysis for federally funded roadway projects valued at over $25 million in federal matching funds. So, most transportation departments are familiar with this analysis process and have access to trained staff within each state's transportation network. The Industrial Engineering/Operations Research (IEOR) community typically uses the SA process.)

[6] (Note: The State of California Department of General Services has posted its "Department of Finance" approved LCCA model which California State departments can use to define the viability of energy conservation projects. The web address to the site is http://www.green.ca.gov/LCCA/default.htm).

Life Cycle Analysis (LCA)

LCA should not be confused with LCCA. While LCA has some linkage to LCCA in what the process is trying to accomplish, LCA's focus is on environmental impact or the burden upon the planet caused by the production and use of a product or service. Over recent years, the LCA process has gained increasing support internationally and is being looked at as a way to compare anything to anything, like running a farm or driving a car. Typical results for the LCA process are a set of sliding scale numbers that display a "green-neutral-not green" rating for any process or product. The data behind each element of the data set documents how that rating was developed and can identify those areas of potential green changes. While this analysis process is briefly discussed here, its inclusion is more for information of things to come.[7]

Example
No example is developed in this section.

Data Sources

Energy data used in energy conservation calculations should always be measured in some way. Does a 40-W fluorescent lamp really only draw 40 W? Most of the time, the answer is No. The reason being twofold: firstly, in order for a 40-W lamp to operate, it needs a ballast and the ballast consumes energy; and secondly, since voltage varies from state to state, the "volts X amps X square root of the phase of the power supply" formula will yield slight variations in true wattage of any load. So, when gathering data, for any kind of analysis, meters or some form of monitors should be used. This process is part of the understanding of how the energy system being investigated works. Obtaining measured consumption data from a load also helps to eliminate errors that occur when energy auditors only use "nameplate data" (energy data that is found on all electrical devices sold in the United States.) Typically, nameplate data describes maximum rated consumption of a device. Devices like TVs, computers, printers, motors, and pumps will use less than the rated consumption.

Example
Consider 2.4-ft 40-W fluorescent tubes in a lighting fixture with an iron-core potted ballast will average 100 W and, after 10 h of on time, will consume 1 kilowatt-hour (Kwh). Use LED replacement tubes that do not need the iron-core ballast and the fixture could now only use 57 W, or 1 Kwh will get you about 17.5 h of lighting operation.

As mentioned before, time of operation for any load needs to be defined. If the on/off cycle of the load is automated, then one needs to understand that on/off

---

[7] For more details the Wikipedia online site has some general information: http://en.wikipedia.org/wiki/Life-cycle_assessment.

pattern. Most lighting systems are on some form of patterned use, like traffic signals (signal controller), roadway lighting (photocell/time clock), and yard lighting (photocell/time clock), and then some systems are more random like building lighting systems (mix of building management system-controlled and occupant-controlled fixtures), HVAC systems (thermal/environmental needs), and office equipment (user needs).[8]

Example
For roadway lighting systems, photocells and sometimes time clocks are used to control the on/off cycles of the day. The sensitivity of the photocell control is set so that annual operation of 4,100 h per year occurs.

Research Needs

While many energy conservation opportunities are known, there is still a need to conduct some form of real research on energy system conservation opportunities and how they interact with people, equipment, or the environment.

Example
Back in the late 1970s, people were looking for easy and quick energy conservation actions that would capture "low hanging fruit" savings. One such action was to set building thermostats to 68°F for heating and 78°F for cooling.

In the beginning, those settings were applied to the traditionally coldest and warmest zones in a building or the heating/cooling systems themselves (cooling locked out when external temps go below 68° and heating systems above 78° and use return air as the main temping resource), and today that maxim is applied to all zones in a building.

Yes, energy was saved. However, was occupant performance impacted? And was the value of the saved energy less or greater than productivity losses during times when the space temp is too cold or warm? Here is an opportunity for research. Lots of open windows, space heaters, and desk fans seem to indicate the occupants may have offset the 68/78 settings savings, plus when people are talking about being too cold or warm or feeling sleepy, they are not conducting business, thus impacting

---

[8] (Note: Environmental pollution, like dust or bird-scat, has been known to cause lights to stay on during the day so most transportation departments have maintenance programs that try to correct the situation within a few days if not sooner. A further note: within the last few years, a national dialog has looked at increasing the intelligence of the roadway lighting controls. The end product is a standardized universal plug socket in new approved roadway lighting fixtures. Into that socket, a simple photocell or time clock could be attached or higher level controls including devices that monitor the condition of the lamp and ballast. Should the fixture monitor note changes in the fixture's operation, the device signals a central control center (via power line carrier signal, WiFi, or cell phone signal) of the condition of the fixture. The control center would then route maintenance crews to service the fixture. This system eliminates nightly driving routes by crews looking for burned out lights, thus reducing fuel and staff maintenance resource costs.)

productivity costs (another indication that there is a real need to understand all the aspects of a proposed project).

A Project Analysis Process

This section of the chapter focuses on a suggested sequence of how to set up a project analysis process. Following this section is an example of the discussed analysis path, and its use to develop the LED roadway and maintenance facilities LCCA model for the California Department of Transportation (Caltrans).

Project Statement

In setting up the analysis model, it is often best to use a spreadsheet application like Excel, Numbers, or some form of free spreadsheet application rather than a database program like Access or FileMaker. The example following this discussion section was originally developed in Excel where each phase of the analysis has its own worksheet with each worksheet dependent upon the previous worksheet. One of the powers of a spreadsheet application is that data cells in one worksheet can be linked to other cells in other worksheets within the same file (in Excel a file is also known as a "Workbook"). The power to this feature is that, if correctly set up, only one of the early worksheets can be the primary data input sheet and the data flows from page to page automatically into the analysis.

Typically, the problem statement groups the projects in order of system group like all the lighting projects, HVAC, building systems, and control systems with a larger grouping of interior then external system packages.

From this point on in this chapter, it is assumed that the reader has a working knowledge of spreadsheet and database computer applications and their differences. There will be a focus on the use of spreadsheets, with a reference to databases.

A version of the project statement would best be located on the first worksheet of the spreadsheet file. It is always just a click away and easy to amend as the analysis progresses. This sheet can also contain project notes and assumptions as they relate to data or ideas. If needed, use a second or a third worksheet for this section of the file is your idea center. And for discussion in this chapter, it is assumed that only one page/worksheet is needed, also known as "Worksheet 1."[9]

---

[9] (Note: A copy of the energy analysis package can also be used to track the implemented project(s) performance over time adjustments to utility costs and other rates and other variables. The power of the spreadsheet would then automatically adjust future year forecasted numbers along with any running total calculations included in the model.)

Base Case Project Setup

As described above, the list of projects master group would be interior then exterior project groups. The interior group would list lighting projects first, then HAVC, then process-related projects, with complete building control systems being the last group of projects on the list. There is a reason for this sequence; lighting systems generate heat in a building that impacts heating and cooling system operations. The HVAC systems will need to address these interactions.[10]

This section would identify current loads being addressed by the projects list. The number of involved lamps/fixtures, hours of operation per year (The calculations for these numbers can be included either in text or in calculation), current energy costs, and green house gas values (Typically in pounds of CO_2 per Kwh or a metric ton factor). Kwh to CO_2 conversion factors will vary from state or region depending on electric power generation sources. Also, over time/years as power production "greens," the Kwh/CO_2 footprint factor will shrink).

Also included in this worksheet section are stated assumptions for inflation factors that will be applied to energy, labor, maintenance materials, or any cost or benefit data element in future years.[11]

Defining the Proposed Action Changes

This section of the project analysis process uses the base case data and assumptions to define the proposed change.

As discussed earlier, these changes will normally impact rate of energy use, hours of operation, replacing utility-supplied energy with another source, cost of energy or maintenance, or some combination.

This worksheet section could also be used to examine various ways to fund a proposed project.

Example
The first project proposal would just state the project proposal in its simplest cost to deploy (assume "free money" for now), including how much time it will take to make the changes and phased in results. (Remember, no projects requiring physical changes happen instantly.) Variations of the proposed project like cost of money, staged deployments, and fund borrowing can then follow by duplicating the alternate cost analysis and layering on the additional data that reflect the focus of the variation.

---

[10] (Note: Process equipment projects may be placed before HVAC projects depending upon the scale of the process-related loads versus HVAC.)

[11] (Note: When developing the LCCA analysis model over time, the use of these adjustment factors can be used to determine forecasted cash flow, net present value of money, and sensitivity analysis factors that would look at what future conditions may impact the proposed project cash flow. By assuming that both base case and project options use the same factors, then changing the variations for all yields a consistent comparison under any scenario.)

Set Up the Analysis Spreadsheet (a.k.a. the "Analysis Engine")

Visualize that each LCCA data set would be a thick band of data stretched across the spreadsheet. The first column (Column A) describes the data contained in the row; the second column (Column B) would contain core data, like number of fixtures, or wattage per fixture, hours of operation, current costs of energy, CO_2 conversion factor, and maintenance cost factors per fixture. Continuing down, Column B would be the results of calculations, like total energy consumed, may be electric grid demand impact, annual value of energy consumed, annual CO_2 footprint, and project maintenance costs, along with other project baseline calculations. This worksheet only contains all the key core data and annual calculations needed for the base case at this point in the workbook development. This page of information will be referred to as "Worksheet 2" for the rest of this element of this chapter.

Insert Real World Data for Base Case

On the next worksheet (Worksheet 3) of the workbook, the two columns of data are copied and pasted from Worksheet 2. Then link the data cell on this worksheet to the previous sheet. And repeat the links to all the other lines of data in the Column B. Doing this allows the user to update the engine by only entering data changes once in the Worksheet 2 and linked data-fields auto calculates into Worksheet 3 (Comment: one of the powers of a spreadsheet-based economic model).

Back in Worksheet 3, add to Column B running totals for those years of data over the years of the analysis engine (and related descriptions for the data in the Column A). The third column (Column C) of data reflects the relevant data of the first year of the analysis. In most cases, "year 1" has no cost inflation factors. The inflation factors only start in those years beyond the first year. Yearly data columns contain adjusted data for those future years. (Comment: If 20 years of LCCA were to be performed, the list of columns would be from year 1 to year 20.) The base case model can reflect various forecasted maintenance patterns or other nonlinear costs that would normally occur over a 20-year timeframe. Ways to automate some of the data fields would include multiplying the previous year's data field by key factors like inflation or some other factor identified in the project statement.

For the running totals in the bottom rows of data, let each year represent that year's totals and only see the full totals be reflected in Column B. Depending on the analysis needs, like a many project program, the analysis engine may require running total data that reflects past years of data plus the current year being viewed. This would be the time to set up this calculation feature. (Comment: This feature might be convenient if looking for total costs and savings less than the full term of the LCCA model.)

At this point in development, the base case analysis engine should reflect all cost elements that reflect maintenance and operational costs of the base case system or process to be studied.

Economic Options: A Discussion

This section is a bit easier to develop than the base case analysis engine. Go back to Worksheet 2 and copy the base case data, then skip down five rows and paste the copied data in align with the base case data on Worksheet 2.

Title this new section "Alternative 1" or something similar; just be consistent for any additional copy and pasted data sets for additional alternatives and their related variation sets.

For the Alternative 1 pasted data, it is time to modify key data elements that reflect changes, like a lower rate of energy use or reduced hours or other conditional changes that occur when compared to the base case. Use the description field (Column A) to note the changes and insert the data changes in the respective cell in Column B. Note that the automatic recalculations occur in the lower Column B data fields that represent annual consumption and other related values. At this point, if additional rows of data are needed (i.e., cost to deploy project, or cost of labor and materials, other costs that relate to the alternative), it is best to include in the basic data section or annual calculations section of Column B at this point. The amount of added rows of data will vary by project. In some cases adjustments to some of the calculated cells in Column B may have to be made.

Copy the Columns A and B data in Worksheet 2 and open Worksheet 3. Paste the data into that worksheet, five rows below the base case data set. Now link the data sets in the Alternative 1 section of Worksheet 3 to their counterparts in Worksheet 2. (Comment: a way to do that would be to copy any cell in the base case section of Column B and paste it into all the Column B cells of the Alternative 1 section. The numbers should not change; however, if you examine any of the pasted cells, you will see text that notes the cell is accessing on the previous worksheet.) Similar copy/paste processes can be used to copy and paste rows of yearly data from the base case into Alternative 1's section of Worksheet 3.

For those new rows of data in Alternative 1 that relate cost or savings, data sets, like utility incentives/rebates, project costs to deploy, or environmental impact information, will need to be added to Column B. These additions would also include and update running or summary totals.

For the yearly data columns, it is suggested to copy and paste similar rows of data from base case to into Alternative 1's data fields. Once Alternative 1's data set seems to be filled, it is time to customize those data cells that reflect a transition from the base case to Alternative 1's operational conditions.

Example
As stated in the project statement, it is expected to take 2 years to fully deploy the project, if that is the case:

- 30% of the project cost may occur in the first year and 70% in the second year.
- Energy savings and utility incentives will have to be offset by time. For the sake of simplicity, the project statement will assume that 15% of the annual savings will occur in year 1, 65% savings in year 2 (30% from year 1 plus 50% of year 2 = 65 %), and 100% in year 3.

- CO_2 reductions follow energy savings patterns, and in this project, the reduction estimates would follow the 15/65/100%pattern.
- Utility incentives are typically awarded after proof of deployment has been documented, and it takes a few months for the checks to be issued, so it is stated in the project statement: assume 0%incentives in year 1, 50%in year 2, and 50%in year 3.
- Annual maintenance costs are also impacted as the new products are deployed, in year 1, 85%of base case costs still occur, year 2 still incurs 35%of base years costs, and year 3 no base case costs occur, and there is zero costs for the remainder of the LCCA model, or if there are identified post project maintenance costs contained in the Alternative 1 project statement, those costs would be entered into the appropriate data cells for identified year(s). Typically, LCCA term of analysis lasts for at least one assumed life span for the newly installed product or service cycle, sometimes longer (a reason might relate to financing requirements). The choice of LCCA timeline should reflect client needs, including the depth of maintenance costs, and these needs should be included in the base case and remain consistent for all the alternative and variation scenarios.

The Alternative 1 analysis model is for the most part complete and is now just a copy/paste and modify data to reflect other alternatives or project funding variations needed to meet the client's needs.[12]

Making a Decision

One of the last worksheets in the Workbook file involves doing the comparison between the base case and each alternative or variation analysis set. For simple projects as in the LED lighting example in the next section of this chapter, the comparison can be placed on the same worksheet as the analysis engine (Worksheet 3). For more involved projects with many options and variations, it may be best to set up a summary worksheet. If the summary worksheet option is used, set it up using the linked cell/copy/paste process discussed earlier in this section. Doing so will allow the summary worksheet to auto-update should any adjustments be made in the base case or alternative scenarios.

As discussed earlier in this chapter, some clients only want to see a simplified version of the economic decision data. Sometimes, the simple cost/benefit analysis will approach the more structured LCCA method; however, most of the time, it will not. However, from the results of the LCCA data set, a simple cost/benefit summary can be generated, and when the client decision maker starts to ask more detailed question, the data paper trail is quickly available.

---

[12] Note: One way to reduce the additions/inserting of rows of data from base case to alternative X variation Z plans is to include every row of data and calculations into the base case model (zeros are okay). Then the copy and paste process only needs to focus on the differences between the various analysis engines in the final workbook file set.

Deployment Options: A Discussion

Sometimes, the decision to proceed is given, and then how to deploy planning begins in earnest. While it would be a good idea to do a few variations of the alternative options based on possible deployment options as part of the original project workbook, sometimes, it is easier to first get the "go" decision, then make a copy of the workbook and remove those alternatives and options that no longer need to be addressed. Now copies of the smaller workbook can be used to illustrate the various "what if" scenarios with a focus on deployment options. Keep one set of the smaller workbook as an original, and then using a copy, just make the changes, then save it as something like "deploy in 18 months" or "deploy in 30 months." These results can be summarized so that management can choose to adjust funding flows to meet an optimum investment goal. There is also a data paper trail to back up the entire decision processes.

Side Bar: A Possible Complexity Issue for Analysis Modeling and a Way to Organize Multiple Projects into an Optimized Investment Plan:

While the processes discussed in this chapter will work for any number of projects (from 1 to 100+ different projects and alternatives and variations), it may be best to develop an Excel workbook file for each grouped or related set of projects, especially if there will be a number of proposed project variation scenarios developed. A way to optimize one's time would be to develop one project workbook with base case, project proposal options, and any variation data sets. Complete the data analysis and then clone that package and adjust data sets that will relate to that next project. (Note: For this approach, assume each project workbook data set is the only project set being implemented. Do not include impacts/relationships with other projects at this stage of development.)

Once all the workbooks are complete, look for the project that yields the fastest payback and rank each workbook. (However, make sure you compare the same funding option and variations for each workbook, if they exist. The fastest payback package should be a consistent first.)

Time to build a summary project package set. First, make a file copy of all the workbooks and put them in a folder marked "First Reduction" and take the original set of workbooks and put them in a file marked "Originals Do Not Modify."

Open the "First Reduction" folder and put a "1" in front of the workbook with the fastest payback, "2" for the second, "3" for the third, and so on for all the projects. Next, assume project #1 has been implemented, then examine all the other workbooks to determine if by implementing project #1, any conditions change in the base case; if so, make note of the changes on the first worksheet and adjust the data in the analysis engine for each project workbook. Once completed, look for the workbook with the next fastest project payback after workbook #1 and note it in some consistent format like 3.1 and add the ranking from .1 to .nx to all the other workbook packages. Then, make a new folder marked "Second Reduction," make a clone of these workbooks, and move them into the new folder. Close the "First Reduction" folder and set it aside. (Comment: If no projects were impacted, and that can happen, then still follow the instruction for this, following the folder instructions).

Open the "Second Reduction" folder, toss the fastest payback project, and focus on the project with #.1, and then assume it is implemented, and adjust the remaining workbooks if needed, then relabel 4.2.1, and so forth. Create a new folder named "Third Reduction" clone, the newly relabeled files, move into the new folder, and repeat the processes until all projects have been reduced as far as they can go. (Comment: This process might end up with some projects no longer being cost-effective, and that is one of the reasons for such an extended process because there is now an "analysis trail" to document why a project(s) is no longer viable.)

Once the reduction process is completed, the resulting list of projects yields the optimum investment strategy and sequence of project implementation. While this process works best for energy programs that have more projects to implement than available funding, this process also works for programs where funding is available and the need for investment optimization is required. This process will identify those areas when the same or partial saving might be claimed by multiple projects if those projects were only viewed on a "stand alone" basis and not taking into account impacts on related conservation opportunities.

Example

Roadway and Maintenance Facility LED Lighting Upgrade Project for All California-Owned and California-Maintained Roadways and Maintenance Center Yards.

Background

As discussed in the early pages of this chapter, the State of California Department of Transportation (Caltrans) spent years developing data and working with lighting industry manufacturers and national standards institutions that resulted in the decision to deploy statewide LED technology upgrades for existing roadways, bridges and maintenance yard lighting systems. High-pressure sodium (HPS) fixtures have been used in the past and currently approve performance standards for LED fixtures.[13]

Here is a sample of the types of HPS and LED fixtures being discussed in this section.

As discussed earlier in this chapter, the LCCA process can involve many alternative options and variations that eventually lead to a decision by department senior management to deploy the upgrade and identified funding sources in February of

---

[13] For copies of the most current specifications and qualified products list, please contact gonzalo_gomez@dot.ca.gov.

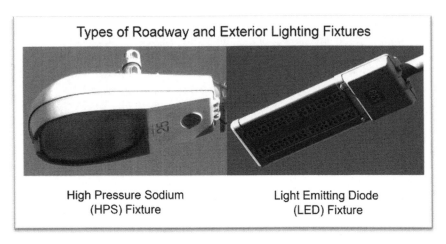

High Pressure Sodium (HPS) Fixture
Light Emitting Diode (LED) Fixture

Images and graphics by Stephen Prey June 20, 2012

2012. Alternative options included internal project funding, construction bond funding, grant funding, utility incentive program funding and combinations of grant internal and external funding in combination with current utility incentives, and 12-, 24-, 36-, 48-, or up to 60-month options for upgrade deployment.

The LCCA analysis model for this statewide project was updated in March 2012 and will be the source of the illustrations in this example. The LCCA model is a public document and can be requested through contact with this book's author.

Project Statement

This project statement has been extracted from the "LED Roadway Lighting Analysis Simple v1" Excel file that is being used as a project tracking document for the Caltrans LED lighting project upgrade for roadway and maintenance facility yard lighting. Funding for the project came from redirected departmental funds and utility incentive funds received as the projects are deployed within electric utility service districts that offer conservation incentive programs.

> **Problem Statement:** Calculate costs savings to convert Department roadway and Maintenance facility lighting systems. Project approved 2/2012 with redirected internal funding for FY 2011/2012. Purchase min. of 40,000 fixtures with encumbered funds by 6/30/2012, and deploy fixtures to districts by late summer 2012 for installation.

For greenhouse gas (GHG) calculations, the following conversion factors were used, and the source was 2007 data set from the State of California Environmental Protection Agency (EPA):

Data sets assumptions, info from CalEPA/CARB in 2007	961 pounds of CO_2/million watt hours	
Ditto	.436 metric ton of CO2/million watt hours	
Or	0.961	Pounds of CO_2/Kwh
Or	0.000436	Metric tons of CO_2/Kwh

Base Case Project Setup

The base case data illustrated here was extracted from Worksheet 2 of the Excel workbook. There are five lighting projects defined in the base case section: roadway, maintenance center yard lighting, maintenance center facility interior lighting, and sand/salt shed facility lighting (interior and exterior). Labor costs for lamp replacement of HPS lamps are documented along with setting up safety barrier setup costs when roadway lighting maintenance is performed. Unlike the column descriptions discussed above (Column A, B, C), that format was only used in Worksheet 3 where the analysis engine was set up. Here, the complexity of the base case assumptions and calculations took up additional columns. It should be noted that the right three columns contain the live data cells that feed the analysis engine data cells.

Defining the Proposed Action Changes

Alternative 1 data assumptions and calculations reflect the flow of assumptions for the five projects. While there is a notation that the interior lighting project for the maintenance facilities has been dropped due to very long project payback, there is a possibility that extra funds may be made available (or operational savings from the installed LED projects could be redirected) at a future date, and the analysis could be included within the overall project package.

Like the base case data sets above, the right three columns contain actual data that feed the Alternative 1 section of the analysis engine in Worksheet 3.

The key data that will change in this section will be the installed cost elements for the LED fixtures, specifically the cost of the LED fixtures. When the LCCA package for this project was first laid out, LED fixture prices being quoted were as high as $1,200 each. As the market matured and volume-pricing (30,000–70,000 units) data was discussed with various vendors, the unit price dropped (as anticipated due to a similar price decline for LED traffic signals back in the early 2000s.) LED prices used in this analysis reflect the range of pricing discussed with qualified vendors. Final revision of that data occurred after the bid openings in July 2011.

Baseline conditions		Line key	Base case line item descriptions	Baseline data	Data extension	Units
240 W (200 w HPS roadway) 34,975 units	34975	Fx ct.	240 w HPS	34,975	8,394	kw
360 watt (310 w HPS roadway) 28,621 units	28621	Fx ct.	360 w HPS	28,621	10,304	kw
450 watt (400 w HPS roadway) 1,903 units	1,903	Fx ct.	450 w HPS	1903	856	kw
4,100 h of on time per year	4,100	Op Hours	Hours of operation	4,100	4,100	Hours
2011 road lighting average energy cost is $0.09 or 9 cents/Kwh	0.09					
HPS average life cycle 3 to 4 years (changeout cycle every 4th year)	16400	Kwh	Annual Kwh		80,171,031	Kwh
Relamp cost per 240 w HPS $100	20	Energy value	Statewide average cost of a Kwh	$0.09	7,215,393	$
Relamp cost per 360 w HPS $300	120	changeout fact costs	Service frequency (SF)	25 %		
Relamp cost per 450 w HPS $300	120		Relamp cost per 240 w HPS $100	20	699,500	$
Fixture/Relamp cost per 240 w HPS $280	200	costs	Relamp cost per 360 w HPS $300	120	3,434,520	$
Fixture/Relamp cost per 360 w HPS $480	300	cost	Relamp cost per 450 w HPS $300	120	228,360	$
Fixture/Relamp cost per 450 w HPS $480	300	1 %	Fixture/Relamp cost epr 240 w HPS $280	200	6,995	$
Annual replacement info: 25% relamp and 1%fixture/lamp replacement		1 %	Fixture/Relamp cost per 360 w HPS $480	300	8,586	$
Average labor (per fixture) 2 people 30 min: $80/h	80	1 %	Fixture/Relamp cost per 450w HPS $480	300	571	$
Traffic control labor factor (1.25 h)	100	Maintenance costs	Annual maintenance costs		1,106,747	$
Labor to service fixture: 1 h		Energy consumption	Est. total energy used		80,171,031	Kwh
Average hours = one PY	2000	Energy value	Est. total value of energy		7,215,393	$

Description	Value	Labor factor	Jobs	1 h/fix/year/SF	Value	Pys Eq.
335 maintenance facilities data for one avg. facility)	335	Labor factor	Jobs			
10 260 w HPS pole mounted fixtures at 4,100 h/year		Grid	Grid demand impact 24/7 (Kw)		19,554	Kw
25 210 w MV/MH wall pack fixture at 4,100 h/year		GHG	Energy carbon factor (.961# of CO_2/Kwh)		77,044,361	Pounds CO_2
Average service cycle 5.5 years for both external fixtures types	5.5	GHG	Energy carbon factor (.000436 MT of CO_2/Kwh)		34,955	Metric tons CO_2
Labor: $70 loaded rate per hour	70					
Wallpack relamp $20/lamp + $46.67/labor	20.00	Facility cound	Number of MS	335		
Pole light servicing (lamp plus install labor)	20	Ext Fx Ct	260 w HPS count	3,350	871	Kw
Fixture/lamp replacement cost (1%failure per year)	120	Ext Fx Ct	210 w HPS/MV/MH count	8,375	1,759	Kw
Office lighting: 40 lamps @ 90 w for 780 h/year		Op Hours	Annual hours of operation	4,100	4,100	Hours
Storage bays: 17 lamps @ 240 w for 260 h		ExLtg	Exterior lighting Kwh		10,781,975	Kwh
Repair bays: 20 lamps @ 240 w for 1,040 h		ExLtg	Value of Kwh @ $0.15		1,617,296	$
Estimated hours of fluorescent lamp life 10,000 h		ExLtg	Service frequency .20	0.2		
Cost of lamps and Ballasts ($10 for 4", $20 for HO/VHO; $20/ballast)		ExLtg	Annual Maintenance costs @ $20/Fx*20%fix …	$49,245	49,245	$
Labor costs: (20 min for lamps, 40 min for ballast failure)/year		Int Ltg	90wFl @ 780 h	No interior	0	Kwh
Average facilities cost per Kwh	0.15	Int Ltg	240wFl @ 260 h	Lighting	0	Kwh
62 sand/salt shed sites	62	Int Ltg	240 wFl @ 1,040 h	Project	0	Kwh
2 260 HPS polelights per site 4,100 h w/ TC and PC controles		Int Ltg	Interior lighting Kwh	Being done	0	Kwh

(continued)

Baseline conditions					
2 210 HPS/MV/MH wall pack per site @ 4,100 h/year	Int Ltg	Value of Kwh @ $0.15/Kwh		0	$
Same cost and field life for exterior and interior lighting	Int Ltg	Service frequency .1	0.1		
8,210 MV interior fixture @ 24/7 h Nov1 through May 1 (4,380 h)	Int Ltg	Int. lighting maintenance costs	$0		
	Int Ltg	Jobs	1 h/fix/year/SF	0.000	Pys Eq.
1,000 watts = 1 kW	Int Ltg	Grid demand impact 24/7 (Kw)		2,630	Kw
Assume annual inflation rate of 4 %	Int Ltg	Energy carbon factor (.961# of CO_2/Kwh)		10,361,478	Pounds CO_2
		Energy carbon factor (.000436 MT of CO_2/Kwh)		4,701	Metric tons CO_2
		Sand?Salt sheds	62		
		260 w HPS pole @ 4,100 h	124	132,184	Kwh
		210w entry light @ 4,100 h	124	106,764	Kwh
		210 w interior lighting (4380 h)	496	456,221	Kwh
		Total Kwh		695,169	Kwh
		Value at $0.15/Kwh		104,275	$
		Service frequency .25	0.25		
		Int. lighting maintenance costs$100/fix	$3,720		
		Jobs	1 h/fix/year/ SF/2000	0.093	Pys Eq.
		Grid demand impact 24/7 (Kw)		162	Kw
		Energy carbon factor (.961# of CO_2/Kwh)		668,057	Pounds CO_2
		Energy carbon factor (.000436 MT of CO_2/Kwh)		303	Metric tons CO2

10 Energy Conservation for Optimum Economic Analysis

	Line key	Alternative 1 line item descriptions	Baseline data	Data extention	Units
	Fx ct.	165 w LED	34,975	5,771	kw
	Fx ct.	235 w LED	28,621	6,726	kw
	Fx ct.	300 w LED	1,903	571	kw
	Op hours	Hrs of operation	4,100	4,100	h
	Kwh	Annual Kwh		53,577,611	Kwh
	Energy value	Statewide average cost of a Kwh	$0.09	4,821,985	$
	costs	165 w LED install cost	884	30,917,900	$
	costs	235 w LED install cost	1,534	43,904,614	$
	costs	300 w LED install cost	1,794	3,413,982	$
	Utility incentives	165 w LED incentive (83 %)	125	3,628,656	$
	Utility incentives	235 w LED incentive (75 %)	175	3,756,506	$
	Utility incentives	300 w LED incentive (75 %)	200	285,450	$
	Utility incentives	Total install cost		78,236,496	$
	Maintenance costs	Total utility incentives		7,670,613	$
	Energy consumption	Est. Total energy used		53,577,611	Kwh
Conversion data for alternative project					
165 w LED replaces 240 w 34,975 units	34,975				
235 w LED replaces (360 w HPS roadway) 28,621 units	28,621				
300 w LED replaces (450 w HPS roadway) 1,903 units	1,903				
4100h of on time per year	4,100				
2011 Road lighting average energy cost is $0.09 or 9 cents/kwh	0.09				
15+ year expected life for LED	61,500				
165 w LED install cost	474				
235 w LED install cost	799				
300 w LED install cost	864	Energy value	Est. Total value of energy		4,821,985	$
165w LED incentive	125	Labor factor	Jobs	1 h/fix/year/SF	13	Pys Eq.
235 w LED incentive	175	Grid	Grid demand Impact 24/7 (Kw)		13,068	Kw
300 w Led incentive	200	GHG	Energy carbon factor (.961# of CO_2/kwh)		51,488,084	Pounds CO_2
% of incentive factors by wattage: 83/75/75					
Annual replacement info: 25%relamp and 1%Fixture/Lamp replacement		GHG	Energy carbon factor (.000436 MT of CO_2/kwh)		23,360	Metric tons CO_2

(continued)

Description		Line key	Alternative 1 line item descriptions	Baseline data	Data extention	Units
Average labor (per fixture) 2 people 30 min; $80	80					
Traffic control labor factor (1.25h)						
Labor to service fixture: 1 h.		Facility cound	**Number of MS**	335		
		Ext Fx Ct	10 165 w LED pole mounted fixtures at 1,000 h/year	3,350	553	Kw
About 10%utility savings to locals, 100%maint. offset to CT		Ext Fx Ct	25 112 w LED wall pack fixtures at 1,000 h/year	8,375	938	Kw
Average hours = one PY	2,000	Op Hours	Annual hours of operation	400	1,000	Hours
335 Maintenance facilities data for one avg. facility)	335	ExLtg	Exterior lighting Kwh		1,490,750	Kwh
10 165 w LED pole mounted fixtures at 1,000 h/ year		ExLtg	Value of Kwh @ $0.15		223,613	$
25 112 w LED wall pack fixtures at 1,000 h/year		ExLtg	LED install costs	$13,684,750		
35 + yars life	5.5	ExLtg	Jobs		5.863	Pys Eq.
Labor: $70 loaded rate per hour	70	Int Ltg	60 wFl @ 780 h	No interior Lighting Project Being done	0	Kw
Wallpack @ 1,000 plus 200 for ms controls	1200.00	Int Ltg	150 wFl @ 260 h		0	Kw
Pole light 885 plus 200 ms control	1,085	Int Ltg	150 wFl @ 1,040 h		0	Kw
Ih per pole or pack to convert (included in price)	200	Int Ltg	Interior lighting Kwh		0	Kwh
Office lighting: 40 lamps @ 60 w for 780 h/year		Int Ltg	Value of Kwh @ $0.15/Kwh		$0	$
Storage bays: 17 lamps @ 150 w for 260 h		Int Ltg	Utility incentives	$0		
Repair bays: 20 lamps @ 150 w for 1,040 h		Int Ltg	Int. lighting maintenance costs	$0		
Estimated hours of fluorescent lamp life 10,000 h		Int Ltg	Jobs	1 h/fix/year	0.000	Pys Eq.
Cost of lamps and Ballasts ($20 for 4", $40 for HO/VHO; $20/ballast)		Int Ltg	Grid demand impact 24/7 (Kw)		1.491	Kw
Labor costs: (20 min for lamps, 40 min for ballast failure) no change		Int Ltg	Energy carbon factor (.961# of CO_2/kwh)		1,432,611	Pounds CO_2
Utility rebates = (4385+2000)*0.79 per site	5044.15	Int Ltg	Energy carbon factor (.000436 MT of CO_2/kwh)		650	Metric tons CO_2

Average facilities cost per Kwh	0.15		**Sand?Salt sheds**	62		
62 sand/salt shed sites	62	LTG	2 165 w LED pole @ 1,500 h	124	30,690	Kwh
2 165 w LED polelights per site @ 1,500 h w/TC and PC MS controles		LTG	2 75 w Entry light @ 1,100 h	124	10,230	Kwh
LED pole light costs: 2*300+400	1,000	LTG	8 112 w interior lighting (910 h)	496	50,552	Kwh
Estimated life od pole and wallpacks @ 1,605 h/year operations	35 years	LTG	Total Kwh		91,472	Kwh
8,112 w LED interior fixtures @ 5*7*26 h Nov1 through May 1 (910 h)		LTG	Value at $0.15/Kwh		$13,721	$
Cost interior LED plus 3 controls=8*400+3*200	3,800	LTG	Utility incentives	$73,470		
8 h labor per interior lighting upgrade		LTG	Costs to install	$799,800		
Cost for wallpack 2*400+200 (includes controls)	1,000	LTG	Jobs	1 h/fix/year/2000	0.372	Pys Eq.
2 75 w wallpacks w/TC-PC-Motion control @ 1,100 h/year		LTG	Grid demand impact 24/7 (Kw)		85	Kw
4 h to upgrade exterior fixtures		LTG	Energy carbon factor (.961# of CO_2/kwh)		87,905	Pounds CO_2
Utility rebates=(12*125*.79)	1,185	LTG	Energy carbon factor (.000436 MT of CO_2/kwh)		40	Metric tons CO_2
1,000 watts=1 kw						
Assume annual inflation rate of 4 %						

Set Up the Analysis Spreadsheet (a.k.a. the "Analysis Engine")

The analysis engine set up on Worksheet 3 has the line description in Column A with the Worksheet 2 linked data appearing in Column B.

Both the base case and Alternative 1 setups are listed below, including their respective 20-year total calculations.

There are numbers contained in the Column B section of the screen shots of Worksheet 3. In the setup process, these data fields would show "0"; however, these screen shots are from a fully populated analysis engine. Also, the line items for roadways and maintenance facility projects have combined respective data from Worksheet 2. Doing this reduces the height of the data set for base case and alternatives, thus allowing for more bands of data sets on a single worksheet.

Baseline	Data
Annual inflation rate	4 %
Annual roadway lighting energy (Kwh)	80,171,031
Value of energy for roadway lighting	$7,215,393
Annual facility lighting energy (Kwh)	11,477,144
Value of energy for facility lighting	$1,721,572
Annual roadway lighting maint. costs	$1,106,747
Annual facilities yard lighting maint. costs	$40,000
Total energy use (Kwh)	91,648,175
Total energy costs	$8,936,964
Total lighting maint. costs	$1,146,747
Total annual system Ops costs	$10,083,712
Mtric tons of CO_2 from electricity use	39,959
20 year totals	**Summary totals**
Total energy use (Kwh)	1,832,963,496
Value of energy	$266,125,627
Maintenance costs	$34,147,928
Total system costs	$300,273,555
Total metric tons of CO_2	799,172

Alternative 1 (2-year implement with redirected internal funding)	Data
Annual inflation rate	4 %
Annual roadway lighting energy (Kwh)	39,283,805
Value of energy for roadway lighting	$3,535,542
Annual facility lighting energy (Kwh)	1,721,572
Value of energy for facility lighting	$258,236
Annual roadway lighting maint. costs (varies after construction)	$1,106,747
Annual facilities lighting maint. costs (Yard lights to LED)	
Cost to upgrade roadway lighting (2 years 25/75 on) Kwh savings	$31,294,598

(continued)

10 Energy Conservation for Optimum Economic Analysis 233

Alternative 1 (2-year implement with redirected internal funding)	Data
Cost of upgrade facilities lighting (2 years 25/75 on) Kwh savings	$5,793,820
Rebates, given after project is on line	$10,855,839
Total energy use (Kwh)	41,005,377
Total energy costs	$3,793,778
Total lighting maint. costs	$1,106,747
Total annual system Ops costs	$4,900,525
Metric tons of CO_2 from electricity use	24,050
20 year totals	**Summary totals**
Total energy use (Kwh)	870,750,333
Value of energy	$118,166,043
Maintenance costs	$1,146,747
Project costs	$26,232,579
Total system costs	$145,545,370
Total metric tons of CO_2	379,647

Insert Real World Data for Base Case

The base case data set build out is shown here. There are two screen captures noted here, the first being for years 1–3 and the second being years 18–20. The years in between follow a similar pattern. Annual maintenance costs are included as well as HPS energy costs.

Baseline	Data	Year 1	Year 20
Annual inflation rate	4 %	1.040	1.040
Annual roadway lighting energy (Kwh)	80,171,031	80,171,031	80,171,031
Value of energy for roadway lighting	$7,215,393	$7,215,393	$15,201,744
Annual facility lighting energy (Kwh)	11,477,144	11,477,144	11,477,144
Value of energy for facility lighting	$1,721,572	$1,721,572	$3,637,092
Annual roadway lighting maint. costs	$1,106,747	$1,106,747	$2,331,749
Annual facilities yard lighting maint. costs	$40,000	$40,000	$84,274
Total energy use (Kwh)	91,648,175	91,648,175	91,648,175
Total energy costs	$8,936,964	$8,936,964	$18,828,836
Total lighting maint. costs	$1,146,747	$1,146,747	$2,416,023
Total annual system Ops costs	$10,083,712	$10,083,712	$21,244,859
Metric tons of CO2 from electricity use	39,959	39,959	39,959
20 year totals	**Summary totals**		
Total energy use (Kwh)	1,832,963,496	91,648,175	91,648,175
Value of energy	$266,125,627	$8,936,964	$18,828,836
Maintenance costs	$34,147,928	$1,146,747	$2,416,023
Total system costs	$300,273,555	$10,083,712	$21,244,859
Total metric tons of CO_2	799,172	39,959	39,959

Economic Options: A Discussion

This section, like the one above contains the fleshed out data set for Alternative 1 for years 1–3 and 18–20. Notice how annual costs to maintain and install the LED fixtures appear in the early years, and after year 3 out to year 20, those values drop to "0." Also, the energy values drop to full impact after year 3.

Alternative 1 (2-yearr implement with redirected internal funding)	Data	Year 1	Year 20
Annual inflation rate	4 %	1.040	1.040
Annual roadway lighting energy (Kwh)	39,283,805	69,949,225	39,283,805
Value of energy for roadway lighting	$3,535,542	$6,295,430	$7,448,855
Annual facility lighting energy (Kwh)	1,721,572	9,038,251	1,721,572
Value of energy for facility lighting	$258,236	$1,355,738	$544,064
Annual roadway lighting maint. costs (veries after construction)	$1,106,747	$553,374	$0
Annual facilities lighting maint. costs (yard lights to LED)		$30,000	$0
Cost to upgrade roadway lighting (2 years 25/75 on) Kwh savings	$31,294,598	$15,647,299	$0
Cost to upgrade facilities lighting (2 years 25/75 on) Kwh savings	$5,793,820	$2,896,910	$0
Rebates, given after project is on line	$10,855,839	$2,171,168	$0
Total energy use (Kwh)	41,005,377	78,987,475	41,005,377
Total energy costs	$3,793,778	$3,793,778	$7,992,918
Total lighting maint. costs	$1,106,747	$583,374	$0
Total annual system Ops costs	$4,900,525	$4,900,525	$7,992,918
Metric tons of CO_2 from electricity use	24,050	34,439	17,878
20 year totals	**Summary totals**		
Total energy use (Kwh)	870,750,333	78,987,475	41,005,377
Value of energy	$118,166,043	$7,651,168	$7,992,918
Maintenance costs	$1,146,747	$583,374	$0
Project costs	$26,232,579	$16,373,041	$0
Total system costs	$145,545,370	$24,607,583	$7,992,918
Total metric tons of CO_2	379,647	34,439	17,878

Making a Decision

In this section, the screen shots depict the 20-year difference between the base case and Alternative 1. Years 1–5 and 16–20 are shown here. The choice to redirect internal funding was based upon two factors: one being the department had a cost savings due to a poor snow season and, more importantly, internal funding the project allowed all the savings to come to the department rather than using a majority of those savings to pay interest costs on borrowed funding. 154.7 million dollars of state funds over the 20 years could be used to match Federal Highway Dollars and leverage up to 1.24 billion in federal funding over the same 20 years (for a total of about 1.4 billion in state and federal project dollars).

Basecase versus A/t. 1	Summary totals 20 year	Year 1	Year 20
Annual energy differential	962,213,163	12,660,700	50,642,798
Annual cost differential	154,728,186	−$14,523,871	$13,251,941
Total metric ton CO_2 differential	419,525	5,520	22,080
Running total cash flow for the project		−$14,523,871	154,728,186

Deployment Options: A Discussion

As to the final deployment decision by Caltrans management, the 2011/2012 fiscal year available funds were used to buy the fixtures now and schedule deliveries over the next few years, as the 12 districts within the department deployed the products. Coordination with the various electric utilities within the state and LED fixture installs has occurred by the fall of 2012 and incentive funds plan to be used to buy additional fixtures from the state contracts awarded in July of 2012. District deployment will utilize a combination of existing programmed roadway lighting maintenance resources and in some cases augmentation of existing resource with private sector service contracts. The incentive to the districts is that as soon as the LEDs are installed, the sooner electrical maintenance staff can redirect their effort to other projects like loop detectors, weigh-in-motion systems, traffic flow monitoring cameras, and traffic accident repairs to damaged roadway infrastructure. Like most government organizations these days, there are too many projects, but not enough staff resources.

Conclusion

This chapter identified and demonstrated the analysis processes that need to occur when deciding to implement an energy conservation and efficiency opportunities. The basic approach can be applied to any and all conservation and efficiency activities, be it energy, water, air quality, or other environmental projects that require economic analysis. The better the data going into the analysis along with best case understanding or the system under study, the better the results. Also, a lesson learned in construct of the analysis engine would be to build as much flexibility into the base case engine to leverage into the alternative and variation portion of the model. Remember, sometimes "zero/nada/zip" is a good thing!

Side Bar 2: Other Energy Opportunities for economic change, technology shifts, job creation, and optimization of resource use:

- Other benefits and opportunities for LED roadway and exterior surface lighting projects: Once the LED lighting systems are used to upgrade the base lighting systems, question can then be asked as to why and when do the lights really need to be used. As Caltrans staff gain knowledge about controlling LED maintenance yard lighting, they can then focus on roadway lighting systems and use the value engineering functional analysis to process to find additional saving opportunities, like do the lights have to be on all the time, what functions can be modified or improved?
- Air condition paradigm shifts: HVAC can include shifting air conditioning system from current high electricity-based systems back to thermal-based absorption technologies (that have been around for hundreds of years), then construct solar-thermal pre-temp systems to reduce the need for carbon-based fuel usage; the hotter the outside air temperatures, the more effective the solar-thermal systems and the lower the carbon-fuel consumption. Reducing the need for peaking electricity needs due to the use of thermal systems, the more electrical energy is available for electric-based transportation. Long-term energy planning for both traditional and renewable energy systems would gain optimum levels if the seasonal use swings could become more level.
- Hydrogen and transportation, a re-think: Ammonia compounds used to transport hydrogen for use in hydrogen energy centers and vehicles. In 2006 the University of California at Berkeley was contracted by Caltrans to look at alternatives to compressed or liquid hydrogen, with the first focus on ammonia-based compounds. The feasibility study was to identify barriers or technology issues and define status of this option.

Chapter 11
Blue-Green Agricultural Revolution

Daniel Nuckols

Abstract New networks of cross-movement coalitions are creating alignments of political, social, and economic action around the issues of sustainable development and renewable food systems. These diverse and transformative coalitions cross boundaries to come together and address the integration of three spheres–environmental (natural resource use and environmental management), economic (cost/benefit calculations and research and development), and social (community-based dialogue and action that addresses standard of living, education, and human rights). To succeed, this chapter argues that effective sustained coalitions will need to improve upon their record of citizenship and governance. Coalitions will need to demand that governments put forth macro-policies that address such issues as intergenerational inequity, market prices that do not reflect ecological damage, and human rights violations. A pre-analytic is called for; one that admits that the crucial ecological issue of successful sustainable planning means that transcendence (compassion, wisdom, understanding, and empathy) will be needed. Tensions between competing perspectives cannot always be solved by the logic and method of neoclassical economics.

Overview: Conflict and Cooperation in Renewable Food System Management

Cross-movement coalitions are one of the more exciting social movements of recent years and provide both opportunities and challenges for those implementing such partnerships and, of course, research opportunities for social scientists. The interactions of organizations across boundaries can become exceedingly complex,

---

D. Nuckols, Ph.D. (✉)
Austin College, Austin, TX, USA
e-mail: DNuckols@austincollege.edu

especially when their policies and agendas are seemingly at cross-purposes, ideologically inconsistent, and too measured when their mission includes a call for strongly relaying their values and ethics. This can be observed when coalitions heartily promote sustainable economic development, and renewable agriculture in particular, but seem relatively withdrawn when cultivating and proselytizing the ethical acumen needed to revive humans' healthy relationship with nature. One can also attest to the concern that these coalitions are not fully addressing the negative consequences that can emanate from a market system nor fully questioning the assumptions and normatively constructed models supported by mainstream neo-classical economists. These coalitions are also not taking full responsibility for their lack of effort in cultivating the art of citizenry and governance.

Coalitions involving initially unexpected allies have centered upon, among other things, energy conservation, regenerative land stewardship, and environmental regulations. Further, as the global economy matures, more and more labor dislocation is manifested, as international competition reduces the bargaining leverage once held by labor unions. This has lead labor unions to seek new allies. With similar concerns to those of labor unions over the ecological impact of free trade and labor displacing technologies, many environmentalists and labor leaders have grasped the opportunity to find common ground, one that appeals to working-class voters.[1]

The Apollo Alliance, founded in 2003, is one such coalition, bringing together what has come to be called blue-green partnerships. Traditional labor unions such as the United Mine Workers, United Auto Workers, and United Steelworkers have strategically aligned with the Sierra Club and National Wildlife Federation. Their mandate was to promote progressive polices toward "green" jobs in the areas of alternative energies, hybrid automobiles, and more efficient transportation networks. Similarly, another coalition, the Blue-Green Alliance, has reframed the issue of job creation by linking it with global warming. Its coalition consists of an alliance of the Sierra Club and Natural Resources Defense Council with the United Steelworkers and Communications Workers of America. Formed in 2006, this grassroots movement began in the relatively strong union states of Michigan, Minnesota, Ohio, Pennsylvania, Washington, and Wisconsin.[2]

In an effort to support a clean energy economy—one that generates sustainable jobs—the Apollo Alliance merged in 2011 with the Blue-Green Alliance and is presently known as the BlueGreen Alliance.[3] Currently, with a coalition of 12 of the biggest and most well-known environmental organizations and labor unions, they continue to partner in battling for jobs in the green economy. Recently, the coalition testified before the US Environmental Protection Agency to limit greenhouse gas emission from new power plants.[4]

---

[1] Mayer (2009).

[2] "Blue/Green Alliance Project" http://www.bluegreenalliance.org.

[3] "Blue/Green Alliance Project" http://www.bluegreenalliance.org.

[4] "Blue/Green Alliance Testimony at EPA Greenhouse Gas Rule Hearings" http://www.bluegreenalliance.org/news/publications/bluegreen-alliance-testimony.

With protecting health being a common ground, labor groups and environmentalists have strategically merged their given perspectives and agendas on job creation, occupational health, and ecological sustainability. One could rightly argue that this speaks to the creativity and compromise needed toward possible solutions, when addressing the issues of broad-base worksite health and the large-scale degradation of the environment. It remains true that most of the country still carries the belief that there is a trade-off between job growth and a healthy environment. The centerpiece of the BlueGreen Alliance political agenda is, however, to generate cooperative dialogue that leads to a reduction in the exposure of toxic elements, both in global communities and particular worksites. These activists feel strongly that a healthy community and workplace sets the stage for resilient job creation and sustained economic growth.

In his book *Blue-Green Coalitions: Fighting for Safe Workplaces and Healthy Communities*, Brian Mayer offers a detailed treatment of the unusual formation of labor-environmental alliances to address health issues. He states:

> Unions are often interested primarily in protecting what remains of organized jobs and preventing further layoffs to maintain a basic standard of living. Environmental protection, which can act as a limit on economic growth, is therefore perceived as a direct threat to jobs—driving the labor movement to ally with industry in opposition to environmental organizations. But, as more in-depth analysis suggests, externalities such as environmental pollution and occupational health hazards disproportionately affect those at the lower end of the socioeconomic structure, the working class, which would theoretically create allies between environmentalists and organizations like unions that tend to represent working class individuals.[5]

A moment's reflection, argues Mayer, reveals why one could initially doubt the existence of such a broad partnership. Unions are built instrumentally, with a hierarchical structure, starting with a national confederacy and ending with member workers of a local union. Mayer maintains that laborers benefit from their membership fees being used for collective representation, resulting in a functional and instrumental relationship with private enterprises. Horizontal membership, however, characterizes the environmental movement, and its success is dependent upon volunteers. A challenge for the environmental movement is to channel the normative and personal values of its participants toward a collective action that cannot, necessarily, be monetized. These two structural types of collective action, horizontal and vertical, would naturally be thought to conflict, given their different and singular modes of internal collaboration and possible class differences. However, as Mayer argues, "when crises occur and disrupt the status quo, unique opportunities to work across class identity divides arise. These moments of opportunity are essential in building blue-green coalitions."[6]

---

[5] Mayer 4.

[6] Mayer 4.

Cooperation in Agriculture

There is also potential for both conflict and cooperation between those private and public activists addressing land management and food production. Coalitions to address partnerships in agriculture are relatively new, and the challenges are daunting. However, the Quivira Coalition is one success story, bringing together private landowners, conservation groups, public land managers, public agencies, and scientists to promote overall conservation, and in particular the healthy stewardship of the land and its fertility. Recognizing that these groups have incentives that many times work at cross-purposes, the Quivira Coalition's initial mission was to show "how sensitive ranch management and economically robust ranches can be compatible" and how partnerships can bring forth "an emerging progressive ranching movement that operates on the principle that the natural processes that sustain wildlife habitat, biological diversity and functioning watersheds are the same processes that make land productive for livestock."[7] The Quivira Coalition treats with askew any hint of it participating in lawsuits, legislation, and arguing of positions. Rather, acting on the belief that the need for sustainable ranch management is a given, their preferred mission is to bring together economic and political actors to address conservation concerns. These include their newer initiatives pertaining to "the accelerating loss of open space to sprawl (often on former ranch lands), the treat of noxious species to native biodiversity, the rise of recreational damage on public land, and the spread of nature deficit disorder" and to "embrace a more 'holistic' vision of land health and restoration, involving grass, water, cattle, and people."[8]

As one member of the Quivira Coalition, the nonprofit Holistic Management International (HMI) represents a successful example of constructing a formal alliance. HMI has brought together communities—environmentalists, ranchers, farmers, educators, and public agencies—that were heretofore relatively unproductive, when looking for ways to systematically coalesce to find common ground in restoring the health of degraded private, public, and communal grasslands worldwide. Formed in 1985, HMI's mission is one of global education, "providing training, courses and consulting services to stewards of large landscapes, including ranchers, farmers, pastoral communities, government agencies, NGOs, environmental advocacy groups and other non-profits."[9] It has over 60 educators and thousands of land stewards who use holistic management strategies to manage more than 30 million acres around the globe.

With a pragmatic "show me the money" philosophy, HMI concentrates on relaying to agricultural producers—in an easy to understand, jargon-free fashion—strategies that will generate the fulfillment of a "triple bottom line" pertaining to sustainable financial, environmental, and social benefits. HMI engages previously perceived

---

[7] "The Quivira Coalition: About Us" http://www.quiviracoalition.org/About.

[8] "The Quivira Coalition: About Us" http://www.quiviracoalition.org/About.

[9] "Holistic Management International: State of Qualifications (2010)."

disparate actors to show how each can be part of "increasing annual profits and enhancing livelihoods, improving soil health and biodiversity of rangelands and pastures, increasing grazing and wildlife capacity, optimally using rainfall and conserving water, growing healthier crops and achieving higher yields, enhancing family relationships, and resisting and positively affecting global climate change."[10] As a specific example pertaining to Texas, HMI has aligned several agricultural groups to work together in reclaiming distressed land and reducing or eliminating the use of heavy equipment, hormones, medication, and insecticides. "One of our primary goals is to *integrate* (emphasis added) holistic management strategies and processes with those of other agricultural colleagues, such as AgriLife Extension, National Resource Conservation Service, Texas Parks and Wildlife, the Texas Section of Society for Range Management, and the Texas Nature Conservancy, to ensure that programs provided for agricultural families and communities are supportive of sustainable land-based enterprises."[11]

Critical Issues Remaining for Coalition Management of Renewable Food Systems

With the overall ecosystem being of paramount concern, the above coalitions are in essence addressing the need to put forth a composite understanding of how nature functions simultaneously with profit-motivated business enterprises, labor groups, communities, educators, public advocacy agencies, public/private partnerships, and any collective responsible for what at the end of the day is ultimately being addressed—a *well-functioning biosphere, restorative land practices, and personal nutritional health, all in the context of sustainable production at a living wage.* One crucial dependent variable here is *food*. A critical trait of humans and their food consumption is that the food web boundary is very indistinct and extremely difficult to map. Now obvious and somewhat acknowledged by the above coalitions, successful supervision must supply strategies that are sustainable, and this can only be achieved if decision makers navigate *within* the systems to be overseen, not against them.[12]

Even Aldo Leopold knew that land and food systems had to be economically expedient to satisfy the market's "bottom-line" profit barometer, but he also insisted that the stability, integrity, and beauty of the biotic community be maintained. Leopold knew that one could not ignore economic factors, especially those behavioral components emanating from individual and collective self-interest. As he states in *The Sand County Almanac*, "It of course goes without saying that economic feasibility limits the tether of what can or cannot be done for land. It always has and always will."[13] The Quivira and HMI coalitions are, to their credit, acknowledging

---

[10] "Holistic Management International: State of Qualifications" 1–10.
[11] Normand (2010, p. 1).
[12] Allen and Hoekstra (1992, pp. 269–277).
[13] Quote taken from Varner (1998, p. 129).

this reality and hence, attempting to operate within its framework. But is such environmental democracy enough — with its acknowledgement of economic realities, movements toward sustainable agricultural coalitions, community-supported agriculture, and public-interest scientists? What is missing?

The arrival of the renewable organic food system movement has become a major and contested arena for the alignment of consumer and producer groups. Obviously, as Marsden, Sonnino, and Morgan admit, "It is clear that the development of alternative and re-localized networks creates new alignments of political, social and economic action around which potentially more sustainable forms of rural development can take hold… this is a highly problematized (sic) area where the construction of new markets can lead to forms of social and economic exclusion being reproduced and questions need to be raised as to the long-term sustainability of these developments."[14] As it applies to the renewable (BlueGreen) food movement, what are these new forms of economic and social exclusion, and what stance should be taken?

To start, renewable food movement activists and coalitions need a focused, coordinated, and sustained confrontation with mainstream neoclassical economics, as it relates to its approach to agriculture. A strong example of a typical orthodox economic description for why agriculture is in trouble can be seen in Steven C. Blank's *The Economics of American Agriculture*, where he relates the increased international competition in commodity markets to the fact that "there are fewer than two million American producers, and in the long run they cannot win any political battles against the 300 million American consumers of the cheap food being provided by the global market."[15] The author goes on to state that land, labor, and capital prices are higher in the USA, and thus, less-developed countries have an absolute cost advantage, enabling them to underprice American agricultural commodities. The author of this diagnosis believes an optimistic note is called for "The difficulty American agriculture has in fighting these trends has this bottom line: *everything that is happening in this development of a global market is good for U.S. agribusiness firms and American consumers* (emphasis in original). The fact that now both domestic and international producers are willing to provide the United States with products at the same or lower prices means that Americans are eating better and prices are not going up."[16] Such a blinding short-sighted description has coalitions needing to ask and answer the following: What are some of the issues neglected in this one-dimensional summary?

To truly understand the sustainable food network and the plight of farmers, one must go beyond neoclassical economic reductionist and mechanical presumptions that these concerns can be reduced to the "workings of a machine and neatly molded to suit the demands of the market"; especially when the proponents of renewable agricultural accepts the scientifically supported claim that "causal forces operate at

---

[14] Marsden et al. (2008).
[15] Blank (2008, p. 129).
[16] Blank 129.

different levels of aggregation and that a comprehensive causal explanation cannot be reduced to a single level",[17] such as "free" international trade. Another specific way mainstream economic analysis clearly ignores the philosophy behind sustainable land production—that encompasses interdependent and synergistic parts—deals with how it normatively frames what is to be measured, studied, and acted upon. In his text, *Civic Agriculture: Reconnecting Farm, Food, and Community*, Thomas Lyson pursues, among other topics, the issue of the *social construction* of modern economic categories. He states that:

> …in the realm of farming, agricultural economists have focused virtually all their attention on the "economically efficient" production and marketing of selected "standard" commodities….those commodities that can be "mass produced" in accordance with the precepts put forth by the neoclassical production function and that articulate with standardized mass markets have garnered most of the attention. Thus, for example, there are detailed econometric analyses of the production practices for all the major market-oriented commodities such as corn, wheat, soybeans, and considerable research time and money are devoted to fine-tuning these models. Non standard varieties or commodities that have not achieved "economies of scale" because they are too embedded in household or community relations to get an "economically unencumbered" reading, have been largely ignored by the conventional agricultural community.[18]

Lyson goes on to explain how such locally supplied products as maple sugar, cedar oil, fruits, and vegetables are deeply embedded in the economic and social fabric of the region—farm communities and specific households of the northeast United States. But these commodities, says Lyson, are approached by mainstream agricultural economics as "marginal" or "peripheral" farm enterprises and are hard to quantify and categorize,[19] even though such agricultural production constitute a livable income for thousands of farm families across America. This lays bare the shortcomings of the modern economic approach to agriculture. Those who supply locally produced organic commodities—on small-scale farms that practice healthy soil management—carry the philosophy that they are not "above nature," or that they can "master nature."

In contrast to large-scale agribusinesses that emphasize high productivity per acre and quarterly dividend payments to stockholders, small-scale sustainable farmers want to know the same as what ecologists desire to know, "how ecosystems function, how they are sustained by sunlight, how species interact and coexist, and how energy and materials circulate within and between adjacent ecosystems."[20] Nature is not an obstacle to be overcome nor can it be used as an "input" in an econometric equation for purposes of adequately explaining the feedback loops of the biosphere, "where wholes cannot be understood by reduction to their component parts, and living things interact in ways that can never be understood fully."[21]

---

[17] Foster et al. (2010, p. 263).
[18] Lyson (1992, p. 80).
[19] Lyson 23.
[20] Soule and Piper (1992, p. 80).
[21] Dryzek (2005, p. 217).

In some respects, farmers and ranchers suffer from what almost all small businesses suffer: a small fraction of the population having inordinate control of banking, industry, and commercial institutions. Likewise, producers of agricultural commodities and livestock also bear the burden of fewer and fewer individuals and corporations controlling more and more land, credit, water, and marketing channels that are adversely affecting sustainable organic food production, distribution, and consumption.[22] Coalitions supporting regenerative farm and ranch practices must concentrate on, and counteract, the origin and distribution of power; they must study and act upon contemporary political economy, as well as the narrative emanating from shallow Newtonian-influenced mainstream economics. Consider the example of the changing technology that is being pushed by some firms, especially biotechnology, that is leading to:

> ...the proletarianization of the farmer, and the appropriation of ownership and control of indigenous plants and animals in third world countries. Subsistence farming is in decline in the third world while the production of luxury crops for export to the rich countries is being expanded as never before. The result is a rise in world food supplies, together with an increase in world hunger. So sharp are these contradictions that hunger is expanding in the United States itself, at the very heart of the system, where it is no longer surprising to see food lines and soup kitchens even during economic expansions. The growth of agribusiness has also generated more and more ecological problems through the subdivision of traditional diversified farming into specialized production, the break in the soil nutrient cycle, the pollution of land and water (and food itself) with chemicals, soil erosion and other forms of destruction of agricultural ecosystems and so on."[23]

The above quotation sheds partial light on the second general concern pertaining to the effectiveness of coalitions; they need to admit to their relative lack of robust participation in political discourse, when it comes to addressing all aspects of sustainable development. Alliances must aim their mission toward the derived consequences of corporate decision making replacing, in many important respects, democratic decision making.

For instance, consider the singular concern over the engineering of genetically modified organisms (GMOs). See below table by Soule and Piper for other non-democratically determined results flowing from modern agribusiness practices. In essence, genetic engineering is the patenting of life. Little democratic discourse—even coalition discourse—has been directed toward the fact that major food and medicinal crop genomes, and their application, are now in the domain of multinational firms. This has taken place because legislative and judicial branches have sanctioned corporate private property rights over democratically induced communal and environmental rights.[24] Almost any anxiety springing from corporation's control and influence of nature, of which GMOs being one of many, but must be faced by environmental activists in the context of the entire political process and policy cycle. Environmental coalitions cannot ignore government and the courts.

---

[22] Lappe et al. (1998, p. 99).

[23] Magdoff et al. (2000, p. 8).

[24] Henson (2002, pp. 227–228).

Modern Agricultural Practices that Have Contributed to the Current Ecological and Economic Crises

Practice	Problems addressed	Problems created
Mechanization	Labor inefficiency	Erosion, energy dependency, capital expenses, interest payments larger farms, and fewer farmers
Inorganic nitrogenous fertilizer	Crop yield	Groundwater contamination, farm specialization, pests, erosion, energy dependency, high input expenses, and less economic resilience
Pesticides	Crop loss to pests (success doubtful)	New pests, resistant pests, water pollution, human poisoning, energy dependency, and high input expenses
Hybrids and genetically narrow varieties	Crop yield and nonuniform traits	Aggravated pest problems, loss of local adaptations, chemical dependency, and high input expenses

In essence, effective sustained coalitions will only continue to progressively evolve by improving upon their record in the skills of citizenship and governance.[25] They must recognize that any successful attempt at sustainability will warrant constant citizen participation in the building and monitoring of democratically controlled governments. Only governments can stop such neoliberal policies of, for example, counting the consumption of natural capital as income and placing taxes on labor and income instead of resource consumption.[26] Moreover, sustainable development discourse should see coalitions demanding that governments put forth macro-policies that acknowledge and correct for the following: (1) both distributional and intergenerational inequity; (2) market prices that do not properly reflect the spillover external social and ecological damage of private-firm production; (3) capital flows to desperate Third World and postcommunist regions, when the result is child labor, little or no environmental and health compliance, human rights violations, and below-the-table financial exchanges between the public and private sector; and (4) the belief in a philosophy of continued and infinite growth in population and GDP that is not sustainable on a bounded planet—with coalitions vigorously arguing that more and more "rational" technology will never deflect from these facts.[27] David Orr best describes what the transition to sustainability will require:

> Only governments moved by an ethically robust and organized citizenry can act to ensure the fair distribution of wealth within and between generations. Only governments prodded by their citizens can act to limit risks posed by technology or clean up the mess afterward. Only governments acting on a public mandate can license corporations and control their

---

[25] Soule and Piper 52.
[26] Daly (2002, pp. 210–212).
[27] Carruthers (2006, pp. 293–94).

activities for the public benefit over the long-term. Only governments can create the financial wherewithal to rebuild ecologically sound cities and dependable public transportation system. Only governments acting with an informed public can set standards for the use of common property resources including the air, waters, wildlife, and soils. And only governments can implement strategies of resilience that enable the society to withstand unexpected disturbances.[28]

Inherently, Orr's message challenges the so-called invisible hand of the market and how it hinges on a philosophy of people pursuing their individual self-interest, with the supposed result of an aggregate economy spontaneously benefitting in a free, nondirective, manner. Orr's quote clearly speaks to this folly. Granted, coalitions are an important start toward an active citizenry, but at the very least, we need even stronger collective action in the form of *coalitions partnering with other coalitions*, in order to form mass social movements that passionately plead and publicly petition for global sustainability.

There is much work to be done toward the promotion and successful implementation of global sustainability, with special emphasis on land use and organic food production. Besides the issues offered above, other reasons include the fact that there is "wide diversity and malleability of the practices and meaning of the organic" and "the range of farming 'styles' implies a related ambivalence toward the philosophy and ethics of sustainable agriculture."[29] Because of this, there will need to be a third directive for coalitions and other collectives promoting sustainability, one with less techno-centric flare, that is, scientific claims that simply highlight the bottom-line logic of sustainable land management, with data regarding soil erosion rates, nutritional benefits of organic food, saved energy costs, etc.

Attention must also be given to programs and policies that explicitly draw out and communicate the importance of social values and relationships—specifically, those deriving from the ethical dimensions of sustainability. As Goodman correctly argues, "The transition to more environmentally sound management hinges on the success of proselytizing campaigns to win 'converts' among researchers, farmers, and consumers."[30] Coalitions, along with sustainable ranching and farming communities, need to communicate more forcefully the moral and ethical dimensions of managing nature. This calls for living in a more reflexive modernity,[31] one that not only concentrates, for instance, on consumer's interests toward food safety and nutritional issues, "but reach beyond these household concerns to embrace an *idea* (emphasis in original) of the good life or the good society that includes some notion of the right relationship to nature."[32]

The coalitions behind the sustainable agricultural movement will ultimately fail if they do not include a "social space" where focus is extended to the complex web

---

[28] Orr (2003).
[29] Goodman (2000, p. 216).
[30] Goodman 218.
[31] Kaltoft (2001, p. 157).
[32] Vos (2000, p. 253).

of natural social interactions. There may be some truth, at least in the short run, to Allen and Kovach's remarks that "ultimately, the dynamics of the capitalist market will consume improvements in ecological sustainability. The question then becomes, does the market for organic agriculture have the potential to instigate these larger changes, possibly fueling a vital social movement around organic food and agriculture."[33] However, for such an international social movement to take place, coalitions and others must speak globally to the ethical relationships surrounding resource stewardship and to the social relationships that holistically encompass the joy and quality of life of all who work in, and benefit from, agriculture.

Coalitions must confront economic inequality and injustice from the perspective "divergence," not "convergence." Some confrontations cannot be won in the usual sense, but only transcended. These are what E. F. Schumacher called, "'divergent' problems formed out of the tensions between competing perspectives that cannot be solved, but can be transcended. In contrast to, 'convergent' problems that can be solved by logic and method, 'divergent' problems can only be resolved by higher forces of wisdom, love, compassion, understanding, and empathy."[34] David Orr calls for a higher level of spiritual awareness, proclaiming that something akin to spiritual renewal is the *sine qua non* of the transition to sustainability. "Scientists in a secular culture are often uneasy about matter of spirit, but science on its own can give no reason for sustaining humankind."[35] Finally, it is not enough for coalitions to be carrying labels promoting themselves to be in favor of agricultural sustainability and renewable food production. As Marx was to reflect, "Value does not stalk about with a label describing what it is."[36]

Conclusion

Much of this chapter has focused on how coalitions, comprised, at times, of cross-functional groups, can come together to creatively address green strategies for sustainability. Effectively working outside of traditional and more mainstream environmental movements, the coordination emanating from these diverse and transformative groups can bring forth the enhanced focus and clarity needed for effective environmental stewardship. However, these coalitions must not take a one-dimensional stance toward sustainability but rather look upon sustainable development as the integration of three spheres—environmental (natural resource use and environmental management), economic (cost/benefit calculations and research and development), and social (community based dialogue and action that

---

[33] Allen and Kovach (2000, p. 230).
[34] Quote taken from Orr (2003).
[35] Quote taken from Orr (2003).
[36] Quote taken from Allen and Kovach 230.

addresses standard of living, education, and human rights). The coming together of cross-purpose coalitions provides unique opportunities to integrate these three spheres. The environmental-economic intersection can address energy efficiency, food security, life-cycle management, and incentives for sustainable resource use.

The social-environmental overlap is best suited to reach the issues pertaining to climate change, environmental regulations, environmental justice, and the stewardship of natural resources, at both the local and global level. A policy marriage of the economic-social spheres can bring together wide-ranging discussions over business ethics, fair trade, social investment, job creation, skill enhancement, and worker's rights.[37] Undoubtedly, green coalitions must bring together the leadership needed to acknowledge these interconnections of successful sustainable planning, all within a viable and environmentally healthy business context. Such a coalescence of leadership will need first, however, to admit that such crucial ecological issues will have to have the perspective of transcendence (compassion, wisdom, understanding, and empathy) and not always look toward logic and method.

References

Holter P (2010) Holistic management international: state of qualifications. Holistic Management International, Albuquerque

Chase G (2012) Workshop for integrating sustainability in education. Association for the advancement of sustainability in higher education, San Diego

Allen TFH, Hoekstra TW (1992) Toward a unified ecology. Columbia University Press, New York

Allen P, Kovach M (2000) The capitalist composition of organic: the potential of markets in fulfilling the promise of organic agriculture. Agric Hum Values 17:221–232

Blank SC (2008) The economics of American agriculture: evolution and global development. M.E. Sharpe, New York

Blue/Green Alliance Project. http://www.bluegreenalliance.org

Blue/Green Alliance Testimony at EPA greenhouse gas rule hearings. http://www.bluegreenalliane.org/news/publications/bluegreen-alliance-testimony

Carruthers D (2006) From opposition to orthodoxy: the remaking of sustainable development. In: Dryzek JS, Schlosberg D (eds) Debating the earth: the environmental politics reader. Oxford University Press, New York, pp 285–300

Daly H (2002) Five policy recommendations for a sustainable planet. In: Schor JB, Taylor B (eds) Sustainable planet: solutions for the twenty-first century. Beacon, Boston, pp 209–221

Dryzek JS (2005) The politics of the earth: environmental discourses, 2nd edn. Oxford University Press, New York

Foster JB, Clark B, York R (2010) The ecological rift: capitalism's war on the earth. Monthly Review Press, New York

Goodman D (2000) Organic and conventional agriculture: materializing discourse and agroecological managerialism. Agric Hum Values 17:215–219

---

[37] Workshop for Integrating Sustainability in Education (2012).

Henson D (1992) The end of agribusiness: dismantling the mechanisms of corporate rule. In: Kimbrell A (ed) The fatal harvest reader: the tragedy of industrial agriculture. Island Press, Washington, DC, pp 225–239

Kaltoft P (2001) Organic farming in late modernity: at the frontier of modernity or opposing modernity? Sociol Ruralis 41(1):146–158

Lappe FM, Collins J, Rosset P (1998) World hunger: twelve myths. Grove, New York

Lyson TA (1992) Civic agriculture: reconnecting farm, food, and community Medford. Tufts University Press, Medford

Magdoff F, Foster JB, Buttel FH (2000) An overview. In: Magdoff F, Foster JB, Buttel FH (eds) Hungry for profit: the agribusiness threat to farmers, food, and the environment. Monthly Review Press, New York, pp 7–21

Marsden T, Sonnino R, Morgan K (2008) Alternative food networks in comparative perspective: exploring their contribution in creating sustainable spaces. In: Marsden T (ed) Sustainable communities: new spaces for planning, participation and engagement. Elsevier, Amsterdam, pp 255–274

Mayer B (2009) Blue-green coalitions. ILR Press, Ithaca/New York

Normand A (2010) Holistic management international. Holistic Management International, Johnson

Orr DW (2003) Four challenges of sustainability. Oberlin College, Spring Seminar Series, Oberlin. http://www.uvm.edu/gice/SNR_seminar/Readings/CB-42

Quivira Coalition: About Us. http://www.quivaracoalition.org/About

Soule JD, Piper JK (1992) Farming in nature's image: an ecological approach to agriculture. Island Press, Washington, DC

Varner GE (1998) In nature's interest? Oxford University Press, New York

Vos T (2000) Visions of the middle landscape: organic farming and the politics of nature. Agric Hum Values 17:245–256

Chapter 12
Going Beyond Growth: The Green Economy as a Sustainable Economic Development Strategy

Laurie Kaye Nijaki

Introduction

Planners and policymakers in the urban context often face very difficult decisions around the development process. Namely, what should the future development of a city be and according to what metrics should this outcome be evaluated? Given the economic trajectory forward, planners and policymakers often must consider growth in terms of production, skills, wealth, and how the pursuit of progress improves or impedes the "true" quality of life on the ground. And, they often face trade-offs between policies that favor aggregate growth over equity considerations that evaluate the way in which this aggregate growth is distributed. They must often define the path forward based upon an evaluative metric that incorporates values of both efficiency and equity for any number of potential stakeholder groups in the pursuit of a better quality of life.

Traditionally, economic development strategies and estimations of economic success were largely concentrated on achievement of aggregate growth as reflected in expanded GDP within a certain geographical area. The goal of development was an aggregate increase in GDP and planners and policymakers could seek to engage in strategies that most effectively led to increased GDP. However, aggregate economic growth alone in terms of measures of productivity, skills, and wealth may be an oversimplification of true economic development. And, in fact, economic growth may sometimes be in conflict with measures that protect equity considerations and quality of life goods.

L.K. Nijaki, Ph.D. (✉)
School of Natural Resources and Environment/Stephen M. Ross
School of Business, Erb Institute for Global Sustainable Enterprise,
University of Michigan Ann Arbor, Ann Arbor, USA
e-mail: lnijakikaye@yahoo.com

Specifically, the definition of growth can be expanded and redefined in two predominant and related veins. First, there is the distributional question in terms of the distribution of both the benefits and costs of growth. This amounts to a question of growth for whom? Who in a city is given the opportunity to benefit from growth, who will benefit from the spoils from development, and who will suffer from the harms from development? Second, there is a widening of the very aims of growth to incorporate the notion of increasing the quality of life—how can different populations use their different economic resources differently in order to achieve varying amounts of success in bolstering quality of life. Increasing quality of life is a much broader estimation of growth than simply economic growth as measured by expanded GDP and one that significantly widens the metrics for evaluating the success of policies for growth including the negative externalities of development. This can be incorporated through the shift from simply growth or development, to the notion of "green" growth.

This chapter briefly discusses the role of growth as a critical value governing decision-making in the urban context and in the development of the ideal city. As a part of the broader paradigm, this chapter seeks to focus on the role of the "next economics" of green growth in fostering new approaches to economic growth and environmental preservation at the locally driven level. Section "Introduction" provides a traditional, growth-oriented, approach toward economics and toward economic development more specifically. First, I will examine some traditional approaches to understanding economic development as a focus on dimensions of a betterment of the quality of life through the pursuit of increased aggregate economic growth. Second, I examine the tensions between economic growth in aggregate and equity through an examination of several key contextual elements including economic restructuring, globalization, and the political/institutional dimension of growth discussions. What may define the "next economics"? In section "Traditional Economics: Understanding Growth and Green" of this chapter, sustainability is examined as a new approach to economic growth for communities. As a fundamental aspect of sustainability, the green jobs movement is discussed as a significant, relatively novel component of the discourse around economic development that addresses concepts of both quality of life increased through development and the role of equity in sharing the spoils of economic growth. This approach provides a unique application of this new paradigm to urban development decisions and locally driven policymaking at the nexus of economic and environmental concerns.

At the end of the day, the discussion around economic growth and development is fundamentally a discussion of progress—or the ideal path forward for urban development that may incorporate both considerations of growth and equity. A deepened discussion around economic development must include a broad swath of questions: Development may be a laudable end, but in what constellation should this be achieved? How should the outcome of development decisions be evaluated? And, equally importantly, how should the costs and benefits of development and growth be distributed? A true estimation of economic development will not define progress as "paving paradise to put up a parking lot," and will come with an assurance that said paving is not merely benefiting a select few.

Traditional Economics: Understanding Growth and Green

The first section of this chapter provides an overview of traditional economic theory. The approach seeks to understand the role of economic development in cities and the way in which growth was privileged by economic thought. First, I discuss the role of traditional growth-driven economics through a brief overview of the literature. Second, I provide an overview of the way in which such growth was contrasted with other dynamic values, such as equity and environmental quality. In the end, traditional economics provides a definitive flavor of policy and planning mechanisms that focus on growth over broader quality of life considerations.

Development, Growth, and Progress: Traditional Economics

Understanding, evaluating, and spurring economic growth is a common theme in planning and policy literature generally and in economic development literature more specifically. In fact, growth has often been a favored end above all others in the decision-making process. How is a better quality of life in the ideal city achieved in this model? In the "classical sense," economic development is hinged on economic growth. Often termed capital fundamentalism, economic growth and development was originally pinned to the investment in capital resources. Economists such as Sir Arthur Lewis assumed that economic growth was reliant on capital accumulation. In the 1950s, Robert Solow revisited notions of economic growth (Blakley and Leigh 2010). As noted by Easterly (2002), "his conclusion surprised many, and still surprises many today: investment in machinery cannot be a source of growth in the long run. Solow argued that the only possible source of growth in the long run is technological change." Thus, economic growth and development is fueled by technological innovation including new products, development of human capital, and new production methods.

A wide array of economic development strategies were and continue to be used that aim to bolster levels of economic growth. For example, economic base theory asserts that economic growth is directly related to the external demand for its goods, services, and products. Economic growth, and thus job generation and a "better quality of life," is created by utilizing local resources to produce exportable goods and services. In order to achieve this, economic developers must bolster growth through attracting businesses such as free-trade zones and tax relief. The economic development processes can be further strengthened by identifying opportunities for cooperative advantage and fostering entrepreneurialism (Blakley and Leigh 2010). Human capital development can also facilitate the process toward economic growth (Mather 1999). Taken together, these strategies will lead to economic growth and consequently to a better quality of life in the aggregate.

What about equity? The resultant growth in this vision has been connected to not only with an increase in the quality of life, but also with the eventual demise of inequality. When understanding the temporal relationship between equity and economic development, the relationship between economy and inequity over time was famously elaborated by the Kuznets curve. According to this theory, the relationship between economic benefit and inequity varies along the temporal trajectory of development. Specifically, at early stages of development, investment in economic growth is primarily focused on investments in physical capital. Inequity is seen as fostering growth by allocating resources toward relatively higher resource holders. These individuals are most likely to invest these resources to their fullest capacity. At later stages of economic maturity, the game changes. Here, an estimate of future costs becomes a more important determinant than the accumulation of physical capital. Inequity at this stage in the game slows growth by lowering educational standards. Thus, when inequality is plotted along the y-axis and income per capita is plotted as the independent variable, an inverted u curve emerges. Thus, the Kuznets curve suggests that more inequitable results will be automatically reached as economic growth continues—equity in this vision becoming a by-product of a normal development trajectory. Given this argument, a focus on economic growth in terms of traditional measures will invariably lead to a decrease in inequality. Economic developers, according to this logic, should consequently focus on improving economic growth in aggregate at the national level and perhaps at smaller geographic scope and scales in the same vein (PERC Research 2003).

Understanding the Economic Reality: Further Theories of Economic Growth

An inherent conflict may exist between aggregate, narrow measures of economic growth and of economic equity. A dwindling degree of economic inequity from aggregate economic growth may in fact not be the case, as envisioned by an examination of the historical trajectory of economic growth. Many scholars assert that the role of economics, equality, and quality of life including environment preservation has changed since the turn of the last century as a part of the trajectory of economic growth. The following section contrasts the focus on growth as the desirable path of progress, through a very brief examination of several contextual elements of the discourse around economic growth: the transition from Fordism to post-Fordism, globalization, and the sociopolitical aspects of development through aggregate and largely inequitable growth.

First, many scholars have envisioned the trajectory of aggregate economic growth and the limitations on reaching equity, through examining the restructuring of industry and its impact on human capital needs and the benefits that result. On the ground, economic growth in the last two centuries and the tensions between equity and growth have been specifically depicted as a transition from the "Fordist" to the "post-Fordist" production era. The Fordist era beginning in the 1920s is the poster child of a policy-oriented focus on aggregate economic growth. It was characterized by economic gain through mass production strategies. Corporate production strategies operated along the maxim of the bigger the better. Products were undifferentiated and available in mass quantities to the public at prices far cheaper than before (Soja 2000). Moreover, the employer/employee relationship was also standardized and was characterized by union-regulated full-time employment. This innovation in economic strategies profoundly influenced space through the resultant distribution of populations and the usage of space throughout the urban landscape. Products were now made available for mass consumption. Increased demand for newly affordable commodities led to economic gains and "economic development" (Soja 2000). And, at the same time, gains in profit margins also seemed to be correlated with increased environmental degradation as unbridled production fueled negative environmental externalities at a higher rate (Graham and Marvin 2001). And in fact, the cheap prices for consumer goods that result from the process further hurt the environment by creating perverse consumer incentives and magnifying the entire process (Soja 2000). Although aggregate growth may have increased, quality of life may not have.

By the 1970s, Edward Soja and other members of the "Los Angeles School" believed that the Fordist era began to wane and point to a major economic restructuring process that they term to be the transition to the post-Fordist state (Soja 2000). In the post-Fordist era, the rigidity of the production processes germane to the Fordist period did not allow for the flexibility of skills, human capital development, and production processes needed in the new globalized, high-tech, service-oriented economy. In contrast, companies in the postmodern era were now operating with smaller profit margins, increasing trade imbalances, raw material shortages, and stagflation. Widespread mass consumption was no longer available to effectively utilize the economies of agglomeration that were so fruitful in the Fordist era, and the environmental impacts were heightened (Soja 2000). The situation was epitomized in Los Angeles, with communities suffering from the environmental impacts of Fordist era production and a decline in manufacturing (see Arvinson 1999; Barbour 2001; Carney 1964; Soja 1986). Thus, economic growth arguably did not ultimately lead to the enhancement of a quality of life and increased equality as could be suggested by traditional approaches to economic growth. Adding this type of context may provide nuanced complexities to an analysis of whether or not a rising tide inevitably leads to a rising of all boats.

Focusing specifically on the new globalized economy and somewhat building on the above approach, many scholars focus on the role of globalization in impacting economic growth and other measures of quality of life. The changing aspect of economic growth along the temporal trajectory can be understood through Thomas Friedman's three phases of globalization. According to Blakely and Leigh's (2010) description of Friedman's theory, he begins with his notion of Globalization 1.0 that "brought significant prosperity to many local economies making manufactured goods, growing crops, and producing services and entertainment." This is followed by Globalization 2.0 where "assessing overseas markets and employing labor to other nations, U.S. local economies experienced plant closing on an accelerating scale. Local economies overly dependent on the industry sectors that found it most profitable to move their operations overseas experienced the greatest devastation. Many of the companies in these industry sectors simply were unable to compete with cheaper imports, and thus moved overseas to take advantage of lower labor and other production costs." Today, we sit within Friedman's third stage of globalization or Globalization 3.0. This phase is characterized by a shift toward non-western nations as drivers for economic growth. In order to meet the demands of economic development in this period, we must adopt "an orientation away from traditional business development and recruitment toward ensuring all participants in a local economy have adequate preparation to make maximum contributions. Recovering from the global recession and creating a new path for prosperity clearly means a shift from business as usual. It also requires an economy focused on reinventing itself though new technologies, innovations, and renewed commitments to ethical leadership…." Perhaps most importantly, economic growth through changing needs for the development of human capital must be done in the context of skills sets demanded by economic forces at the international scale. This will increase the need for competitive advantage derived from specialization in a new economy where "it is it no longer feasible for a firm located in one place to be unconcerned with the network of institutions and suppliers that can provide in materials and talent…" (Blakley and Leigh 2010).

Finally, providing further contextual meat to the analyses and perspectives above, the process of favoring growth through measures of productivity and wealth in terms of achieving an expanded quality of life, and reevaluation of its impacts on the equitable and contextual distribution of goods, has also been described by a wide array of urban theorists from the sociological, political perspective. Providing one of the more influential analyses of the role of growth and inequity in an urban setting, the process of favoring growth was described by Harvey Moltoch and others as one that was fueled by the development of "growth coalitions." In this vision of economic development at predominately the local scale, the most influential actors in fostering urban development were so-called place entrepreneurs who are defined as individuals who profit from renting out real estate.

Thus, according to Altshuler and Luberoff (2003) in Mega-Projects: The Changing Urban Politics of Urban Public Investment, "What most distinguishes place entrepreneurs from ordinary citizens is that they value land for its 'exchange value' (its capacity to generate profit) rather than its 'use value' (as a locus for social interaction, the enjoyment of nature, and ecological health). Their unswerving aim is growth-which, above all, means real estate development—regardless of the negative consequences it may entail for current land users, such as the ordinary residents of established neighborhoods. And they routinely seek government action to facilitate their endeavors. To secure such action they organize local "growth machines"— that is alliances of those in the community who stand to profit from development. These alliances include not just place entrepreneurs themselves but also their contractors, bankers, architects, engineers, and advertising firms; the employees of such enterprises and their labor unions; local media, utilities, and retailers who think that growth will bring them more business; and politicians who recognize that growth-oriented interests are the largest contributors to local campaigns." Thus, through the development of an effective regime perhaps based on an understanding of regime theory (see Stone 2006), the growth coalition brings together a broad swatch of powerful interest that were firmly entrenched in fostering the largest amount of growth possible in the urban setting.

Further defining the reality of the interface between growth quality of life, and the equitable enjoyment of progress, the process of favoring growth can likewise be understood and played out by Allan Schnaiberg's "treadmill of production." Corporations and individuals invest and reinvest capital to seek maximum economic returns on the treadmill of economic production. This behavior resultantly strains the ecosystem and social systems as a continual withdrawal of resources is fueled by economic growth. As described by David Pellow in Garbage Wars (2002), both private consumption and industrial production create a variety of negative by-products or externalities that range from "effluents, waste, and other forms of ecological disorganization." Moreover, the most profitable strategy for industries becomes capital intensification and a resulting gain in worker displacement and unemployment. In the end, Pellow (2002) asserts that, "these ecological and social strains place pressure on the state, communities, workers, and corporations and address these ills—often, ironically, through future pro-growth policies."

In conclusion, given the discussion above, then, we can understand how growth often became the singularly favored (and often incompatible with equity considerations) urban strategy for planners and policymakers. Growth then becomes a very convenient approach to conceptualizing the decision-making process. As noted by Altschuler and Luberhoff (2003), the emergence of the growth machine fosters the development of what they and Logan and Moltoch term "value free development." They note that, "the essence of this ideology is that while people disagree about values, there is no serious reason to disagree about growth. Growth means prosperity, and with much more money everyone in the community can pursue his or her own values better." Thus, the notion of growth becomes a politically powerful force in mobilizing interests and in reaching development decisions moving forward. Equity may be left off the list of key considerations of development as a consequence. With growth increased further in aggregate and a simultaneous increase in wealth concentration in the hands of elite, the process of economic restructuring, of globalization, and of institutional and political realities of growth coalitions demonstrates the complex and often contradictory relationship between aggregate growth and the equitable distribution of development's benefits. It highlights the related tensions between economic growth and true progress through the "true" improvement of a quality of life.

Toward the Next Vision of Economics: The Green Economy as Sustainable Economic Development Strategy

The previous section sought to provide an overview of economic thinking in the context of development decisions and the broader pursuit of a better quality of life along traditional, neoclassical, institutional economics terms. Simply put, growth matters and defines the right path forward. As mentioned above, the pursuit of growth has guided thought throughout the twentieth century.

The "next" vision of economics may begin to reorient the discussion at the locally driven level through the rise of the concept of sustainability. This chapter's second section, contrasting with the first, turns toward sustainable economic development as a new way of understanding the nexus of economic and environmental values as one which provides a textured and, at the same time, symbiotic pursuit of such goals. First, a brief overview of sustainability is provided. Second, the green economy movement is examined as an outcome of this new approach. Thinking and action around the green economy may be most indicative of a new and improve visioning of economic solutions at the local level.

Sustainability and Sustainable Development: Going Beyond Growth

As indicated in the discussion above, economic growth in aggregate may not actually fuel the optimum end we seek and/or may only do so for a small segment of the population. Growth may be inequitable and incomplete. Amartya Sen (1999) provides an in-depth analysis of the complexities inherent to the growth process through a consideration of quality of life factors and the ultimate benefits of development. He begins his text with the central thesis of his work—an expanded definition of growth and of development. He notes that, "Development can be seen, it is argued here, as a process of expanding the real freedoms that people enjoy. Focusing on human freedoms contrasts with narrower views of development, such as identifying development with the growth of gross national product, or with the rise of personal incomes, with industrialization, or with technological advance, or with modernization. Growth of the GNP or of individual incentives can, of course, be very important as the means to expanding the freedoms enjoyed by members of the society. But freedom depends also on other determinants, such as social and economic arrangements (for example, facilities for education and health care) as well as political and civil rights (for example the liberty to participate in public discussion and scrutiny)." In this way, Sen (1999) asserts that, "viewing development in terms of expanding freedoms directs attention to the ends that make development important, rather than merely to some of the means that inter alia, play a prominent part of the process." Thus, Sen (1999) asserts that, "development requires the removal of major sources of unfreedom: poverty as well as tyranny, poor economic opportunities as well as systematic social deprivation, neglect of public facilities as well as intolerance or overactivity of repressive states." Thus, aggregate growth is not everything. Sen's plea is for a redefinition of growth and of development that fundamentally goes beyond the accumulation of resources (Sen 1999).

Further emphasized by Blakely and Leigh (2010), much like Sen (1999), the complexities in measuring and fostering economic growth are multiple. A variety of scholars envision a different approach to economic development that specifically focuses on rising economic inequity (or in Sen's case capability). Blakely and Leigh note (2010) that, "There is nothing wrong with creating wealth and jobs and increasing the tax base. But it is a great mistake to equate economic growth with economic development. The blind pursuit of economic growth can destroy the foundation of economic development. For example, if an economy's growth is based on an exhaustible natural resource supply (e.g., timber, seafood, coal), then it will eventually come to a halt. The workers will be unemployed and, without proper attention to the education and skill development of the labor force, or to the development of a more diversified industry structure, the community can enter a death spiral." This process, they argue, is similar in towns with singular industries that are vulnerable to global economic shifts. In this case, "The industry may move, or its owners may exit the industry and the town, taking their capital worth with them. These are the simplest of examples and it should be understood that a town with more than one industry, but with a narrow industrial base, can be just as vulnerable." And in fact, they assert that economic growth may in fact directly lead to increases in income equality and further marginalize already marginalized populations. Blakely and Leigh thus argue that an economic development approach should focus on raising the basic quality of life for everyone (Blakley and Leigh 2010).

As noted at the outset, the goals of economic development can be expanded by looking beyond the aggregate level of growth to incorporate measures of equity and then to a broader conception of growth in itself in terms of the desired ends of progress. Given the conflicts between growth and equity, how can progress be reframed through the infusion of extended quality of life and equity considerations into economic development. How can environmental considerations similarly be balanced through a revisioning of the relationship between the environment and the economy?

The emergence of sustainability has most significantly reenvisioned the balancing of growth with these other competing values. Sustainable development seeks to unite these disparate interests and to overcome these difficult contradictions between competing values in the urban context. Sustainability is defined by the integration of economic and environmental goals in economic development decisions. Higgens (1996) refined the definition and notes that: "Sustainable development is a concept which encourages both economic growth and a healthy environment. It recognizes the desirability of economic growth and change and acknowledges the right of individuals and organizations to pursue economic goals, including sales and profits." Although the complexity of the topic makes defining and then realizing the ultimate end goal of sustainability difficult, scholars have nonetheless continued to redefine "sustainable economic development," including the operationalization of the concept through sustainability programs, plans, and evaluative metrics. Roberts (2004), for example, notes that, "Sustainable economic development strategies promote mutually beneficial environmental, social and economic processes....Sustainable economic development are ideas related to: The effective and efficient use and management of natural resources; The promotion of a hierarchy of waste solutions that places the avoidance of waste at the top of the list and the disposal of unsorted waste at the bottom of the list of options; The introduction of new methods and techniques for design, production, distribution and end-of-life management, which emphasize the avoidance or minimization of waste and environmental damage; The establishment of new economic activities based on opportunities for the production of environmental goods and services and for the distribution, maintenance and eventual disposal of such products; The promotion of high standards of environmental management and performance in all aspect of economic development and in all aspects of economic development and in all business activities, including energy conservation environmental sound construction, green transport and a wide range of other occupational areas; The establishment of new and collective and collaborative institutional structures that can assist in the introduction and management of sustainable economic development." Similarly, as noted by Reinhardt (2000), "macroeconomic definitions of sustainability focus on the need to maintain aggregate stocks of natural and manufactured capital constant over time so that future generations have consumption possibilities similar to those of the current generation." Thus, sustainable economic development provides a descriptive, albeit elusive, conception of a true balancing of the multifaceted three "Es."

Green Economy as the "Next Economics": Revising and Revising the Idea of Growth and Equity Through Sustainable Economic Development

Sustainability can become a new aim for planners, as well as more broadly among policymakers who aim to improve conditions in their community. This may represent a fundamental shift in addressing development conflicts and forging decisions aimed at fostering improved quality of life. And, as some scholars assert, the purported "new green economy" or a new "environmental epoch" (Mazmanian and Kraft 2009) may provide a nexus between environmental preservation and economic development through the creation of a new "paradigm" of sustainable economic growth (Roberts 2004). The green economy movement seeks to change the way that development is viewed by reenvisioning two primary conflicts inherent to discussions around progress—economic development through a better quality of life and then a more equitable distribution of this quality of life.

Green jobs, as the outcome of this movement, are a critical vehicle of achieving the sustainability ideals as described above. Simply put, green jobs provide employment opportunities that will lead to the environmental preservation dictated in the sustainability vision described above. A green economy includes a wide spectrum of products that do not adversely affect the environment while providing economic benefits in the form of revenue and job creation (OCED 1999). The Organization of Economic Cooperation and Development provided the first conception of the green economy as an economic development strategy by defining the green economy in terms of the environmental goods and service industry consisting of: "… activities which produce goods and services to measure, prevent, limit, minimize or correct environmental damage to water, air and soil, as well as problems related to waste, noise and eco-systems. This includes cleaner technologies, products and services that reduce environmental risk and minimize pollution and resource use." Although definitions of specific industries vary, the green economy can be conceptually seen as the overall framework in which economy activity can occur. In practice, for example, the Occupational Information Network (Rivkin et al. 2010) defines the green economy as follows: "The green economy encompasses the economic activity related to reducing the use of fossil fuels, decreasing pollution and greenhouse gas emissions, increasing the efficiency of energy usage, recycling materials, and developing and adopting renewable sources of energy." New classes of "environmentally friendly" businesses are now considering sustainability in their corporate choices as the cost of pollution is increasing while the marginal cost of pollution reduction is dropping and the consumer demand for green products is increasing. Government procurement polities around green products are also fueling niche markets and altering corporate behavior (Nijaki and Worrel 2012). And perhaps offering the starkest example, environmental regulations are creating business opportunities in a subset of innovation-driven, new green industries. New efficiencies, regulatory opportunities, and consumer responses are making sustainable development an economic possibility for some communities (Wasik 1996; Roberts 2004; Lopez et al. 2007).

The "new green economy" may provide a new way of envisioning economic growth that is more along the lines of development as proposed by Sen (1999) and others. The green economy aims to widen the view of economic growth or progress through an integration of environmental considerations in the development process. It reframes growth as "green growth" and thus limits development by taking into account quality of life considerations that are hinged on environmental quality today and into the future. In this way, the metrics for evaluating development choices and their successes is changed to one that seeks to reference the long-run environmental effects of economic action and inaction.

And, in fact, this metric seeks to completely change the equations from which these discussions are judged. Traditionally, the differentiation was seen as a zero sum game between the environment and the economy (Andrews 1999). In the green economy literature, the key difference here is that there is a stated attempt to mediate the gap between the environment and economic development. And, the starkest example may be in deriving economic development opportunities directly from environmental preservation and environmental cleanup. A recent University of California Study (Chapple et al. 2009), for example, posits that "at its most basic level, the green economy consists of economic activity that reduces energy use and/or improves environmental quality…The green economy is not just about the ability to produce clean energy, but also the growing market for products that consume less energy, from fluorescent light bulbs to organic and locally produced food. It also encompasses economic sectors that improve the environment, for instance, through remediation of toxic signs or the design of a more compact city." Although not all economic activity in the "new green economy" is imagined to be environmentally beneficial, the overriding conception of environmental growth is that there will be a net increase in economic opportunities that are not generally harmful to the environment and that may provide sources of environmental remediation as a viable business case leading to aggregate growth.

Measuring and analyzing the green economy has remained a complex task. Although sustainability and related concepts around green jobs growth are increasingly used, there is not a widely accepted definition of the green economy across regions and states domestically and internationally. Figure 12.1 summarizes a few

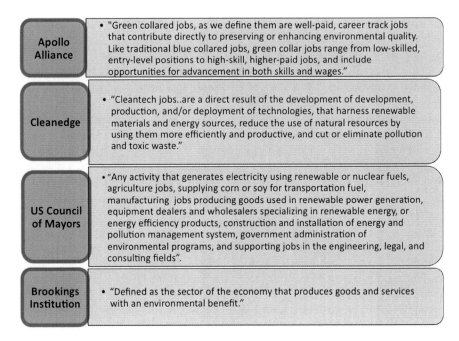

Fig. 12.1 Defining green jobs differently

varying approaches. It is difficult to determine what types of sustainable practices, and at what threshold, define an environmentally friendly business. Moreover, there is no agreed upon metric to determine what sectors of industry qualify as a green goods and service.

Just as rigor is important in developing this definition, so is flexibility. Green jobs are designed around environmental preservation. Understanding the green economy is challenging because of the diverse and disparate spectrum of products produced and used. Although the green economy is sometimes coined the "green sector," this conception in not an accurate portrayal of the sweeping constellation of firms within the green economy. The green economy is not a traditional industry sector. It includes a wide array of industry sectors and subsectors. Potential candidates include anything from large-scale producers of solar panels, to small-medium environmental consulting firms. Both a manufacturing and a service economy can be potentially expanded and refashioned in the new green view. The common uniting force between disparate industries is the environmental benefit of the product that is being consumed or produced. A variety of new industries, and the greening of existing ones, consequently fit the bill—all united under the notion of environmental sustainability.

A useful conceptual framework can be constructed in order to facilitate an understanding of the patchwork of activities around sustainable economic development. Specifically, green jobs analysis can occur at several different levels: the green economy, the green industry, the green firm, and the green job or occupation.

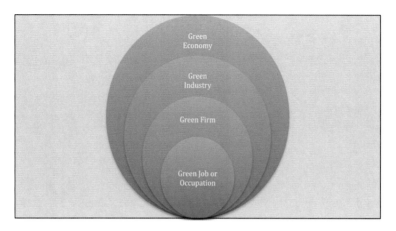

Fig. 12.2 Defining the green economy

As summarized by the component parts of corresponding Fig. 12.2 above, we can drill down through these different layers of analysis in order to understand the opportunities and obstacles around green jobs development and proliferation.

1. First, the *green economy* is the overall framework in which economic activity can occur. ONET (see Rivkin et al. 2010), a national agency in the United States, defines the green economy as follows: "The green economy encompasses the economic activity related to reducing the use of fossil fuels, decreasing pollution and greenhouse gas emissions, increasing the efficiency of energy usage, recycling materials, and developing and adopting renewable sources of energy." Again, although not all economic activity may be environmentally beneficial in the "new green economy," the overriding idea is that there will be a net increase in economic opportunities that are not generally harmful to the environment. And, it is the overriding commitment to the ideas of sustainability and sustainable economic development as indicated above.
2. Second, the green economy consists of *green firms*. These are the businesses that make up the green economy. According to ONET, again, "A green firm is an organization that provides products and/or services that are aimed at utilizing resources more efficiently providing renewable sources of energy, lowering greenhouse gas emissions, or otherwise minimizing environmental impact. Green firms with similar activities, production value chains and/or products can form a green industry, sub-sector or sector." These firms run the gamut in terms of potential products and services used and consumed. Examples include broad classifications such as the following: renewable energy, energy generation, systems installation and storage, green building and energy efficiency, biofuels production and farming, transportation and alternative fuels, water, wastewater and waste management, environmental compliance, and sustainability planning.

    Green firms can be "green" at a variety of different levels as summarized by the Fig. 12.3 below. First, firms can be identified as "green" because of their end

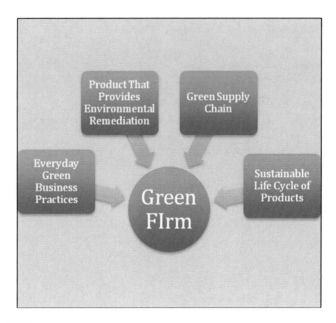

Fig. 12.3 Defining a green firm

product or service. These are the firms that are most clearly identified as green producers because their products or services are directly aimed at some aspect of environmental remediation or promotion. Second, and in a related vein, firms can be "green" in terms of the life cycle of their product. Although the particular good or service is not directly being utilized for environmental remediation or promotion, it may be sustainable over the lifetime of the good and therefore may be a green product. The firm is green by this definition, then, because it is a producer of sustainable products that indirectly benefit the environment throughout their life cycle. Third, a firm can be "green" according to the environmental effects of its supply chain. Fourth and finally, a firm can be "green" according to its everyday business practices including its use of green practices, services, and products (i.e., the use of alternative fuel vehicles, energy efficiency/conservation, sustainable farming, recycled products or recycling, water conservation, or pollution reduction). These green practices can be implemented across most industry categories and likewise spans a multitude of potential occupational categories. More broadly, in terms of these firms within the green economy, green firms can be understood as falling into two broad areas: the consumers or producers of green technology.

(a) *Green producers*—include any company that directly producers green goods and services. This includes both the manufacturers and service providers of green goods and services. Examples of green producers include solar panel manufacturers and manufacturers of diesel particulate filters and heavy-duty vehicles.

(b) *Green consumers*—include any company that integrates sustainability practices within their business practices. A range of sources and certification programs define businesses as "green." For example, the study by Chapple et al. (2009) defined green practices as "activities that reduce energy consumption and/or improve environmental quality into their operations." Thus, these companies can be categorized as "environmentally friendly companies." Examples include green restaurant owners, workers who utilized recycle products, or installers of particulate filters and owners of the trucks on which these devices are installed.

These two groups are interconnected to one another. Green producers can, and perhaps are more likely to, also be categorized as green consumers. However, the connection between green goods consumer and producers are conceptually and practically linked at another level. Namely, green consumers become the source of demand for the green producers. Thus, "environmentally friendly" companies are directly using environmental products. For example, in the diesel truck example briefly alluded to above, the green consumer that is putting diesel particulate filters on their truck *may* qualify as a part of the green economy. The green producer here is the company that produces diesel particulate filters—an employment function that directly reduces the environmental hazards of diesel truck operations. Thus, the green consumer (the diesel truck operators or owners) is consuming the environmental product (the DPF technology) from the green producer. The green consumer in this example is an important part of the green economy because the truck owner/operator is a necessary component of the green producer or DPF technology's demand. It is important to note that production processes are complex and multilayered. The demarcation between consumers and producers is not mutually exclusive. A producer, thus, can be a consumer of green goods and services as components within a layer of their wider production processes.

3. Third and finally, green firms are made up of *green jobs or green occupations.* According to ONET, for example, "the greening of occupations refers to the extent to which green economy activities and technologies increase the demand for existing occupations, shape the work and workers requirements needed for occupational performance, or generate unique work or worker requirements needed for occupational performance, or generate unique work and worker requirements." Moreover, according to ONET, a green job is an application that (1) directly works with policies, information, materials, and/or technologies that contribute to minimizing environmental impact and (2) requires specialized knowledge, skills, training, and experience. They thus provide an individual-level of green economic opportunities.

New categories of goods and services related to environmental protection are emerging and are therefore creating opportunities for jobs and economic development in communities. Once the green economy is defined, implementing governance strategies is the next step toward realizing the paradigm. Green jobs have

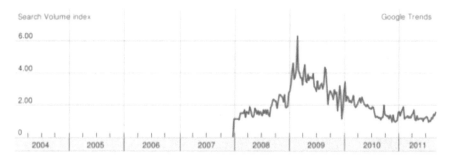

Fig. 12.4 Green economy references by Google search volume

become increasingly popularized in policy circles and in the mainstream media. This has operated concurrently with the increasing impetus to redress environmental pollution. Google trends, as a rudimentary manner to pick up trends in key terms across search data, clearly illuminates an uptick in this focus. As summarized in the Fig. 12.4 below, the term was increasingly utilized since the end of 2007, with a significant peak in 2009. The results are similarly corroborated through Proquest searches of the term "green jobs." From 2000 to 2005, the term was referenced 26 times. From 2000 to 2010, the term appeared with a frequency of 1,074. A search in the date range of 2008–2010 yields 1,066 references. Thus, "green jobs" may be an increasingly utilized rhetorical hook for both economic development planning and environmental preservation efforts domestically. States, regions, and cities have increasingly taken prominent roles in this discussion and have provided leadership beyond nation-state action.

The interest is there. The political reality in developing such strategies is likewise complex. The complexity in terms of integrating multiple objectives can be called new in the sense that it will likely necessitate institutional changes. Such changes will necessitate economic development strategies that include an integration of environmental expertise in economic development planning and policy efforts. The specific economic development strategies that are used to cultivate economic growth in green industries may not be all that much different, particularly in terms of operationalizing the concept through traditional business attraction and innovation-driven industry strategies, once the desired green industries are selected.

In fact, much of the focus in terms of aggregate job growth and economic development has been focused in the green jobs realm on higher-skilled, higher-wage employment. This has been done through a focus on innovation. Drawing back to theories of growth toggled to innovation from Solow onward, this strategy toward a green economy has sought to focus on opportunities engendering from technological advancement in green technology. Economic growth through these strategies will occur through the innovation-driven process of coming up with new technological fixes to environmental problems. Additionally, market opportunities may increasingly exist in marketing companies' practices through the new move toward corporate social responsibility (see Hardjono and de Klein 2005). Thus, in evaluating green jobs as a novel economic development strategy, it is important to keep in mind that much of the implementation of green economy strategies can be seen as not dramatically different than any other economic development strategy in the actual implementation, once the desirable list of green industries are targeted for growth. A better quality of life may be fostered through these efforts, expanding the values and aims of aggregate environmental growth, with little attention to equity in terms of distributional considerations.

How can a green jobs strategy address distributional issues and thus expand the vision of growth and progress beyond aggregate growth? Distributional equity considerations can directly be addressed through a green economy strategy. As explained above, the green economy movement seeks to integrate economic and environmental goals. However, who should benefit from the economic gains of cleaning up environmental blight, for example? As noted by Bullard (1980), "Industrial flight from central cities has left behind a deteriorating urban infrastructure, poverty, and pollution. What kind of replacement industry can these communities attract? Economically depressed communities do not have a lot of choices available to them. Some workers have become so desperate that they see even a low-paying job as better than no job at all. These workers are forced to choose between unemployment and a job that may result in risks to their health, their family's health, and the health of their community. This practice amounts to 'economic blackmail'." If green jobs create economic development opportunities, the next question to ask is who will benefit from environmentally preservative growth? Will the same narrow groups of individuals who historically benefited from the production that led to environmental externalities suffered in environmental justice communities now benefit economically once more in finding the next best way to clean up said externalities?

In fact, no part of the movement may epitomize the tensions between equity and growth, given the green economy lens and the reality of urban development as described by Bullard, than the differentiation between the green jobs movement generally and the emergence of the green-collared jobs movement in particular. Green-collared jobs seek to directly give economic opportunities to those populations that fit this description (Apollo Alliance 2007; Bowen et al. 2007). Green-collared jobs, at the operational level, seek to specifically target employment opportunities to populations that are most in need, perhaps using Rawlsian (1971) logic. The green-collared job movement seeks to remake the movement toward a green economy through the integration of growth and equity that is specifically targeted at those individuals that are blue collared and that are widely identified as often "left out" of the economic opportunities along the historical trajectory toward greater levels of economic growth. The pursuit of quality of life improvements and the degree of equity that is added along with the merging of environmental and economic desires will alter the types of industries that may be targeted and the types of companies that may be courted. Jobs in this sector may be specifically targeted at cleaning up the disproportionate environmental blight in poorer communities and thus can increase both the equity of economic opportunities and the closely related environmental blight attributable to production. Thus, rather than the situation of "economic blackmail" described by Bullard, the green-collared jobs movement can focus on the pursuit of opportunities targeted to those communities that have disproportionately suffered from past environmental blight.

Will the green jobs movement provide a harmonic integration of economic growth and equity considerations? If the green economy is about finding employment opportunities in environmental remediation, then environmental justice communities will assuredly have their share of economic opportunities. As an example, Agyeman and Evans (2003) describe the process by which the primary tools of land use planning via zoning have purportedly resulted in widespread geographic segregation of people and land utility within the urban context. And, as a result, this land use policy created the overburdening of cumulative effects from environmental hazards in low-income minority communities. Planners can resultantly utilize both the views of sustainability and of environmental justice to identify the historical flaws in such land use planning. These flaws can include the separation of uses and low-density development and thus can fuel urban sprawl and auto-dependent transportation. As a result, things are beginning to progress with recent changes in urban planning and in public policy aimed at accommodating the green economy through "more efficient land development, mixed-use and mixed-income developments, and the reuse of former industrial sites." Through the reenvisioning of economic development, green jobs development around the reuse of this land and may provide fruitful ground to operationalize environmental justice and sustainability at the same time—really amounting to a economic development strategy that is preservative of current and future environmental considerations and thus looping together components of inter- and intragenerational equity and justice. In a perfect theoretical world, environmental blight can provide economic opportunities for those who have disproportionately suffered from the externalities attributable to production in an environmentally preservative economic development strategy drawing together the three "E"s.

Although it is possible in theory, mixed results will likely persist in practice, with particular difficulty in addressing equity issues. As noted above, the implementation process is a specifically important area in terms of targeting growth versus equity. Whether or not it is important to see growth in aggregate, or more important to focus on areas where the distribution of economic spoils from green jobs programs could be distributed, may dictate the path of green jobs economic development strategies. A strategy that targets employment in high-technology industries that require highly skilled, specialized human capital will incorporate a different vision of progress than a green-collared job strategy aimed specifically at workforce development at the lower-skilled end of the spectrum.

Keeping the discussion around green-collared jobs in mind, it seems to follow that, at the end of the day, the devil is in the details in terms of whether or not equity plays as a guiding role or will just remain another aggregate-focused economic development strategy leaving issues of equity and justice by the wayside. Political dynamics will likely be a critical factor determining the actual implementation and achievements of green jobs strategies as opposed to traditional economic development approaches. Time will only tell whether or not the green jobs movement and the green-collared jobs movement will actually increase equity in practice, or just remain a utopian effort to draw together economic growth and equity in a new approach to merging environmental and economic goals.

Conclusion: Toward the Next Economy of Green Growth and Green Jobs

The "next economics" may in fact be green economics! This chapter sought to examine sustainability and green jobs at the localized level as a new aspect of both economic development and economic thought in general. First, traditional growth-centric theories as a focus of economics and of economic development were covered. Second, this chapter pointed toward the "next economics" of sustainable economic development. The green economy can provide a unique extension of such reenvisioning of economic principals and the pursuit of progress through economic growth *through* environmental preservation. Rather than environmental externalities from economic growth, economic growth can occur through mitigating previous environmental externalities.

This provides a useful contrast, but how can traditional and the "next," "green" economics be weaved together into a coherent conceptualization? How can the new paradigm grow from the old, to reform economic thought to reach beyond growth alone in an incorporation of richer sustainability principles? In the end, traditional economic thought and the "next economic" thought around green growth can be drawn together into a coherent, complex story. Although future research is needed in order to define a path forward, the picture liking traditional economics to "next economics" can be preliminary painted in two ways—first through theory and then through practice.

Traditional to the "Next Economics" of Green Growth: In Theory

Sustainability, or a lack of sustainable economic development focused on green jobs, can be understood in traditional economic, institutionally driven terms. Environmental problems offer a plethora of economic and political challenges. A brief retreat into the institutional economics perspective can aid in understanding the concept of "eco-efficiency" as a critical related concept. This thereby provides new and often controversial assumptions around the nexus of environmental and economic concerns and outcomes. In the neoclassical, institutional perspective, environmental goods are often underprovided in the marketplace relative to their pareto-optimal level or equilibrium level. This leads to market failures (Perloff 2004). Most natural resources are club goods or commons goods. This, according to neoclassical economists, is because most natural resources are both non-exclusionary (the cost of excluding individuals from use is high), and rivalrous (the use of the good precludes others use of it). It is unrealistic or too costly to limit the enjoyment of these goods. Moreover, each person's enjoyment depletes the resources and consequently depletes the potential utility that can be obtained through other individual's enjoyment of the good (Perloff 2004).

This often fuels a market failure, or the tragedy of the commons as identified by Hardin (1998). Much of the problem is also attributable to an inability to appropriately gauge the market value of environmental resources, in the first place, and the social marginal costs to the environment generated from industrial production. When polluters pollute, for example, they do not fully take into account the cost that they are exerting on other individuals. The social marginal cost of resource depletion is not fully considered, and their activity is guided by comparisons of their marginal costs and benefits. The social marginal costs are greater than the private marginal costs of their action, or lack thereof, and the resource is consequently depleted and underprovided in the market (Hardin 1998).

However, the institutional argument may be changing. The nexus between the environment and economics, cast from distinctive assumptions about a burgeoning economic epoch, has been reenvisioned. Using environmental indicators, analysts can now determine how efficiently wealth is generated in environmental terms. Several empirically based studies have extrapolated on the notion of eco-efficiency. Results have indicated that industrial economies are increasing efficiency when measured by resource use per dollar earned in a country's GDP. Thus, the pursuit of economic goals can also preserve the environment at some level of general development within a country. As often depicted by the Kuznets curve, the relationship between environmental degradation and environmental preservation is curved whereby at some level of economic growth, environmental preservation can be mutually preserved. Sustainable development is bolstered, moreover, by an overall restructuring of the world's economy to service industry domination and the mostly abandonment of many of the Fordist mass production tendencies (Bleischwitz and Hennicke 2004).

Many manufacturers are adopting sustainability concerns within their operations due to the reduction in energy, and materials reduce production costs. At some point, some researchers nevertheless have indicated that countries may achieve a level of eco-efficiency. Sustainability and eco-efficiency can be realized when efficiency gains are reached through actions that will not limit the overall prosperity in our society. Raimund Bleischwitz and Peter Hennicke (ibid. 2004) argue in Eco-Efficiency, Regulation and Sustainable Business: Towards a Governance Structure for Sustainable Development. "…Thus there is a so-called double efficiency built into the triple bottom line of sustainability theory. Societies can produce wealth from nature more, or less, efficiently and can produce well-being from prosperity more, or less, effectively…." Thus, in countries that have achieved eco-efficiency, both an increasing level of economic development (often measured in GDP per capita or in more complicated indexes of development) and an increasing level of sustainability will result. In addition, corporations may be voluntarily adopting sustainability standards. Increasing economic freedom correlated with an increasing level of sustainability may be exhibited within a country or region (ibid., 2004). Eco-efficiency is predicated on the notion of developing what has been termed a "green economy." Specifically, new categories of goods and services related to environmental protection are emerging—creating opportunities for employment specifically and economic development generally, in a cornucopia of communities.

The argument follows that businesses are increasingly demanding these green goods and services as a part of their operations. Theoretically, strategies for sustainable business are becoming more prominent at both the international and national level. Businesses themselves can get a bigger bang for their buck by engaging in environmentally friendly practices and utilizing green purchasing strategies. According to Bleischwitz and Hennicke (ibid., 2004):

> In this context, proactive strategies for sustainable business development are of increasing relevance for companies and markets, and for future business opportunities, as with, for example, the adoption of environmental management (sustainability) objectives and even though there is no apparent attempt by governments to regulate the issues under consideration. A company may act proactively because, for example, it wishes to position itself as environmentally friendly or more broadly sustainable on the market, or because it has realized that achieving environmental objectives is linked to economic gains (a win-win solution).

A company's green procurement policies can be widely varied. They commonly amount to the purchasing of eco-labeled products or services, in-house or third-party evaluations of the product, or supply chain initiatives that improve efficiency along the supply chain (Young 1994). Green purchasing then not only provides savings for the company itself but also may improve profit margins for the consumer (Bleischwitz and Hennicke 2004). And the entire system loops around to create demand for the green goods and services employed by these green companies. Intuitively and grounded on these assumptions, this will create increased numbers of sustainable businesses and a heightened overall level of sustainability within a country, state, or city.

Moreover, this framework asserts that changing technology is enabling business to clean up in a cost-effective manner. The cost of pollution prevention, in board averages, has decreased in recent years. This has helped to give businesses further incentives to clean up and tighten the differential between private marginal cost and social marginal costs of production activities. Sustainability, for this reason, has been seen as a technological fix. According to the assumptions, eco-efficiency may now be possible, and even profitable, in the new economy (Bleischwitz and Hennicke 2004).

Thus, through the eco-efficiency lens and the wider institutional perspective, the set of incentives have been restructured within the institutional framework in the networked, service economy in such a way so that entrepreneurs can pursue sustainable operations within their corporate model. It is not about the "warm and fuzzies" but also about profits and possibilities. This will help to ensure the sustainability of the sustainability movement because it means that, once again, sustainability will not be categorized as a luxury good. Even in harder economic times, corporations will still be motivated to pursue sustainable objectives. The resultant producers of green goods and services (theoretically, at least) will remain a viable part of the market. Thus, eco-efficiency is established as a possible forgoing goal of the wider ontological framework of the sustainability discourse and of the "next economics" of green economic growth. Simply put, the "next economics" of sustainability is guided by eco-efficiency goals.

Traditional to the "Next Economics" of Green Growth: In Practice

One key aim that underlies much of the analysis and theoretical work around sustainable economic development is how to appropriately move forward with workforce and economic development for the green economy. Thus, what does the "next economics" of green growth look like in practice? This question is governed by the basic, yet difficult, question: How is workforce development or economic development different for green jobs and green firms? Although when reduced to specific green industries, a variety of traditional strategies remain viable, several key differences that build upon traditional economic strategies are critical to keep in mind. Areas of difference between the "next economics" of green economy-focused growth and traditional economic development strategies seeking industry and occupational growth include:

1. *Diversity in sectors.* We can think of green business as divided between the larger green economy, the potential green industries that lead to green producers or green practicers, green firms that make up the green economy, and the green jobs or occupations that fall within these larger firms. We should understand these green occupations and firms in the context of the larger green economy. A vast array of industries, in fact almost all, can potentially fall within the green economy as green practices (environmentally friendly companies). Although much fewer in number, a wide diversity of different industries are direct producers or service providers. Therefore, there is no one green industry or green sector within which it is appropriate to train a workforce within. Local or regional governments cannot pursue an overall "green sector" or singular "green cluster" strategy. Sector strategies within the "green" portions in identified industries can be effectively pursued from both the economic development and workforce development perspective. However, sector strategies can only deal with one sector of the green economy at a time. They will likely not address the full diversity of green firms within the green economy.
2. *"Practicers"* versus *"producers."* Developing a green economy can be bolstered by strategic consideration beyond simply the green product or service production. Across many industry categories, firms are thinking about ways to green up production processes and to integrate sustainability principles into their everyday business practices. This shift is arguably often fueled by two principal factors. First, increased consumer demand for green goods and services may lead to business opportunities for environmentally friendly companies. And, second, companies often must change operations in order to meet current or pending environmental regulations. Training and job opportunities are located beyond the green producers to the need for a workforce that is knowledgeable about everything from chemical reduction to energy efficiency.
3. *Regulation centric.* In understanding and unlocking needs and opportunities around the green economy, it is important to consider the role of environmental regulations in the equation because green jobs are often in response to current or pending regulations. Many new products and market opportunities, subsequently

leading to new market needs, are incentivized in response to regulations. For example, Chapple et al. found that such a case applied to recent climate change regulation at the statewide level. They note that, "firms whose operations are affected by California Assembly Bill 32 (AB 32), which establishes the first comprehensive program of regulatory and market mechanisms to reduce green house gasses, also are more likely to innovate new processes." Thus, future economics-driven research should seek to consider environmental remediation and associated regulations in order to locate potential opportunities and needs for green workforce development.

4. *Small and start-ups.* May green producers are thought to be small and/or start-up companies. Much of the industry is new and reactive to changing environmental conditions and regulations. Innovation is central in the development of opportunities in the green jobs sector. This may provide significant growth opportunities — this may provide opportunities in the "next economics" of green growth. But, at the same time, this also presents challenges in terms of understanding future trends and opportunities in the context of this theorized, and increasingly realized, next economies of green growth. It may be particularly important to understand venture capital trends in unlocking the future potential of green jobs.

In the end, not all growth will be green growth; not all jobs will be green jobs per se. But, some economic opportunities can be generated in the new economy, and the new direction of economics must include considerations of green economic opportunities and sustainability more broadly in order to offer a comprehensive pursuit of greater quality of life and a multifaceted pursuit of progress for local communities. Local and regional governments can play a key role in fostering sustainable economic development opportunities. The institutional structure will play a key role moving forward, and future research will seek to apply academic rigor to an understanding of the purported paradigm shift around sustainable economic development domestically and internationally.

References

Agyeman J, Evans T (2003) Toward just sustainability in urban communities: building equity rights with sustainable solutions. Ann Am Acad 590:35–53

Altschuler A, Luberhoff D (2003) Mega-projects: the changing politics of urban public investment. Brookings Institution Press, Washington, DC

Andrews RM (1999) Managing the environment, managing ourselves: a history of American environmental policy. Yale University Press, New Haven

Apollo Alliance (2007) New energy for cities: energy-saving and job creation policies for local governments. Apollo Alliance, Washington, DC

Arvinson E (1999) Remapping Los Angeles, or, taking the risk of class postmodern urban theory. Econ Geogr 75:134–156

Barbour E (2001) Metropolitan growth. Public Policy Institute of Los Angeles, Los Angeles

Blakley EJ, Leigh NG (2010) Planning local economic development: theory and practice. Sage, Los Angeles

Bleischwitz R, Hennicke P (2004) Eco-efficiency, regulation and sustainable business: towards a governance structure for sustainable development. Edward Elgar Publishing, Northampton

Bowen A, Lee J, Ito J (2007) Green cities: green jobs. Los Angeles Apollo Alliance/Strategic Concepts in Organizing SCOPE, Los Angeles

Bullard RD (1980) Environmental justice in the 21st century: race still matters. Phylon 40:151–171

Carney FM (1964) The decentralized politics of Los Angeles. Am Acad Pol Sci 353:107–121

Chapple K et al (2009) Innovating the green economy in California's regions. University of California Berkeley, Berkeley

Easterly W (2002) The elusive quest for growth: economists' adventures and misadventures in the tropics. MIT Press, Cambridge, MA

Graham S, Marvin S (2001) Splintering urbanism: networked infrastructures, technological mobilities, and the urban condition. Routledge, New York

Hardin G (1998) The tragedy of the commons. Science 162:1243–1248

Hardjono T, de Klein P (2005) Introduction on the European corporate sustainability framework. J Bus Ethics 55:99–113

Higgens J (1996) Canadian perspectives on the world environmental industry. Environmental Technologies Development Corporation, Toronto

Lopez MV, Arminda G, Rodriguez L (2007) Sustainable development and corporate performance: a study based on the Dow Joes sustainability index. J Bus Ethics 75:283–300

Mather V (1999) Human capital-based strategy for regional economic development. Econ Dev Q 13:203–216

Mazmanian D, Kraft ME (2009) The three epochs of the environmental movement. In: Mazmanian DA, Kraft ME (eds) Towards sustainable communities: transition and transformation in environmental policy. MIT Press, Cambridge, MA

Nijaki LK, Worrel G (2012) Procurement for sustainable local economic development. Int J Public Sect Manage 25:133–154

Organization Economic Cooperation Development (OCED) (1999) The DAC guidelines for sustainable development. OCED, France

Pellow DN (2002) Garbage wars: the struggle for environmental justice in Chicago. MIT Press, Cambridge, MA

Property and Environment Research Center (PERC) (2003) The environmental Kuznets curve. PERC, Bozeman

Perloff J (2004) Microeconomics. Person Publishing, New York

Rawls J (1971) A theory of justice. Harvard University Press, Cambridge, MA, p 1971

Reinhardt F (2000) Sustainability and the firm. Interfaces 30:26–41

Rivkin D et al (2010) Greening the world of work: implications for ONET SOC and new and emerging occupations. Bureau of Labor Statistics, Washington, DC

Roberts P (2004) Wealth from waste: local and regional economic development and the environment. Geogr J 170:126–134

Sen A (1999) Development as freedom. Random House, New York

Soja EW (2000) Postmetropolis: critical studies of cities and regions. Blackwell, Oxford

Soja EW (1986) Taking Los Angeles apart: some fragments of a critical human geography. Environ Plann D Soc Space 4:255–272

Stone C (2006) Power, reform, and urban regime analysis. City Community 5:23–38

Wasik JF (1996) Green marketing and management: a global perspective. Blackwell, Oxford

Young JE (1994) The next efficiency revolution: creating a sustainable materials economy. Worldwatch Institute, London

Chapter 13
Conclusions: The Science of Economics

Woodrow W. Clark II

Abstract The purpose of this chapter is to stimulate discussion and concerns over how economics is both studied and used. The chapter reviews some of the history of economics with a strong focus on the neoclassical economics from Adam Smith and how it has performed or, probably better said, failed to perform today. Economics must be a science, and there are a number of ways to approach that issue. Linguistics is one as well as the use of science for filing patents which can set a new standard for economics, as illustrated in this chapter.

The Science of Economics

The last four decades or so of supply side economics with even the revisionists ranging from Thomas Friedman (the world is flat) to Jeremy Rifkin (the world entropic) fail to consider the basic problem with economics: it is not a science. The Next Economics is about how economics can become a science.

While many economists, including those in this book, make the point that economics can or should include environmental externalities and even concerns on how energy is measured and used, they need to note the basic problem: economics is not a science.

W.W. Clark II, Ph.D. (✉)
Qualitative Economist, Academic Specialist, Cross-Disciplinary Scholars in Science and Technology, UCLA and Managing Director, Clark Strategic Partners, California, USA
e-mail: www.clarkstrategicpartners.net

This was brought home to me personally in June (2012) at a UCLA meeting for the Cross-disciplinary Scholars in Science and Technology (CSST) in which I am involved as a teacher and doing administrative work. We had a UCLA pre-summer school meeting of over 40 mentors (all scientists from fields ranging from physics to chemistry, biology, medicine, and engineering with combinations thereof) to the CSST students. I was the only economist along with one sociologist. After I was introduced and later in the meeting during a discussion, I made the comment: "I am here as an economist. And economics needs to become a science." The entire room laughed and applauded my presence there. Later, several people got into conversations with me and hence gave talks in my class on sustainable communities at UCLA. Their goal and mine is to teach social scientists what science is and for economists in particular to become scientists.

In The Next Economics, that was one of the key goals for all the chapters. But even more significant, underlying that goal has been the need for economists to understand what science is about and how it works. Every scientist is trained the same way in terms of thinking, creating, and proving hypothesis so that they become repeatable theories for application and use in the future. To say that economics is scientific because Adam Smith followed Sir Isaac Newton misses the point.

Science has a set of processes and procedures to which every scientist follows. The conclusions of every scientific study must be subject to challenge and replication. That is why climate change issues, without a doubt, are now significant and proven theories. The evidence and replicable data are now daily events. But that was only after over two decades of work by the UN Intergovernmental Panel on Climate Change (UN IPCC) and UN Framework Convention on Climate Change (UN FCCC). I worked in both of these groups, starting almost 20 years ago.

The basic conclusions from The Next Economics are how society (primarily political leaders and decision makers) along with academics need to look into the definition and meaning of "science" itself. The review of modern economic theory with its failure to be a science means that the field of economics must find a new paradigm and philosophical roots. For one thing, in order to address climate change, communities and nations must set plans. They need to state visions, which have measurable objectives as well as frequent updates and revisions. On the business level, the same criteria and needs must take place. However, whereas a community or government entity is accountable at election time, businesses need to be so on monthly, quarterly, and annually in verified accounting reports. Nonetheless, all organizations must have leadership that is held accountable.

Consider physics: "The 'rules' of physics are also of fundamental help in understanding the associated sciences of chemistry, biology, metallurgy, astronomy, etc. For example, before we understood the physics that govern the behavior of electrons in an atom, chemistry was purely a phenomenological science; that is, we knew we could repeat what we had done or seen before, but we didn't know why, nor could we predict what would happen in a new situation. Now, because we understand the physics of the atom, we understand fully why there are around one hundred basic chemical elements and around a few million chemical compounds." (Perkins 1996:1)

The Economics of Finance

On a recent trip to east Africa to talk about climate change for the US State Department, I met a group there from the World Health Organization (WHO) in London. Over several meals, this issue and need was discussed. Yet, there was little or no data, and certainly, none of us had been able to get either funding or commitments to do the scientific research and data gathering necessary for creating sustainable communities there.

Consider some cases where renewable energy for designated areas has worked or is being considered. Again, the basic issue is economics. How to get renewable energy installed through long-term financing? For example, in 2008, the City of Berkeley, California, created the PACE (Property Assessment for Clean Energy) program for renewable energy systems through additional payments on local city tax bills. The program meant that an owner of a building could chose to tax themselves in order to pay for the renewable energy systems to be installed. The monthly tax bill would be the way to pay off the funds needed to buy and install the renewable energy system. Should there be a sale or transfer of the building then the added tax would also be part of the sale or transfer.

Herein lay the problem for the program. In the summer of 2010, Fannie May and Freddi Mac ruled against the PACE program for individual homes: if there was a default on the home mortgage or taxes and hence a lien for those funds, Fannie May or Freedi Mac as the financial owner would be placed in second position. The situation for commercial properties was different and depended on a variety of other variables. But PACE was stopped for residential homes due to that decision by the key economic source for financing homes.

However, that meant that other creative economic mechanisms needed to be found for financing renewable energy and sustainable communities. One example that worked in other areas in order to fund new, but higher costs technologies was through government purchase programs or other nonprofits who could pay over long periods of time. Some of these programs would establish competitively bid master contracts for vehicles, college campus and government buildings, infrastructures, and other public needs.

The strategy was to establish a master contract with one company who successfully won the bid. From that one contract, other customers could buy the same products. This was done in some cities and states to reduce carbon emission, since natural gas cars polluted less that gasoline fueled vehicles. But as the master contract took effect, it then allowed other local government entities to purchase off the same contract and save large sums of money. The key was that bulk purchasing can greatly reduce the cost of innovative technologies to mitigate against climate change.

Applications of this financing mechanism could be used for other nonprofit organizations like religious groups, hospitals fire stations, and public schools among others. Additionally, some renewable energy companies have provided a very interesting economic addition: provide employees, staff, students, and administrators or even subcontracts the ability to buy off the master contract for a lot less money than if they were purchasing the systems on their own. Group and large-scale economic contracts yield lower costs than single-item purchases.

In more conventional economic areas also, here are creative approaches to financing renewable energy and technologies to make communities sustainable. One is the mortgage for any building or complex. The other is a lease. In both cases, the cost for a building must be seen as to include renewable energy much as the plumbing, lighting, heating, and cooling systems are all today. Not more than 50–60 years ago, after WWII, mortgages for homes and buildings did not include bathrooms let alone air-conditioning.

These "modern" technologies over time were then included in the mortgage or cost for a home or building. The logic seems apparent to do for renewable energy. Instead of considering solar panels, for example, as an add on cost, they could be made part of the mortgage cost for a new or refinanced home or building. The impact on reducing climate change and establishing buildings and communities sustainable would be a large step in reducing reliance on fossil fuels and make communities and regions energy independent and carbon neutral. The new economic paradigm that includes long-term cost reductions can occur and provide solutions to climate change.

Hypotheses, Plans, Rules, Standards, Measurements, and Accountable Results

Economics must become a science:

> One very important job in physics is to determine the fundamental constants of nature that are needed to understand the "rules" and apply them to predict and calculate things. A good example of such a constant is the speed of light (approximately 186,000 miles a second). Typically experiments are used to measure these physical constants. Theoretical studies can also be used to infer their values but an experimental test is usually ultimately required. A remarkable achievement of physics is that we know the value of virtually all the constants of nature that we are aware must exist for our present understanding of the "rules." (Perkins, op.cit. p. 3)

Consider a very recent example of science at work. In early July 2012, it was reported in the newspaper that "Physicists are celebrating their Higgs boson 'triumph'" (Brown 2012a). Over the entire twentieth century, scientists have tried to learn more about the standard model and thus answer one important question, "why does matter exist?" (ibid., p. 8) For physicists, the answer was similar to "landing on the moon or the discovery of DNA" (ibid. p. 1). The basic question was posed by British physicist Peter Higgs who heard the scientific beginning of the answer to his question from over 48 years ago, at the European Organization for Nuclear Research (CERN) in late June 2012 by two independent research teams (CMS and ATLAS). Both teams at the CERN conference independently reported data from December 2011 that "uncovered 'tantalizing hints' of a Higgs boson with the a mass of about 125 billion electron volts" (ibid. p. 8).

However, as with all scientific research, this is not the end. The next step all of the physicists stated "would be to figure out whether the particle is indeed the single Higgs boson described by the Standard Model or some exotic variant." (ibid., p. 8).

Some physicists believe that there are multiple Higgs bosons that require more research, data, and evidence. Hence, the need for a massive proton collider that needs to be updated in the next two years (Brown 2012b, p. 9). Scientific research never ends.

What does it then mean for economics becoming a science? One basic concern is even with the report on the Higgs boson. Aside from what is Higgs boson in terms of both definition and meaning (smallest atom that could be the basis for all matter). But then, consider the some of the discussion reported above: what does "tantalizing" and even "hints" mean in science? Clearly, not something that is proven beyond a doubt; thus, not subject to further investigation. Just the opposite, as the last quotation indicates, "figure out" the next steps and how that relates to the "standard model" used in physics. In short, science is a never ending search for truth through theories, data collection that have conclusions, yet with a continuing need to investigate, valid, and double check. Economics needs the same set of rules.

A very simple application of economics is with everyday life. All forms of groups, families, businesses, and governments need sets of rules. People and families need rules as much as all businesses and governments. While some people complain about government rules and codes, the fact is that they have the same thing for their families and children. Science follows that same line of thinking. In order to discover something or investigate an idea, there must be some set of hypotheses which turn into a plan with rules and standards that are measureable. If not, then science tries another set of hypotheses with rules. Even when the results are proven, it is also critical to replicate those hypotheses and test them again and again and again.

For example, I teach entrepreneurship in the USA, China and EU since the term was still new in the early 1990s. One of my first published articles was about entrepreneurship in Silicon Valley and how it worked. And my reason for teaching entrepreneurship was so that people who want to start businesses need to understand that there are sets of rules to follow from other NewCo experiences. I knew from own personal experiences because I had started my first business, a landscaping company with my younger brother (Wayne) in the late 1950s called "Wayne-Wood Nurseries" when we were in not even in high school yet, but that paid our way through college. We sold the company after we both went on to graduate school outside of Connecticut, where we grew up. We were both eager to move "west," gradually as we both spent 5–6 years in Illinois before going to the University of California. Berkeley. Neither of us had loans or had to pay back for our undergraduate and graduate education. But we wanted to continue our education.

The best example of where an area of social sciences or field of study like economics can be seen as a science is linguistics. In the field of economics, as described and outlined in several chapters in this book as well as Qualitative Economics (Clark and Fast 2008), linguistics changed into a science under the leadership of Noam Chomsky with what he calls "The Galilean Style" which comes from the natural sciences and is the process to construct "abstract mathematical models of the universe to which at least the physicists give a higher degree of reality than they accord the ordinary world of sensations" (Chomsky 1975: 28). I had read and followed Chomsky's work before and while working on my PhD degree in Anthropology at the University of Illinois, Urbana, in the early 1970s which I decided not to earn, and settled for a MA degree instead. Chomsky at MIT had became a long-distant mentor.

Later in his book on Reflections on Language (1975) and then in far more detail with Rules and Representations (1980) as well as a number of books since then, Chomsky states that language (a field of study that needed to become a science) is seen as "A comparable approach (which) is particularly appropriate in the study of an organism whose behavior, we have every reason to believe, is determined by the interaction of numerous internal systems operating under conditions of great variety and complexity" (ibid., 218). Chomsky called all of his work in linguistics "transformational grammar which led an intellectual revolution, while at MIT in the early 1960s." What Chomsky did decades ago, continues to this day.

In short, language moved from being a field of study to a science. Therefore:

> Creative aspect of language is a characteristic species property of humans. Language serves as an instrument for free expression of thought, unbounded in scope, uncontrolled by stimulus conditions though appropriate to situations, available for use in whatever contingencies our thought processes can comprehend. (ibid., 222)

The challenge for linguists is to approach language and its study as a science. Therefore, what language means to the linguist, grammar (as distinct from speaker-hearer's grammar) is "a scientific theory, correct insofar as it corresponds to the internally represented grammar ... The grammar of the language determines the properties of each of the sentences of the language. For each sentence, the grammar determines aspects of its phonetic form, its meaning, and perhaps more" (ibid., 220). And therein became a challenge to linguistics, not as a science but as needing more than just structure that determines meanings.

The biggest challenger was George Lakoff, whom I had for classes while at the University of Illinois, Urbana, and then again when he came to California for a summer school program at the University of California, Santa Cruz. Lakoff then went to the University of California, Berkeley, in 1972 where he is still today. I chose to leave Urbana and follow Chomsky to MIT. But he urged me to instead go to Berkeley due not only to the work of Lakoff, but also due to the "politics of Vietnam" at the time. MIT was not the best place to be if as a student or faculty member, if you were interested in politics as I was and as was Chomsky. Lakoff was in a more friendly environment in Berkeley.

My classes and studies with Lakoff and other faculty at Berkeley, however, led me in a different direction. For example, I studied with Herbert Blumer who was a retired professor in sociology and created the Sociology Department at UC Berkeley and the field of "Symbolic Interactionism" (Blumer 1969) which was the beginning of making sociology into a science.Blumer's work was rooted in the philosophy of George Herbert Mead (1932) who at the University of Chicago in the early part of the 20 th Century which took a different philosophical approach to societal issues, which was fast becoming "behaviorialism" from the influence of BF Skinner. Meanwhile in the 1970s, Lakoff became a critic of Chomsky on transformation grammar since from Lakoff's perspective, it only dealt the rules and structures of language. Lakoff and others at UC Berkeley (Lakoff and Núñez 2000) were far more interested in the meanings and definitions of words and even phrases in language. It was with the "deep structure" of language that both the structures for language and their impact had the most significance (Lakoff 1999, among earlier articles and books). What I got from all this was seeing how science worked in linguistics but also another perspective to other social sciences, including anthropology, sociology, and political science.

With these theoretical concepts in place from linguistics, the actual transformation rule-making process can be seen. That is, the business relationship becomes successful or unsuccessful because she/he draws upon the defining characteristics in the deep structures (universals or common properties across cultures) of the new business creation interactive process (surface structure) and applies the proper rules. For example, when a business agreement appears to have been very successful (e.g., material wealth, power, or head of large company), there were many transformational rules that got the business actors to that place (surface structure interactions). The transformational rule-making process is often intuitive and based upon common sense. It is the surface structure in scientific terms.

However, there is also the deep structure of words, concepts, numbers, and phrases that needed to be explored in-depth. I did that since the early 1980s in a number of areas but really decided that understanding deep structures and meanings was a core element needed in the social sciences. That leads me to taking classes from Herbert Blumer, whose concerns were more psychological than mine with his symbolic interactionism. However, the theory and research on understanding how people think and act is critical and becomes a core interest of mine. My book with Professor Michael Fast on Qualitative Economics (2008) reflects those ideas and concerns in practical ways for economics to be a science.

In the end, I changed directions at UC Berkeley and even took law classes at Boalt Hall wanting to understand how law gets involved with determining and defining meanings of words, phases, and numbers. Unfortunately, there were not too many law professors then who cared about science. So I decided to take a Ph.D. in an applied area that I had experiences in as well as a social science. But did continue to take law classes such as Professor Laura Nader in the Department of Anthropology. Hence, I got a combined degree in Higher Education and Anthropology with my thesis on "Conflict in Public Schools." The topic gained a great deal of interests over 30 years ago and still does today. The basic issue is, what causes violence in public schools, then and now today?

The scientific understanding of surface and deep structures helps provide an examination of schools and other institutions which leads to solutions, especially since the problem is repeated and can thus be examined repeated in a scientific manner. What I did however after the Ph.D. was go into business, applying some scientific insights that I had gained over the years while in academics and then practice as a teacher. What I had learned with Wayne-Wood Nurseries was now needed to be put into practice. In the academic world, if you receive your Ph.D. from an institution, you cannot then teach there but must go elsewhere for 6–7 years before coming back. I had offers to go to other universities, but I liked the San Francisco Bay Area. So I stayed.

What I did was start a mass media company in San Francisco that produced dozens of documentary and educational media. My dissertation topic on "School Violence" was a perfect topic to start me producing documentaries and then talk shows for local television. In the 1980s, Clark Communications was earning over $1 million annually, until the early 1990s when the end Cold War came and California was hit particularly hard with the need to downsize, convert and sell-off military bases, and restructure the entire American (global economies). The point is that as an entrepreneur, I knew what it meant and how to teach the subject.

Was starting a business, let alone, economics a science? No. But I had studied chemistry and physics in high school and at my graduate universities to know what the basics of science were. Then later on, I was asked to go to Lawrence Livermore National Laboratory to train scientists in technology transfer and commercialization. While there, I learned a lot more about science from hundreds of scientists there (over 1,300 physicists and over 3,500 engineers were there).

However, it was when I went to Aalborg University (AAU) in Denmark as a Fulbright Fellow in 1994 that I learned even more about the basics of science in the Faculty of Natural Sciences while teaching entrepreneurship in the Economics Faculty. I was asked to start a program at AAU that involved their science park (NOVI) across the street from the university. This was not an incubator and very different from what Americans were doing then (and even now) to commercialize new technologies and ideas. I have written and published many articles on science parks from over a decade ago. But the articles are not as well as received in the USA, as they are in the EU and Asia (Clark 2003a, b). Science parks there have been extremely successful and moved each country into the green industrial revolution (Clark and Cooke, to be published).

The application of my three graduate M.A. degrees with the Ph.D. made me unique. Today, I refer to it as "cross-disciplinary studies" and my focus then and now is on "sustainable communities." And even more so since, I had an entrepreneurial series of businesses. However, even more significant was my working with scientists at LLNL and also at AAU. I became a sort of translator from science to business and then back to science again. My experience and work on science parks in Denmark and then throughout Northern Europe over a decade of research and publications lead to a number of other ways in which economics can become a science in practice.

My last book with Grant Cooke on "Global Energy Innovation" (2011) and our next one on "The Green Industrial Revolution" reflect the concerns with the environment and also the impact and need for a science of economics to do that. It is one thing for people, communities, and governments at all levels to talk about climate change and the environment but very different discussions about what to do. The biggest and most concerning barrier for taking any action is always economics. It is too expensive to save our earth, reverse climate change, and stop pollution. That is foolish and frankly just playing politics with the field of economics—not the science of economics.

Let me end with a personal case that proves economics as a science can help resolve and mitigate environmental (and other) societal issues. Consider energy efficiency and conservation. Today, LED light bulbs are available that take a faction, if any electricity. The issue is that today these LED lights cost too much. Any and every new invention always has a slightly economic cost when first brought out and into any market. However, LED is an example of where that economics when considered in scientific terms makes the costs minimal and the scientific economic process to do so, patentable.

In the attachment below (Appendix 1), that scientific economic process is now as a patent pending (February 2012) done by two electrical engineers and myself through a NewCo, called Nularis. The scientific economic process for LED lights is patented that shows how to conserve and save energy through the purchase of LED bulbs. With this case, example from scientific research the rules and formulas are laid out in detail. The patent sets a new standard for economics to be a science with its application to business for The Next Economics. More and similar ideas for the science of economics will be forthcoming and welcome. It is time that economics becomes a science.

Appendix 1

Patent Invention Disclosure: Applied Dated: February 12, 2012

Economic Efficiency Through Lighting (Nularis Corporation)

Title: Method and design of software and related systems which couple amortized loan terms and payments with predicted and actual energy cost savings.

Inventor(s): Woodrow W. Clark II Ph.D., Wendell Brown, and Jonathan Fram

Background of the Invention

Energy efficiency and conservation are important in order to achieve international goals for reduction of greenhouse gas emissions, fossil fuel usage, grid load strain, costs, and a wide range of other benefits. However, many approaches to energy efficiency and conservation involve significant capital outlays that create financial management risk and provide undetermined return-on-investment rates and payback periods, which often hinder their adoption.

These inventions relate particularly to methods and design of software and systems which run in computing environments (computer hardware, virtual CPU environments, servers, computers, tablets, wireless mobile devices, etc.) that couple and integrate amortized payment terms and amounts with predicted and actual energy cost savings. The inventions are thus novel, innovative, and useful in that they provide a mechanism for financial risk reduction/management and predictable cost outlays (loan repayment terms that are directly linked to energy savings), thus serving as an enabler for the financing of such energy efficiency and conservation projects.

Description of the Inventions

The inventions relate to the methods and design of software and related systems which couple amortized payment amounts with predicted and actual energy cost savings.

Features as Formulas: 1–14

14: The formula for the periodic payment amount A is derived as follows. For an amortization schedule, we can define a function $p(t)$ that represents the principal amount remaining at time t. We can then derive a formula for this function given an unknown payment amount A and $r = 1 + i$.

$$p(0) = P$$

$$p(1) = p(0)r - A = Pr - A$$

$$p(2) = p(1)r - A = Pr^2 - Ar - A$$

$$p(3) = P(2)r - A = Pr^3 - Ar^2 - Ar - A$$

$$p(t) = Pr^t - A\sum_{k=0}^{t-1} r^k$$

Applying the substitution,

$$\sum_{k=0}^{t-1} r^k = 1 + r + r^2 + \ldots + r^{t-1} = \frac{r^t - 1}{r - 1}$$

After substitution and simplification, we get

$$\frac{p(t)}{P} = 1 - \frac{(1+i)^t - 1}{(1+i)^n - 1}$$

The annuity formula is

$$A = P\frac{i(1+i)^n}{(1+i)^n - 1} = \frac{P \times i}{1 - (1+i)^{-n}} = P\left(i + \frac{i}{(1+i)^n - 1}\right)$$

References

Blumer GH (1969) Symbolic interactionism: perspective and method. Prentice-Hall, Englewood Cliffs

Brown E (2012) Physicists are celebrating their Higgs boson 'triumph'. Los Angeles Times, Los Angeles, pp 1 and 8

Brown E (2012) Massive proton collider is a great big hit after all. Los Angeles Times, Los Angeles, p 9

Chomsky N (1975) Reflections on language. Pantheon Books, New York
Chomsky N (1980) Rules and representations. Columbia University Press, New York
Clark WW II (2003a) Science parks (1): the theory. Int J Technol Trans Commercialization 2(2):179–206. Inderscience, London
Clark WW II (2003b) Science parks (2): the practice. Int J Technol Trans Commercialization 2(2):179–206, Inderscience, London
Clark WW II, Cooke G (2013) The Green Industrial Revolution. Publisher to be selected for 2013
Clark WW II, Cooke G (2011) Global energy innovation. Praeger Press, New York, NY
Clark WW II, Fast M (2008) Qualitative economics: toward a science of economics. Coxmoor Press, Oxford
Lakoff G (1999) Philosophy in the flesh: the embodied mind and its challenge to western thought. Basic Books, New York
Lakoff G, Núñez R (2000) Where mathematics comes from: How the embodied mind brings mathematics into being. Basic Books, New York
Mead GH (1932) The philosophy of the present. Prometheus Books, New York, NY
Patent Invention Disclosure. Inventor(s): Clark WW II Ph.D., Brown W, Fram J (2012) Method and design of software and related systems which couple amortized loan terms and payments with predicted and actual energy cost savings. Applied dated: 12 Feb 2012
Perkins LJ, Senior Physicist (1996) What is physics and why is it a 'science'? Lecture at University of California Physics Seminar, University of California, Berkeley, pp 1–3

Index

A

Accountable, 276, 278–283
Action, 11, 13, 17, 18, 31, 44, 45, 48–52, 55, 56, 59, 64, 68, 73–83, 87–90, 98, 128, 129, 133, 134, 140, 166, 186, 216, 218, 225–231, 239, 242, 246, 247, 256, 258, 261, 266, 270, 282
Actors, 73–83, 87–90, 129, 178, 179, 240, 241, 256, 281
Aggregate, 17, 140, 246, 251–255, 257–261, 266–268
Agile energy, 11, 13, 26
Agriculture, 11, 17, 238, 240–243, 246, 247
Alliances, 16, 17, 47, 146, 238–240, 244, 256, 267
Alternative, 5, 45–47, 61, 66, 72, 74, 106, 124, 128, 129, 137–139, 146, 151, 159, 160, 176–178, 210, 214, 220–223, 225, 229–236, 238, 242, 263, 264
America, 23, 25, 46, 47, 67, 147, 152, 153, 161, 238, 243
Analysis, 1–3, 5, 82, 85, 89, 94–113, 116, 117, 125, 135, 138, 170, 176, 179, 187–205, 207–236, 239, 243, 255, 258, 262, 263, 272
Anthropology, 279, 281
Artifacts, 77, 83, 89
Asia, 10, 23–27, 33, 35, 38, 127, 147, 152, 153, 161, 282
Asian Development Bank, 35, 151
Astronomy, 276

B

Balance, 2, 3, 27, 46, 95, 112, 144, 151, 186
Bank, 35, 38, 151, 161, 166, 193

Bay Area (San Francisco), 63, 281
Beyond growth, 251–273
Biology, 47, 276
Blue-green, 16, 22, 237–248
Brazil, India, Russia and China (BRIC), 12, 147
Buildings, 7, 8, 11–16, 34, 37–39, 60, 63, 66, 71, 74, 108, 131, 132, 134, 138–140, 145, 159, 168, 174, 211, 216–218, 239, 245, 255, 263, 277, 278

C

Campus, 35, 94, 277
Canada, 15, 23, 27, 28, 62, 63, 166, 168
Cap and trade (CAT), 8, 45, 50–52, 54–60, 64–68
Capitalism, 2, 22, 144, 146, 155, 161, 175
Capitalist, 35, 144–147, 160, 247
Carbon, 8, 12, 16, 26, 30, 31, 39, 44, 46, 48, 50–54, 57–66, 68, 95, 106, 108, 127, 129, 133, 134, 137, 144, 147, 149, 157, 159, 161, 162, 167, 168, 172, 177, 194, 202, 204, 208, 212, 213, 227–231, 236, 277, 278
Carbon emissions, 12, 30, 45, 46, 50, 59, 61, 63, 147, 149, 157, 277
Central grid, 14, 15, 37
Century, 2–5, 9–12, 15–17, 22–25, 29, 31, 32, 44, 73, 108, 144, 147, 149, 160, 180, 186, 187, 254, 257, 278, 280
Chemistry, 3, 276, 282
China, 2, 8, 9, 12, 22–25, 27, 29, 31, 33, 35, 36, 38, 143–162, 279
City, 29, 63, 134, 188, 208, 251–253, 261, 271, 277
Classical economics, 2, 8, 9, 17

Index

Clean coal, 24, 46, 47, 158
Clean tech, 25, 31, 137, 147
Climate change, 2, 3, 6–11, 13, 15, 17, 22–24, 27, 31, 36, 37, 43–68, 93, 101, 105, 108, 126, 127, 137, 150, 158, 159, 166, 167, 178, 180, 207, 213, 241, 248, 273, 276–278, 282
Coal, 5, 10, 13, 14, 29, 33, 34, 38, 46–50, 63, 65, 93, 95, 98, 101, 105–107, 114, 115, 118, 135, 148–151, 158–160, 259
Codes, 279
Cold War, 23, 25, 146, 147, 281
Common sense, 72, 73, 76, 79, 81–83, 88, 281
Conscious reality, 74, 75
Conservation, 16, 18, 26, 31, 32, 36, 38, 45, 47, 95, 97, 102, 112, 126, 128, 131–133, 138, 139, 141, 156, 158, 161, 207–236, 238, 240, 241, 260, 264, 282, 283
Cooperative, 17, 79, 157, 239, 253
Cost-benefit analysis (BA), 36, 37, 189
Costs, 2, 22, 44, 96, 125, 145, 166, 188, 209, 242, 252, 277
Cross-disciplinary, 276, 282
Culture, 2, 80, 81, 87, 155, 242, 247, 287

D

Deep structures, 4, 85–87, 280, 281
Definitions, 5, 12, 17, 31, 34, 74, 75, 78, 86, 87, 95, 130, 137, 138, 161, 252, 258–262 264, 276, 279, 280
Demand, 2, 10, 14, 15, 17, 24, 26, 27, 30, 33, 34, 36, 45, 52, 53, 62, 72, 77, 88, 93–95, 97, 99, 100, 102, 103, 105, 107, 110, 113, 116, 117, 126–128, 131, 144, 145, 148–150, 152, 153, 156, 158–162, 170, 171, 174–177, 180, 193, 208, 211, 212, 219, 227–231, 242, 245, 253, 255, 256, 261, 265, 271, 272
Denmark, 14, 15, 26, 29, 54, 61, 143, 160, 187, 282
Dependence, 46, 67, 102, 106, 146–148, 155, 160
Design, 12–14, 29, 44, 45, 51, 55, 57, 67, 68, 74, 87, 139, 173, 187, 208, 259, 261, 283
Disease, 46, 52, 54
Distributed, 33–34, 108, 137, 194, 252, 268
Dynamics, 72, 87, 132, 146, 161, 188, 247, 268

E

Economic(s)
 efficiency, 51, 283
 growth, 17, 32, 36, 54, 55, 61, 62, 126, 127, 130, 132, 141, 144, 147–152, 154–160, 174, 177, 239, 251–261, 266–271
 raise, 143–160
 reality, 254–257
Economic-environmental, 15, 185–205
Economist, 2–4, 9, 12, 17, 22, 24, 25, 27, 45, 46, 50, 52–54, 56, 58, 63, 68, 81, 144, 152, 155, 166, 171–173, 177–181, 238, 243, 253, 269, 275, 276
Electrical engineering, 282–283
Electric vehicles, 169, 178
Emerging nations, 12, 147
Emissions, 9, 16, 24, 30–33, 45, 50, 53–68, 105, 106, 128, 129, 132–134, 137, 139, 147, 149, 156, 158, 159, 161, 167, 172, 176, 187, 194, 202–204, 238, 261, 263, 277, 283
Emotions, 89
Energy
 consumption, 12, 36, 52, 62, 93, 95–102, 106, 108, 109, 118, 135, 136, 144–152, 154, 155, 157–159, 226, 229, 265
 dependency, 152–156, 245
 efficiency, 14, 26, 30, 32, 33, 36, 45, 52, 60, 63–66, 95, 100, 108, 126, 128, 130–139, 141, 178, 248, 263, 264, 272, 282, 283
 independence, 61, 64, 96, 128
 security, 2, 6, 147, 152–155, 157, 158, 180
Engineering, 3, 138, 139, 170, 213–214, 244, 276
Entrepreneur, 46, 126, 130, 141, 144, 256, 271, 282
Entrepreneurship, 34, 279, 282
Environmental impact, 94, 107, 117, 127, 189, 215, 220, 263, 265
Environmentally sound, 13, 14, 23, 32, 246
Epistemology, 72
European Union (EU), 12, 14, 15, 22–36, 38, 58, 62, 112, 113, 147, 279, 282
Everyday life, 72, 73, 75, 76, 78, 87, 88, 0, 279
Exports, 53, 107, 148, 155, 156, 159, 161, 244
Externalities, 3, 8, 9, 16, 27, 37, 115, 172, 177, 239, 252, 255, 257, 267–269, 275

F

Feedback, 15, 243
Field of study, 6, 279, 280
Finance, 7, 11, 14, 31, 36–38, 60, 100, 117, 136, 138, 139, 145–147, 161, 169, 173, 174, 178, 214, 277–278
Firm, 32, 74–79, 89, 90, 245, 256, 262–264
Five-Year Plan, 145, 151, 156–159
Foreign policy, 150, 152–154, 158, 160, 161
Fossil fuels, 13, 14, 22–27, 36, 38, 50–52, 58, 60, 66, 68, 94, 96, 98, 101–107, 113–116, 127, 147–151, 167, 168, 172–174, 176–179, 261, 263, 278, 283
Free market, 45, 49
Fuel cell, 23, 26, 95, 106, 113, 138

G

Gas, 10, 13, 14, 23, 24, 26, 29, 30, 34, 37, 38, 44, 47, 50, 52, 61–63, 66, 93–95, 104–107, 112, 117, 118, 133, 146, 150, 153, 158, 167, 173, 178, 212
GDP. *See* Gross Domestic Product (GDP)
Generalized other, 75
Generate, 6, 14, 17, 26, 33, 85, 97, 112, 116, 127, 128, 135, 187, 218, 239, 256, 265
Germany, 15, 25, 27–31, 34, 147, 160
GHG. *See* Greenhouse gases (GHG)
GIR. *See* Green Industrial Revolution (GIR)
Globalization, 17, 147, 252, 254–257
Global warming, 13, 45, 47, 48, 58, 63, 64, 68, 93, 94, 97, 99, 101, 102, 133, 134, 238
Government, 8, 11, 12, 14, 16, 18, 25–33, 35–39, 45–52, 55–61, 64, 65, 67, 68, 96, 113, 130, 145–148, 152, 154–157, 159, 166–168, 171, 173–176, 181, 193, 207, 209, 210, 212, 214, 235, 240, 244, 256, 261, 276, 277, 279
Greenhouse gases (GHG), 30, 45, 50, 52, 56, 61–63, 67, 68, 95, 128, 133, 134, 137, 158, 172, 213, 225, 227, 229, 238, 261, 263, 283
Green Industrial Revolution (GIR), 9, 22–34, 36–39, 147–148, 150, 160, 282
Green jobs, 28, 67, 138, 238, 252, 260–263, 265–273
Green tech, 24, 47, 131, 152, 264, 266
Grid, 13–16, 26, 28, 33, 34, 37, 39, 115, 116, 119, 153, 159, 192, 203, 208, 209, 211, 219, 227–231, 283
Gross Domestic Product (GDP), 17, 44, 62, 100, 106, 135, 144, 148, 149, 154, 156, 158, 245, 251, 252, 270
Growth, 7, 23, 48, 93, 126, 144, 168, 239, 251

H

Health, 1, 6, 9, 17, 21, 27, 37, 46, 52, 54, 62, 64, 126, 132, 172, 238, 239, 241, 243, 245, 248, 256, 258, 259, 267, 277
Higgs boson, 278, 279, 294
Household, 33, 34, 49, 52, 125, 126, 128–130, 159, 243, 246
Human
  capital, 17, 253–256, 268
  construction, 72, 73
  dynamics, 71, 72
Hydrogen, 16, 24, 95–97, 100, 105–116, 138, 236
Hypotheses, 6, 10, 18, 86, 101, 278–283

I

IMAR. *See* Inner mongolia autonomous region (IMAR)
Imports, 51, 53, 56, 60, 96, 97, 106, 107, 109, 118, 149–151, 155, 255
Incentive, 11, 25, 27, 28, 31, 36, 38, 45, 47, 50, 51, 57–59, 64, 67, 68, 114, 126, 130–133, 135, 141, 159, 160, 173, 179, 180, 193, 207, 220, 221, 223, 224, 229–231, 235, 238, 255, 258, 271
Independence, 60, 61, 64, 96, 128, 134, 145, 146, 154
Industrialization, 146, 160, 175, 258
Infrastructure, 9, 11–14, 22, 24, 25, 34, 38, 39, 97, 110, 114, 138–140, 143, 145, 153, 158, 174–176, 180, 235, 267, 277
Inner mongolia autonomous region (IMAR), 35, 151
Innovation, 7, 17, 55, 57, 62, 87, 128–131, 133, 138, 140, 156, 158–160, 180, 253, 255, 265, 266, 273, 282
Integrated systems, 34
Interaction, 71–84, 86–90, 161, 218, 237, 247, 256, 278, 280, 281, 284
Interactionism, 71, 76, 85, 87, 280, 281

International energy agency, 150
Inter-subjective, 72, 77, 78, 81, 88, 89
Invention, 88, 187, 282, 283
Investigation, 74, 83, 87, 189, 279
Investments, 13, 16, 31, 33–38, 51, 59, 61, 67, 101, 114, 115, 125, 128, 130, 137, 139, 144, 147, 148, 151, 153, 155, 156, 159, 160, 168, 177, 178, 180, 193, 199, 205, 209, 223, 248, 253, 254, 256, 283
Italy, 15, 16, 27, 29, 147, 188

J

Japan, 2, 15, 23–25, 29–33, 36, 43, 46, 112, 113, 144, 149

K

Knowledge, 3, 4, 73–82, 84, 86, 87, 89, 158, 160, 181, 217, 236, 241, 242, 245, 259, 265, 272

L

Laws, 3, 7, 8, 71, 95, 98, 159, 169, 193, 240
Leadership in Energy, Environmental Design (LEED), 14
Leapfrog, 22, 24, 25, 147–148, 160
LED. *See* Light-emitting diode (LED)
Lexicon, 86
Lifeworld, 73, 76, 82
Light-emitting diode (LED), 15, 16, 18, 33, 207–209, 212, 214, 215, 217, 221, 223–225, 229–232, 234–236, 282
Lines of action, 74, 75
Linguistics, 4, 5, 10, 73, 85–89, 279–281
Load shifting, 51
Logic, 5, 49, 73–74, 88, 146, 246–248, 254, 267, 278
Lung, 51

M

Macro-economics, 171, 260
Market
　driven, 64, 148, 160, 161
　economy, 25, 154
　forces, 2, 18, 25, 27
　manipulation, 59
　oriented, 144, 146, 243
　solutions, 8, 42–68
Master contract, 277

Matera, 15, 16, 188, 205
Mathematics, 2–5, 10
Meaning, 4, 72–77, 79–81, 83–89, 130, 140, 161, 180, 246, 276, 280, 281
Measurements, 5, 18, 29, 36, 37, 87, 111, 188, 189, 192, 278–283
Mediterranean, 15, 185–205
Methodology, 37, 73, 86, 88, 138, 189–194, 204
Micro-wind, 16
Models, 5, 7, 12, 27, 28, 37, 82, 88, 95, 238, 243, 279
Mortgage, 7, 37, 38, 134, 277, 278

N

National, 2, 24, 43, 93–121, 125, 144, 188, 207, 237, 254, 282,
National economic, 145, 146, 156, 161
National oil companies (NOCs), 153
Natural, 4, 8, 10, 13, 14, 23, 24, 26, 29, 31, 32, 34, 46–48, 62, 66, 73, 76, 83, 85, 87, 93, 95, 98, 99, 101–105, 114, 115, 117, 118, 127–129, 132, 134, 135, 137, 138, 147, 148, 151, 152, 161, 166, 170, 172, 178, 186, 205, 238–240, 245, 247, 259, 269, 277, 279, 282
Natural gas, 10, 13, 14, 23, 24, 29, 34, 47, 63, 93, 95, 98, 101–105, 114–118, 135, 138, 148, 151, 152, 158, 170, 172, 178, 277
Neo-classical economics, 3, 9, 17, 22, 170–176, 242
Neo-colonialism, 154
Netherlands (Holland), 6, 29
NewCos, 279
Nordic countries, 8
Norway, 26, 29, 54, 62
Nuclear power, 14, 16, 22, 24, 26, 29, 46, 61, 108, 148, 152, 157, 169, 204

O

Observation, 4, 18, 24, 72, 79, 83, 86, 102, 113, 172, 194
Oceans, 6, 11, 12, 17, 23, 24, 26, 48, 96, 102, 160
Organization, 12, 25, 29, 32, 46, 49, 65, 67, 74–77, 79, 80, 85, 87, 89, 90, 125, 129, 132, 139, 145, 152, 153, 157, 235, 237–239, 259, 260, 263, 276–278

Organization firm, 74
Organizing, 72, 73, 77–80, 89

P
PACE. *See* Property Assessment for Clean Energy (PACE)
Paradigm, 2, 3, 6, 8, 9, 13, 15, 22, 23, 27, 78, 80, 90, 146, 171, 176, 236, 252, 260, 265, 269, 273, 276, 278
Patents, 18, 131, 282–284
Patents pending, 282
Peoples Republic of China (PRC), 36
Perspective, 2, 8, 9, 11–13, 21, 25, 72, 73, 76–79, 83, 84, 87, 88, 114, 138, 144, 154, 170, 173, 180, 239, 247, 248, 256, 269, 271, 272, 280, 281
Phenomenon, 73–75, 79, 84, 87, 89
Philosophical, 2, 6, 8–10, 23, 73, 74, 276, 280
Photovoltaic (PV), 35, 37, 101, 104, 110, 111, 114, 119, 120, 159, 170, 174
Physics, 2–5, 181, 276, 278, 279, 282
Planners, 93, 95, 96, 251, 257, 260, 268
Plans, 4, 10–12, 15, 22, 35, 36, 38, 45, 52, 55, 66–68, 76, 88, 93, 94, 100, 130, 145–147, 151, 156–159, 178, 221–223, 235, 259, 276, 278–283
Policy legitimacy, 154
Political science, 43, 281
Politicians, 8, 24, 48, 49, 54, 153, 166–168, 176, 180, 256
Pollution, 8, 17, 24, 32, 33, 39, 44, 46, 47, 51, 52, 54, 57, 60, 62, 64, 65, 67, 126, 128, 129, 133, 137, 144, 147, 149, 158, 161, 166, 172, 216, 239, 244, 245, 260, 261, 263, 264, 266, 267, 271, 282
PRC. *See* Peoples Republic of China (PRC)
Prediction, 4–6, 10, 29, 171
Price, 7, 43, 100, 125, 149, 166, 193, 225, 242,
Private, 11–14, 36, 45, 49–51, 81, 112, 130, 133, 139, 145, 208, 235, 239–241, 244, 245, 257, 270, 271
Problematic, 83
Process, 5, 17, 37, 62, 74–79, 81–85, 89, 90, 112, 113, 128–130, 138, 139, 166, 169, 177, 178, 210–215, 217–223, 232, 235, 236, 240, 241, 244, 253, 255–259, 261, 265, 266, 268, 272, 273, 276, 279–283
Productivity, 17, 126–128, 135, 141, 171, 216, 243, 251, 256
Profits, 35, 52, 133, 169, 175, 176, 241, 255, 256, 259, 271

Property assessment for clean energy (PACE), 38, 134, 277
Public–private, 145, 241
Public schools, 277, 281

Q
Qualitative economics (QE), 4, 9, 10, 24, 71–90, 279, 281
Quality of life, 17, 247, 251–261, 267, 273
Quantitative economics, 87

R
Reality, 2, 14, 22, 25, 31, 33, 39, 47, 56, 72–83, 88–90, 126, 141, 146, 161, 170, 173, 174, 179–181, 189, 242, 254–257, 266, 267, 279
Reflecting, 89
Re-generative braking, 26
Region, 6, 10, 11, 13–15, 18, 25, 26, 28, 32, 44, 63–67, 93–121, 126, 128, 131, 133, 134, 145, 147, 151–154, 156, 161, 181, 218, 243, 245, 261, 266, 270, 272
Regulations, 18, 44, 45, 49, 50, 52, 57, 58, 61, 63, 64, 67, 68, 98, 126–128, 131, 140, 156, 159, 166, 176, 238, 248, 261, 270, 272–273
Renewable energy, 8–15, 22, 23, 25, 26, 28–31, 33–38, 61–63, 65, 66, 95, 96, 98, 100–110, 112, 114–118, 128, 130, 131, 133, 134, 138, 139, 141, 145, 147, 148, 151–153, 158–162, 168, 169, 175, 176, 178, 236, 263, 277, 278
Researchers, 45, 48, 54, 73, 87, 88, 105, 246, 270
Residential, 7, 13, 14, 31, 33, 39, 116, 119, 134, 135
Rhetoric, 53, 80
Road maps, 12, 109, 145
Rules, 3–5, 11, 18, 58, 63–67, 75, 80, 85, 86, 151, 276, 278–283

S
Savings, 15, 45, 127, 128, 130, 134, 209, 212–214, 216, 219, 220, 225, 230, 232, 234, 271, 283
Science, 1–11, 17, 18, 24, 25, 35, 36, 48, 71–90, 132, 139, 247, 275–284
Science of economics, 3, 9–12, 89, 275–284

Scientific, 3–6, 10, 11, 18, 38, 72, 74, 80, 82, 83, 86–88, 144, 242, 246, 276, 278, 280, 291–293
Second industrial revolution, 22, 23, 179
Self-reliance, 144–146, 160
Semantics, 85, 86
Shale oil, 23, 47
Shopping malls, 11, 14, 39
Sino, 146
Smart grid, 15, 26, 37, 39, 153, 159
Social capitalism, 9, 12, 35, 143–162
Social development, 95, 145, 154, 156, 157
Social institutions, 80, 81, 86
Socialist, 144–146
Social science, 4, 10, 73, 81, 82, 88, 149, 279, 281
Sociology, 181, 280, 281
Solar, 7, 24, 46, 93, 130, 145, 168, 236, 262, 278,
Solar thermal, 16, 112, 135, 145, 169
Special interests, 25, 59
Standards, 1, 12, 17, 45, 57, 60, 62, 64, 65, 95, 102, 104, 111, 115, 116, 126–135, 137, 140, 141, 154, 156, 157, 161, 191, 192, 208–210, 223, 239, 243, 246, 259, 270, 278–283
State
  business, 8, 12, 22, 23, 25, 27, 33, 36, 137, 140, 256, 271
  capitalism, 12, 22, 23
Strategy, 39, 45, 51, 57, 94–95, 97–99, 110–112, 114, 117, 128, 138, 144, 145, 147, 155, 158–160, 167, 180, 223, 251–272, 277
Streams of consciousness, 84
Structures, 4, 5, 13, 22, 27, 28, 72, 74, 75, 80, 82, 83, 85–88, 110, 134, 141, 145, 170, 174, 176, 187, 212, 239, 257, 260, 270, 273, 280, 281
Subjectivist, 72
Superpowers, 146
Supply, 2, 17, 18, 26, 41, 57, 63, 98, 102, 106, 118, 143, 145, 148–152, 159, 161, 167, 172, 174, 175, 177, 179, 185, 193, 211, 215, 241, 243, 264, 271
Surface structures, 85–87
Sustainability, 12, 31, 94, 95, 97, 129, 151, 155, 182, 239, 240, 245–247, 252, 257–265, 267, 269–273
Sustainable communities, 10, 14, 22, 23, 26, 33–35, 95, 96, 100, 147, 276–278, 282

Sustainable development, 34, 94, 151, 154, 244, 245, 247, 258–260, 270
Sustainable economic development, 94, 238, 251–273
Sweden, 26, 29, 53, 54, 61
Symbolic interactionism, 73, 85, 280, 281

T
Tax, 8, 23, 26–28, 31, 33, 36–38, 50–54, 57–64, 68, 102, 115, 159, 168, 173, 179, 204, 253, 258, 276, 277
Technical analysis, 187
Technology transfer, 282
Thinking, 5, 12, 23, 73–75, 81–85, 88–90, 257, 258, 272, 279
Third industrial revolution, 3, 22, 24
Trade, 8, 24, 30, 43, 45, 49, 51, 52, 54–59, 64, 68, 114, 146, 147, 150, 153–155, 160, 161, 167, 238, 239, 243, 248, 253, 255
Traditional economics, 2, 25, 269
Transformational grammar, 4, 180
Transformational theory, 86

U
UN Framework Convention on Climate Change (UN IFCC), 276
UN Intergovernmental Panel on Climate Change (UN IPCC), 23, 276
United States of America (USA), 6, 8, 11, 13–16, 21–27, 31, 32, 34, 36, 38, 43, 46, 63, 65, 126, 129, 135, 137, 144, 148, 149, 165–181, 279, 282
Universal, 8, 72, 85, 86, 216, 281
US Congress, 14–16, 60, 64, 144, 146, 148, 149, 152, 153, 158
US Environmental Protection Agency (US EPA), 134, 238
US Green Building Council (US GBC), 12

V
Value free development, 257

W
Western nations, 9, 12, 36, 145, 152, 156
Wind energy, 15, 35, 107, 111, 135, 186–188, 192–193, 200
Wind turbine, 15, 144, 160, 168, 185–205

Printed by Publishers' Graphics LLC